Betsy Bowditch   5-

THE DEEPEST WOUNDS

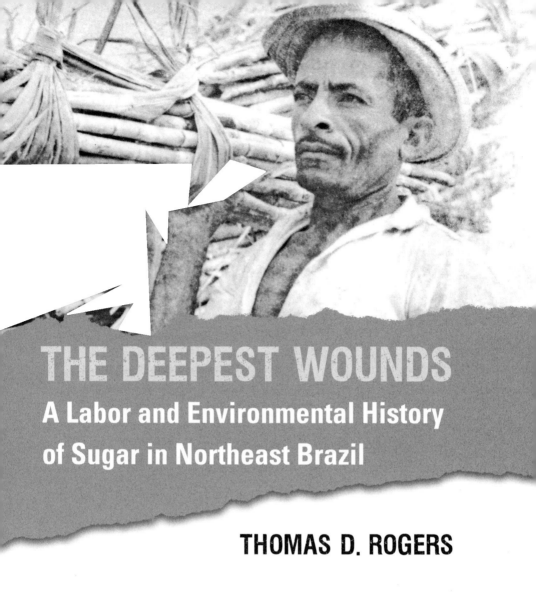

THE DEEPEST WOUNDS
A Labor and Environmental History of Sugar in Northeast Brazil

THOMAS D. ROGERS

The University of North Carolina Press
Chapel Hill

© 2010 The University
of North Carolina Press
All rights reserved

Manufactured in the
United States of America
Set in Charter, Univers, and Viper Nora

The paper in this book meets the guidelines for permanence and durability of the Committee on Production Guidelines for Book Longevity of the Council on Library Resources.

The University of North Carolina Press has been a member of the Green Press Initiative since 2003.

Library of Congress Cataloging-in-Publication Data

Rogers, Thomas D., 1974–
  The deepest wounds : a labor and environmental history of sugar in Northeast Brazil / Thomas D. Rogers.
    p. cm.
Includes bibliographical references and index.

ISBN 978-0-8078-3433-6 (cloth : alk. paper)
ISBN 978-0-8078-7167-6 (pbk. : alk. paper)

1. Sugar trade—Brazil—Pernambuco—History. 2. Sugarcane industry—Environmental aspects—Brazil—Pernambuco—History. 3. Sugarcane industry—Social aspects—Brazil—Pernambuco—History. 4. Pernambuco (Brazil)—History. I. Title.
HD9114.B63P477 2010
338.1'7361098134—dc22

2010018134

Parts of this book have been reprinted with permission in revised form from "Laboring Landscapes: The Environmental, Racial, and Class Worldview of the Brazilian Northeast's Sugar Elite, 1880s–1930s," *Luso-Brazilian Review* 46, no. 2 (December 2009): 22–53, © 2009 by the Board of Regents of the University of Wisconsin System, reproduced courtesy of the University of Wisconsin Press, and "Geneticistas da gramínea doce em campos decadentes: Variedades de cana-de-açúcar, agrônomos e plantadores na abordagem da modernização agrícola (1930–1964)," *Clio: Revista de pesquisa histórica* 26, no. 2 (2009): 161–88.

cloth 14 13 12 11 10 5 4 3 2 1
paper 14 13 12 11 10 5 4 3 2 1

For Hannah

CONTENTS

Acknowledgments xi

Abbreviations xv

Introduction: The Wounds of a People and a Landscape: Labor and Agro-Environmental History 1

PART I: THE LANDSCAPE OF THE *ZONA DA MATA* TO THE 1930S

1 An Eternal Verdure: The *Longue Durée* of the *Zona da Mata* 21

2 A Laboring Landscape: The Environmental Discourse of the Northeast's Sugar Elite, from Nabuco to Freyre 45

3 A Landscape of Captivity: Power and the Definition of Work and Space 71

PART II: OPENING UP THE *ZONA DA MATA*, 1930–1963

4 Modernizing the Sugar Industry: Cane Expansion and the Path toward Rationalization 99

5 The *Zona da Mata* Aflame: Political Upheaval, Strikes, and Fire 125

PART III: THE DICTATORSHIP COMMANDS THE *ZONA DA MATA*, 1964–1979

6 The Only Game in Town: Workers, Planters, and the Dictatorship 157

7 An Agricultural Boom and Its Unexpected Consequences 179

Conclusion: Power, Labor, and the Agro-Environment of Pernambuco's Sugarcane Fields 203

Notes 219

Bibliography 269

Index 297

ILLUSTRATIONS

Engenho Araticuns, 1940s–50s 56

Donkeys carrying cut cane, 1940s–50s 80

Workers plowing with oxen, 1940s–50s 82

Work gang, 1940s–50s 117

Cane cutters and foremen on horseback, 1940s 118

Cane cutter with bundles of cane, 1940s–50s 120

Municipalities of Pernambuco's *zona da mata* 127

Burning cane to prepare the field for harvest 143

ACKNOWLEDGMENTS

This book has emerged as the product of years of intellectual thrashing about, increasing focus, concerted research, and ongoing refinement (roughly in that order). A childhood in Appalachia helped foster a fascination with society's relationship to its surroundings and perhaps led to an intuitive understanding of the paired exploitation of labor and the environment. So the two basic themes that stand at the core of the book have been abiding interests of mine. I would not necessarily have taken the time to examine and reflect on how my thinking had developed over the long term, or on how labor and environmental histories connect, had I not enjoyed the unparalleled mentorship of John D. French in the Ph.D. program in Duke University's Department of History. With him prodding me, I looked back with new eyes on how much I had learned with Shanti Singham and Craig Wilder at Williams College. And I saw that the invaluable period I spent with Tom Hatley at the Southern Appalachian Forest Coalition had pushed me to see landscapes historically. I thank Shanti and Craig for encouraging me to pursue history. I thank Tom for nearly steering me away from academia, and for sharing so much through his thinking, writing, and friendship.

John directs boundless energy and intensity toward his students, and I found in him someone to grow with and draw inspiration from. I am proud and grateful to count myself one of the fortunate beneficiaries of John's intellectual generosity, devotion to learning and teaching, and humane spirit. I grant him substantial credit for my development as a scholar, and I have deep gratitude for his painstaking engagement with this manuscript as I turned it into a book.

I hope other mentors who contributed time and precious energy to improving this work find in it some confirmation of their skills and influence. A revelatory seminar on oral history with Danny James was a fitting beginning to years of feeling repeatedly enlightened by his thinking. He has shared brilliant insight and (probably unwittingly) helped keep me focused on what is really important in this line of work. I am lucky that Wendy Wolford arrived at the University of North Carolina when she did, with her deep knowledge of the *zona da mata* and willingness to share ideas; I have depended heavily on these virtues. In addition to reading my work critically, Jocelyn Olcott offered an example of how to teach, approach diverse literatures, and engage the profession. Barry Gaspar's broad thinking and good humor helped along the way, and his work is a continual enticement back to the Caribbean. And for key advice at different moments thanks to Greg Grandin, Gunther Peck, Susan Thorne, and Rob Anderson.

I formed warm friendships at Duke and UNC that have provided pleasure and sustenance. Thank you Noeleen McIlvenna, Dan Golonka, Jan French, Gabi Lukacs, John Grennan, Mark Healey, Jody Pavilack, Alejandro Velasco, David Sartorius, Bianca Premo, and Kristin Wintersteen. Gonzalo Lamana's calm, incisive thinking serves as a model and a sounding board. Ivonne Wallace Fuentes will always be a dear intellectual sibling. And there are those friends who have had nothing to do with this book but who have shaped me profoundly nonetheless: Gist Croft, Todd Poret, Dan Polsby, Mike Keim, Ron Chowdhury, and John Ochsendorf.

A Fulbright Fellowship funded a year of research between 2002 and 2003, along with an Albert J. Beveridge Grant from the American Historical Association. For other research trips, I thank the Foreign Language and Area Studies program of the U.S. Department of Education, the Duke Center for Latin American and Caribbean Studies, the Lewis Hanke Prize from the Conference on Latin American History, and the University of North Carolina at Charlotte. Duke's Center for Latin American and Caribbean Studies also provided support of a less tangible kind, and I appreciate Natalie Hartman's hard work over the years. For support in writing, I thank the Duke Graduate School and again UNC Charlotte.

In Recife, Christine Rufino Dabat's assistance has been endless, always offered cheerfully and in the same spirit of generosity and seriousness that has made her such an important teacher and mentor to students at UFPE. I cannot thank her enough for the truly extraordinary efforts that she and Espedito made to help my research and to help Hannah and me (and now Dinah and Juno, too) enjoy Recife. Thank you also to Christine's *turma* of excellent and engaged students. Discussions of my work at the second Oficina "Sociedades Açucareiras" in May 2007 were extremely helpful, and I was pleased to attend the fourth in 2010 as well. I am grateful for the help of the Fulbright Commission in Brazil and for the welcoming atmosphere at UFPE, cultivated particularly by Marc Hoffnagel and Joaquim Correia de Andrade. Writing these acknowledgments now, back at UFPE as a Fulbright visiting professor, I have even more reason to thank Christine, the History Department, and the students.

Thanks also to Marcília Gama and the rest of the staff at the Arquivo Público Estadual Emerenciano Jordão and Clóvis Cavalcanti, Osvaldo Henrique Barros, and others at the Fundação Joaquim Nabuco. Augusto Cézar Ramos at the Tribunal Regional de Trabalho archive in Vitória de Santo Antão deserves special mention for his grasp of the archivist's value and responsibility and for his heroic efforts to uphold both. I also acknowledge the assistance of Gerson Quirino Bastos and Francisco de Oliveira at UFRPE, Múcio Wanderley and Roberto Moura at IPA, and João Farias at INCRA. Many thanks to all of those people in the *zona da mata* who generously lent me their time and, not infrequently, intimate details of their life histories: Arnaldo Liberato and the administration at Usina Catende, João Correia de Andrade at Engenho Jundiá, José Guilherme at Usina Cruangí, "Rei" in Jaqueira, Ronivaldo and Benedito in Palmares, Paulo and Severino in Vicência, and so many others. Ana Engrácia worked hard transcribing these interviews; I appreciate her help.

In the time spent writing this book, I incurred more debts as I tried out new ideas and rethought old ones. The Latin American Labor History Conference has been a crucially important place for this sort of work (for me and many others along the way). Thanks of course to John and Danny for starting and sustaining the conference and for allowing me to organize it in 2005. I learned a great deal from Steve Striffler, Myrna Santiago, Chris Boyer, Steve Marquardt, and Gunther Peck. Thanks to Barbara Weinstein and Joan Bak for key comments on my work. In other forums, thank you to Paulo Fontes, Alexandre Fortes, Todd Diacon, Sarah Sarzynksi, Gillian McGillivray, Emily Wakild, Peter Beattie, Erica Windler, David Montgomery, and Kevin Boyle. Lyman Johnson and Jerry Dávila read portions of the text and gave sound and useful advice. And as a close reader and helpful critic John Soluri has been very encouraging. Thank you also to the cogent reports of the University of North Carolina Press's anonymous reviewers, the efficient, intelligent efforts of Elaine Maisner, the careful eye of Alex Martin, and the support of other editors at the press.

I had the great good luck of coming to UNC Charlotte and finding wise, supportive, and savvy colleagues. Rob Smith opened doors of all sorts for me, and for this and his reliable sanity he has my lasting appreciation, respect, and affection. Jerry Dávila, Lyman Johnson, and Jürgen Buchenau have provided abundant advice, fellowship, solidarity, and warmth. I do not take for granted the company of three excellent Latin Americanist historians. Akin Ogundiran has been an exemplary chair, and it has been a pleasure to teach with and learn from my colleagues.

Finally, I thank my family members for their intelligence, good spirits, and grasp of how long a project like this takes. For being such crucial sources of support, encouragement, and most of all love, I thank Hugh and Ruth Blackwell Rogers and Bob and Elen Knott. Omma and Appa, thank you for being who you are and for shaping who I am. Bob and Elen, I am continually grateful to be a part of your family. Dinah MacNeill Rogers and Juno Augustine Rogers are younger than this book, but they have become sources of inspiration and revelation. It is wonderful to see them gamely making Brazil part of their lives. Without the love, encouragement, and unquestionable good sense of Hannah Knott Rogers, neither the research for this book nor its writing would have been possible. Hannah lived with this project at all phases—joining me for interviews, discussing the materials I gathered, and then gently prodding me to move through the chapters over the years of work. She can claim more responsibility for the book's completion than anyone else. For this and for so many other reasons, the book is for her.

ABBREVIATIONS

The following abbreviations are used throughout the book.

ANL	Aliança Nacional Libertadora
AP	Ação Popular
ARENA	Aliança da Renovação Nacional
ASPAN	Associação Pernambucana de Defesa de Natureza
CLT	Consolidação das Leis do Trabalho
CO	Coimbatore (sugarcane varieties from the Sugarcane Breeding Institute, including CO 331, or 3x)
CONTAG	Conferência Nacional dos Trabalhadores na Agricultura
CP	Canal Pointe (sugarcane varieties from Canal Pointe, FL, station)
CPRH	Agência Estadual de Meio Ambiente (Pernambuco) (formerly the Companhia Pernambucana de Controle da Poluição Ambiental e de Administração de Recursos Hídricos)
CTP	Companhia de Tecidos Paulista
DOPS	Departamento de Ordem Político e Social
ELC	Estatuto de Lavoura Canavieira
ETR	Estatuto do Trabalhador Rural
FETAPE	Federação dos Trabalhadores na Agricultura do Estado de Pernambuco
FUNRURAL	Fundo de Assistência ao Trabalhador Rural
GEA	Grupo de Estudo do Açúcar
GERAN	Grupo Especial para a Racionalização Açucareira do Nordeste
GTIA	Grupo de Trabalho para a Indústria Açucareira
IAA	Instituto do Açúcar e do Álcool
IANE	Instituto Agrícola do Nordeste
INCRA	Instituto Nacional de Colonização e Reforma Agrária
INPS	Instituto Nacional de Previdência Social
IPA	Instituto de Pesquisas Agronômicos

JAC	Juventude Agrária Católica
JCJ	Junta de Conciliação e Julgamento
JOC	Juventude Obrera Católica
JUC	Juventude Universitária Católica
MST	Movimento Sem-Terra
PLANALSUCAR	Programa Nacional de Melhoramento da Cana-de-Açucár
PMDB	Partido do Movimento Democrático Brasileiro
POJ	Proofstation Oost Java (sugar cane varieties from Java)
Proálcool	Programa Nacional de Álcool
PROCANOR	Programa Especial de Apoio às Populações Pobres das Zonas Canavieiras do Nordeste
PRORURAL	Programa de Apoio à Pequena Produção Familiar Rural Organizada
PROTERRA	Programa de Redistribuição de Terras e Estimulo à Agroindústria do Norte e do Nordeste
SAPP	Sociedade Agro-Pecuária de Pernambuco
SORPE	Serviço de Orientação Rural de Pernambuco
SUDENE	Superintendência de Desenvolvimento do Nordeste
3X	CO 331, a sugarcane variety from Sugarcane Breeding Institute, India
TRT	Tribunal Regional de Trabalho
TST	Tribunal Superior de Trabalho
UFPE	Universidade Federal de Pernambuco
UFRPE	Universidade Federal Rural de Pernambuco
ULTAB	União de Lavradores e Trabalhadores Agrícolas do Brasil

THE DEEPEST WOUNDS

INTRODUCTION

The Wounds of a People and a Landscape
Labor and Agro-Environmental History

"Monoculture, slavery, and latifundia, but principally monoculture; they opened here, in the life, the landscape, and the character of our people, the deepest wounds."[1] Early in an extended 1937 essay on the Brazilian Northeast, the Pernambucan writer Gilberto Freyre sketches this melodramatic summary of the Pernambuco sugar region's historical inheritance. Freyre indicts the brutal, race-based system of bondage and the concentration of land ownership in the hands of a powerful and avaricious few as destructive forces in his region's past. But he felt that the ills of monoculture—the extensive cultivation of a single crop—exceeded either of these. Sugar, not the people who profited from it, brought slavery and required production on a large scale, foreclosing on the emergence of a balanced, diverse agricultural society. He used similar imagery elsewhere, describing "the two running sores of monoculture and slavery, two wide-open mouths that clamored for money and for blacks."[2] For Freyre, sugarcane and African slavery tore into and lay open the land and left behind an injured society. They polluted rivers with sugar mill waste, destroyed vast forests, and fostered the brutal domination of slaves by masters.

Even as he decried the twinned domination of land and labor, Freyre helped build an elite discourse about the sugar region that naturalized these operations of power. Still, his observations offer revealing insight into the region's past and prompt important questions. What damage did monoculture do, and how did the exploitative agricultural and labor systems that he highlighted for the colonial period remain relevant in the twentieth century? Did the wounds he described remain open, continuing to harm the landscape and the people? Centuries of cane agriculture and repressive labor regimes did shape the landscape of the sugar-growing region of Pernambuco, frequently called the "forest zone" or *zona da mata*. This book examines these legacies, through a narrative of labor and agro-environmental history. This approach highlights new facets of the relations between workers and planters and casts new light on the role of the state in agricultural change. Incorporating the intellectual and cultural as well as material dimensions of the environment, it focuses on those most intimately involved in environmental change: workers. It is a labor history viewed through the lens of agro-environmental change rather than a labor history of a community, union, or productive unit.

Questions of labor and land have animated writing about Pernambuco for more than a century. In 1883 the abolitionist Joaquim Nabuco condemned slavery for setting up a "struggle of men against the land."[3] When Freyre picked up on this theme, lamenting the wounds inflicted by slavery and sugar, he was joined by other intellectuals of the 1930s who pondered similar associations. The nationally renowned novelist José Lins do Rego set a series of five novels in the *zona da mata*, even writing one of them from the perspective of a sugar worker who migrates to the city.[4] Still, the elite discourse that highlighted the importance of the environment also naturalized power relations between planters and workers. The persistence of these themes over the course of this period helps explain why the historiography of Pernambuco lacks extensive work on the transition from slavery to freedom, since the assumption of ongoing planter power attenuated the importance of abolition as a moment of transition.[5]

The transition from slave to free labor did not bring to Pernambuco changes quite as dramatic as those experienced in the economically dynamic coffee-growing regions of southern Brazil, but it did stimulate the emergence of a new set of labor relations revolving around tenancy on plantations. This book first provides a telescoping view of the *zona da mata* over the long term of European occupation, then describes the period from the 1870s to the 1930s, examining the elite discourse about the landscape as well as the discourse of those who worked in the sugar fields. The book then tightens the focus in the 1930s, moving forward to examine new perspectives brought by agronomists, social scientists, reformers, and bureaucrats. These groups' increasing interventions impinged on the basic social structure balanced between planters and workers, introducing new dynamics in this story of labor and agro-environmental change in the *zona da mata*.

This narrative, which captures the full complexity of the region's agroindustrial essence, allows me to add new context to the events that have brought the region notoriety. In November 1963 the descendants of the historical master class faced a vigorous challenge to their dominance from a cane labor force, largely descendants of African slaves, when the workers mounted the biggest strike in the history of Brazil's rural labor movement. During the strike, two hundred thousand rural workers walked off the job, bringing the state's sugar mills to a standstill.[6] The cane workers drew on momentum from eight years of peasant organization followed by rapid rural unionization. National and international attention turned to the struggles of a region where, according to a 1960 *New York Times* article, "physical survival is [the] only concern."[7] The upheaval helped spur broad national debate over agrarian reform, and the region earned the highest priority for aid from John F. Kennedy's newly founded Alliance for Progress.[8] The social mobilization had built on the emergence of trained agronomists, social scientists, activists, and progressive politicians, but the rural workers' actions also were guided by specific trajectories and challenges in the different areas of the cane zone. The history told in this book provides a richer understanding of the tumultuous period that culminated in the 1963 strike.

The movement's full potential went unrealized when the military toppled President João Goulart's democratic government in 1964 and quickly intervened in most of the rural labor unions. The years following the coup brought a sharp increase in interaction between state agencies and worker and planter groups, altering social relations and agricultural practices. The 1970s oil shocks spurred the creation of a massive state support structure for ethanol production, which ushered in a cane boom and accelerated the move away from traditional relationships of tenantry and toward strictly wage-based work. The production and labor management framework overseen by the state became the foundation for a new wave of organization and mobilization at the end of the 1970s when the sugar workers again strode to the forefront of the rural labor movement. In 1979 they organized another strike, their efforts paralleling the actions of striking industrial workers in

São Paulo following the lead of Luiz Inacio "Lula" da Silva, who would be elected president in 2002. Together, the two movements in 1979 accelerated redemocratization.[9]

Studies of these major rural labor movements in Pernambuco have tended to analyze their political dynamics while leaving largely unexamined their agricultural contexts. The influence of reformist and activist elements within the Catholic Church played their roles, as did Communists and technicians from the military regime. But the expansion of cane fields from the 1930s onward, the introduction of a new cane variety in the 1950s, new payment schemes for cane cutting, and other key changes influenced workers' day-to-day lives. This book explores these crucial but overlooked aspects of the period, bearing out Steve Marquardt's argument that "the entanglement of nature and labor can make a difference at the macro-political level of labor movements and even states."[10] Closely examining the broad history of sugar work in Pernambuco, this book reinterprets the episodes of mobilization that reverberated around the country while offering broader lessons about how a history of labor overlaps with the agricultural environment's history. Field workers' stories help illuminate this history. For example, the experiences of Paulo and Severino, whom I introduce in chapter 3, illustrate many aspects of change in the twentieth century.

Despite the clear links between the environment and rural labor, and despite the fact that social relations with the environment have been on the intellectual agenda in Pernambuco for years, scholars have yet to construct a satisfying framework for analyzing the two in tandem. Valuable work on the region has provided suggestive starting points. Early in the 1970s, a group of Brazilian anthropologists began a research project in the *zona da mata* that would result in a set of excellent monographs examining field and mill workers' worldviews, challenges, and relationships to power. This literature offered important insights into labor and hierarchy and sensitively analyzed worker culture and mobilization.[11] Despite the extraordinary richness of this work, which includes evidence directly relevant to landscape change, these scholars never consciously focused on the environment.

The *zona da mata* regained the attention of a wide audience in the early 1990s with the publication of the American anthropologist Nancy Scheper-Hughes's sprawling ethnography *Death without Weeping*. Without the benefit of insights from her Brazilian colleagues, Scheper-Hughes advanced the startling argument that women in the small towns of the cane region faced such desperate poverty that they played more than a passive role in the deaths of some of their children.[12] Though she invokes images of the environment, Scheper-Hughes does not engage substantively with the labor and agro-environmental realities of the *zona da mata*. Drawing on interviews and questionnaires, Christine Rufino Dabat's recent study of sugar workers in Pernambuco provides a detailed portrait of cane workers' experiences over the course of the twentieth century and analyzes their place in debates within leftist politics and the academy about their social roles in the region's history.[13] Dabat dismantles a traditional, elite-centered view of the region's history and reveals that workers exerted a measure of control over their own lives. Concurring with her judgment, I also ground social relations and expert debates in their agro-environmental reality. The pressing materiality of the agro-environment, and of people's ideas about and relations to their landscape, signals that the lens of class relations and social structures does not see everything in history.

Seeing and Thinking Landscapes

Pernambuco's political police visited the Pindobal sugar plantation northwest of Pernambuco's capital, Recife, in June 1960, responding to reports of agitation among tenant workers. Some workers had apparently joined the five-year-old Peasant League movement and wanted their colleagues to follow suit. Investigators reported that the tenants affiliated with the league had prohibited nonaffiliated tenants from planting any crops in the garden areas to which they had access through their tenantry agreements. The leaders declared that the holdouts could resume cultivating only when they joined the league.[14] The use of this tactic reflects the traditional role of land in workers' living and work relationships. Planters' near monopoly on land ownership meant that workers' only "legitimate" access to land lay through tenantry of some sort, which required the regular provision of labor, and planters placed tight restrictions on tenants' use of garden plots. By exerting control over land, the Pindobal workers demonstrated their internalization of prevailing assumptions about power and its relationship to land.[15]

The Pindobal workers' alleged tactics would have been familiar to any sugarcane planter. Indeed, a month before the investigation the owner of Pindobal had spoken to police about the incipient conflict and had explained some of his rules for tenant-employees. "Planting fruit trees of any kind is never permitted, nor root crops such as cane, coffee, etc.," the investigators summarized in their report.[16] During a period of upheaval, these workers attempted to exercise some power themselves. But even as the workers' and planters' views of land influenced each other, important differences separated them. Both planters and workers recognized control over land as a central mode of power in the landscape. But planters saw their own control as essential to their identity; they commanded workers and dictated the use (or manipulation) of the environment. Workers saw themselves as captives of the planters' controlling power, expressed in the control over land, and sought freedom in various ways, including through the cultivation of trusting relationships with employer-patrons. The planter discourse was dominant, which makes sense considering their greater power, but even as the workers' language of captivity reflected this discourse it also opened possibilities to contest the elite view. In addition, the interventionist state of the military dictatorship introduced a new landscape discourse into the region also marked by the assumption of control but reliant on the transforming power of science and technocracy. I call on these schematic outlines of the planter discourse of landscape, the workers' iteration, and the state discourse as explanatory variables at various moments during the narrative of the region's history and describe how they changed. The processes that drove the history of the region were indissolubly bound to discourses and metaphors of the landscape.

The concept of landscape is marked by a fundamental duality: it is at once an idea—a panorama associated with particular meanings—and an environment with a "consuming materiality."[17] It is what we see and the way we see it. This rich "enviro-cultural" quality makes the concept especially valuable for the study of agricultural environments. The landscape of the *zona da mata* has been and continues to be both a place and a diverse set of discourses about that place. Landscape carries the weight of centuries of use in disciplines ranging from art and art history to geography, anthropology, and history. With Renaissance-era European roots, the concept began as a genre of painting that portrayed landforms and the signs of local customs and later referred to aesthetically engineered environments before eventually becoming a unit of geography as a discipline. The use of the term *landscape* to describe the scenery constructed around the country houses of the growing European bourgeoisie associated the word with manipulation and the encoded implication of power (one can think about the term "a commanding view" in this respect).[18]

Landscape played a central role in geography at the beginning of the twentieth century and then declined before resurging in the 1980s.[19] Despite proliferating definitions, scholars consistently treat landscape as a concept that links cultural sensibilities to the material environment.[20] People associate sentiments, experiences, dispositions, traditions, and politics with their surroundings and thus contribute to building the collective landscape. In everyday life the physical apprehension of a place conjoins with the ideas and feelings about that place (themselves linked to the physical experience) and so emerges a landscape—something seen but also felt.[21] When workers describe their fear of a legendary green serpent that grows with maturing cane to protect the crop, they are also communicating their intimidation by planters, who own and lord over the cane. As enviro-cultural forms, landscapes can have social impacts and can change. Because they are visible and persistent, they are readily perpetuated over the course of time, but of course they can also evolve in response to other social dynamics.[22]

The concept of landscape has its skeptics. Some critique historians' use of the idea, charging that they fail to clearly connect it to historical causality.[23] Does a landscape make change, and if not how does it matter to history? Others feel that landscape is only a way of seeing common among tourists or outsiders and that it fails to capture "natives'" apprehension of their surroundings.[24] They suggest that landscape implies a privileged view, a view from a position of power, and therefore only some people "see" landscape. Denis Cosgrove suggests that "outsiders," those in positions of power, may lack the intimate knowledge born of labor but they have the privilege of detachment to see things from a distance, what he calls a landscape view. He argues that other people—the people actually working in constructed landscapes or depicted in landscape painting—do not see landscapes. It is more appropriate to think of them as having, or constructing, recognizing, and associating meanings with "place."[25]

Agricultural workers may come from largely oral cultures, but that does not render them incapable of detachment or abstract thought. As Allan Pred writes, "Local and regional agricultural transformation is generally one with the reconstitution of everyday life," a process that "involve[s] the concrete lives of actual women and men." This thoroughgoing upheaval includes "the emergence of new forms of practice-based individual and collective consciousness."[26] Scholars of countries with largely illiterate rural populations must confront the significant problem of recreating workers' landscape discourses, since they do not produce the abundant letters, memoirs, novels, and other sources that communicate these ideas. They more frequently articulate their views through collective action rather than through individual, reflexive intellectual production. By assiduously pursuing evidence in folklore, oral histories, and the memoirs that have been produced, we can still achieve some understanding of the central metaphors workers have used to describe their environment and how they have mobilized within it. For Pernambuco's cane workers the crucial metaphor was captivity, and it hinged on their long experience of exploitation on plantations.

Raymond Williams suggests that visitors to English country estates look at the land and its productivity and "think it through as labor" to understand the scale of change in the landscape.[27] This book adopts a similar view, attempting to think through Pernambuco's sugarcane landscape as labor—as a system of relationships changing over time that have involved the exploitation of labor and the transformation of the land. I have asked how deeply held views of the agricultural environment interacted with significant agricultural change during the twentieth century to influence the labor history of the region. Labor and landscape have close ties in agricultural settings. Society's incremental construction of landscapes through material processes "make[s] no sense without attention to practice," as the anthropologist Hugh Raffles observes. Labor lies at the heart of landscape construction, even if it is frequently obscured.[28]

The Brazilian Northeast's sugar region was a landscape of labor and a labored landscape. In the minds of the planter class, it was a *laboring landscape*: the landscape itself labored for planters, as they saw no distinction between land and labor.[29] Land and people were bound together in their minds, and not just by the ties of culture, law, and belief to which "modern" societies are accustomed. They were tied by coercion, the sense of entitlement and control passed down across generations by those in power, and a history of slavery and plantation agriculture. Workers reacted to these ideas and to their own realities living amid the cane fields and making a living in relationships negotiated with planters. As the case of the Pindobal workers makes clear, workers internalized the idea that power lay in the control of land. Their consistent use of the metaphor of captivity to describe their work and living conditions directly reflected the planters' discourse of a laboring landscape, and not infrequently contested that discourse. In this book I seek to connect the meanings, feelings, and perceptions about landscape to actions and interactions on the ground in the context of a major agro-industry.

Situating *The Deepest Wounds*

What lies at the center of this book's narrative, workers or their environment? I argue that the answer can be both—that by examining processes of agricultural change, the stories of workers and the environment can be told together. Indeed, it is one story. In human relations to the environment generally, "our own bodies and our labor . . . blur the boundaries between the artificial and the natural," and the history of an agro-ecosystem such as a sugarcane region is "made from and inscribed on the human body."[30] The *zona da mata*'s history is intimately linked to the people who effected its transformation—to generations of hard labor, of people hoeing and plowing and planting and cutting a plant species that hails from the South Pacific. Since these worker's lives were spent in the fields, as in any intensive agricultural region, their story must be told in conjunction with agricultural developments such as crop variety shifts, field expansion, the use of fire, or changing harvest schemes.

Pernambuco's history has been marked by repeated bursts of linked agricultural and social changes, each driven by the crop—sugarcane—that has held the region in thrall for four centuries. While the consecutive centuries of expanding cane fields nurtured some continuity, real agro-spatial changes also took place. For example, cultivation practices, the treatment of soil, irrigation techniques, and labor regimes all changed, and sometimes changed again through history. The choice of a hardier cane variety could mean expanding cultivation into new, less-fertile areas. The expansion of planting, in turn, could extend labor into hillier country than had previously been cultivated. These changes impacted the lives of the workers who populated the cane fields in the *zona da mata*. Working on a steep hill, battling undergrowth and sliding soil, workers have a far different experience than they do moving down lines of cane on a flat valley floor. Few sugar planters or workers saw a clear distinction between agricultural change and social change; they assumed that changing cultivation techniques or innovations implied changes in work responsibilities and demands.

In agricultural settings, shifts in crops (or varieties), technology, or scale catalyze social transformations. In "agricultural capitalism generally," Marquardt observes, "environmental change and change in labor process are inextricably entangled with one another at the point of production."[31] A clear example comes from the massive expansion of straight-line grain crops in the North American plains, which decimated prairie grass at the end of the nineteenth and beginning of the twentieth centuries, producing an explosion in wheat and corn production and then, in the 1930s, the environmental disaster of the Dust Bowl. Even when agricultural change does not have such massive and transparent effects, we should be attuned to the social changes linked to agriculture.

Using the concept of landscape, the first part of this book explores how people thought and wrote about their surroundings and what dispositions toward the natural world they developed over the passage of time. My goal here is to illuminate *why* they may have acted in the ways that they did.[32] Without placing environmental conditions in a deterministic role, I describe the interactions planters and workers had with their surroundings while also analyzing the deep-seated habits, assumptions, and practices that informed those interactions. This approach accounts for what Fernand Braudel called "the necessary reduction of any social reality to the place in which it occurs." Such a perspective demands that the historian recognize long- and short-term processes; continuity and rupture. As Braudel pointed out, we break history into varying lengths and periods, even if time itself is not our own creation. By accommodating these different periodizations, from the *longue durée* to the "history of events," we can resolve stable phenomena such as a class's culturally influenced view of the landscape while also grasping the immediate import of a major event such as a strike.[33] Reinaldo Funes Monzote has provided a masterful example of this method in his recent study of how forests gave way to sugarcane in Cuba. Funes Monzote explicitly acknowledges the role of attitudes and ideologies in structuring various actors' treatment and exploitation of the environment, and he tracks debates over resource use and agricultural development in Cuba across centuries.[34]

The long-term primacy of sugarcane in the *zona da mata* paralleled the centuries-long domination of the cane industry by a small planter class. This stable context helped produce durable regional continuities in power relations and land use. The trend toward local control distinguishes this story from many other histories of tropical commodities in twentieth-century Latin America—sugar in the Caribbean, for instance, or bananas in Central America or northern South America. The companies producing sugar in Pernambuco and employing workers were overwhelmingly locally owned. The Dutch took over the region for a few decades in the seventeenth century, and British and French companies tried unsuccessfully to produce sugar in large mills in the 1870s and 1880s, but aside from these episodes production has generally been in local hands. This is not a story of "penetration" by foreign capital, or of enclave communities.

Though foreign capital did not predominate, Pernambuco's history still has much in common with the histories of other regions that produced tropical goods for export. Monoculture itself, for instance, persisted from the early period of extractive colonies and became an assumed feature of modern agriculture. (The term *monoculture*, however, does frequently obscure a more complex reality in which some combination of fallowing, forest fragments, or workers' subsistence gardening introduces a level of diversity into the agricultural environment.) Tropical commodity regions also found themselves tightly linked to demands that stretched across the globe, and the dynamics of these networks resonated back through to the production zones.[35] I appreciate the importance of these forces and include their relevance in analyses of such moments as the boom in alcohol production associated with the 1970s oil shocks. However, in this study I look the other way in a sense, deeper into the agricultural region itself. Taking this path, I follow scholarly work analyzing the productive worlds of commodity-producing regions. Historically minded anthropologists have made some of the most valuable advances here, paying close attention to the ways production itself is organized and how social relationships grow around those processes.[36]

The second part of this book follows some of the commodity scholarship that has delved into the scientific research that supported particular crops and facilitated their spread. Agricultural science played a crucial transformative role, especially in the twentieth century. At the beginning of the century, major multinational commodity companies spread their influence and crop of choice (the most heavily studied being United Fruit and bananas), and the middle to latter part of the century brought the productive leaps of the Green Revolution. The role of agricultural science has been described in lucid detail by scholars such as Stuart McCook writing about sugar and Steve Marquardt, John Soluri, and Lawrence Grossman writing about bananas. These scientific innovations brought dramatic change, memorably inspiring Edmundo Navarro de Andrade to describe coffee spreading across São Paulo as a "green wave."[37]

Major agro-industries have had a history of close oversight, at times indulgent, by the state. In the case of Brazil, sugar and coffee (along with other products) came under close regulation in the 1930s. In the 1960s and 1970s, the focus of the third part of this book, the state became a dominant force both guiding production and intervening in labor relations. Recent scholarship in tropical commodities and in Latin American history more generally has shown a keen interest in the state, and important insights have emerged. Steve Striffler, for instance, elaborates for Ecuador's south coast a complex picture of the intertwining and frequently opposed forces of capital, state power, and worker or peasant agency. His explanation of the exchanges between peasants and the state, resulting in the transformation of peasants' experience as subjects, parallels my interpretation of the aggressive, technocratic Brazilian military regime's transformation of workers during the 1960s and 1970s, pushing them into roles amenable to state regulation and oversight.[38] In an illustration of the dialectical relationship between state power and individual or group agency, the determined resistance of the cane workers at the end of the 1970s helped topple the dictatorship.

In pursuing the connection between agricultural change and workers' lives, I also draw inspiration from labor and environmental history and the small body of work emerging at the intersection between the two. Brazilian labor history has grown in recent decades, and the field's scope has expanded to address complex new themes.[39] A new generation of scholars, many reflecting the influence of E. P. Thompson, has placed a clear-eyed analysis of working class culture on its agenda alongside careful city-, sector-, or factory-based studies. The field's ambit in Brazil has also expanded to include "unfree labor," or slaves.[40] John D. French's analysis of discursive frameworks in the evolution of labor law in Brazil and how they shaped labor politics enriched my analysis of the important labor court records from the sugar zone.[41] In this book I use two types of sources that have featured in some of this recent work: oral histories and police and court records. The police files come from the archive of the Departamento de Ordem Político e Social (DOPS, the political and social police), which were opened in many Brazilian states only in the late 1990s.[42]

Brazilian environmental history, and the field in Latin America more generally, is at a comparatively early stage of development. Excellent models appeared in the early 1990s, including Warren Dean's masterful "biography" of the Atlantic Forest. With José Augusto Pádua's important work, we have new insight into environmental and conservationist thought during the late colonial period and the nineteenth century. Funes Monzote's long-term recent study of Cuban forests and sugarcane is a welcome environmental perspective into sugar studies. And a new cohort of young environmental history scholars has been energized recently by a series of four international conferences.[43] Much of the work contributing to Latin American environmental history has focused on commodity-specific studies, forest issues, urban environments, and the histories of conservation and environmentalism. In this study, I focus on omissions that remain in the field, including perceptions of the environment and interactions between populations and their states.[44] Using a unique narrative mixing labor and agro-environmental histories, I attempt to shed new light on major political events in the region.

Few English-language scholars have explicitly balanced the explanatory weight of environmental circumstances and human interaction. Walter Rodney's classic history of workers in late-nineteenth-century Guyana includes extensive discussion of the material realities of sugarcane agriculture—

mucking drainage canals, digging cane holes, pulling cane on barges, and more—showing that a history of workers is most effective when it is also a history of the environment they transform.[45] This book follows Rodney's lead, and contributes to a conversation about how to grant analytical weight to environmental factors without allowing them to balloon to deterministic proportions.[46] The U.S. historian Gunther Peck suggests that environmental history tends to elide class and power differences in an effort to grant the environment its own historical agency. But the broader and more synthetic view afforded by this "holism" could grant a novel perspective on labor history. This book seeks to realize the promise Peck envisions by charting the contrasting discourses of landscape between planters and workers.[47]

Agricultural change and its effects played out on varying scales, as did the operations of power in the *zona da mata*. When planters moved to new payment systems—remunerating cane cutting according to large measured areas of field rather than a day's work, for instance—they affected tens of thousands of workers. But the small scale mattered, too. In his discussion with the police in 1960, the owner of the Pindobal plantation noted his concern about Peasant League organizing, but he appeared to indicate that relations with his workers had truly soured when he inadvertently killed a family's chicken. After falling from his horse on a ride around his property, he had thrown a piece of wood in frustration and struck a chicken in the head, killing it. Though he had reimbursed the family, he felt that resentment persisted. That he felt this vignette important enough to detail for the police speaks to the complex relations of patronage and power on the plantations. On a very basic level, the family's bare subsistence needs had been threatened. The boss had also taken his power a step too far by killing the chicken, and his workers used the political forces unfolding around them to seek revenge.[48] The boss understood these dynamics, and he knew that the quotidian affairs of his workers interacted with and informed larger political currents.

Organization of the Book

The Deepest Wounds centers on the sugar-growing region of Pernambuco during the twentieth century. Both archival and ethnographic sources weave into the study, from colonial travelers' accounts and contemporary scientific articles to interviews with workers. The most heavily used archival sources include government and industry documents relating to agricultural research and management of the sugar sector, labor court records, and the DOPS files, the last of which were first opened in Pernambuco in 1999. Between 2002 and 2003 I also conducted forty-nine interviews with sugar workers and foremen, mill managers and owners, union officials, agronomists, and environmentalists. Twenty-seven of the interviews were recorded, amounting to nearly thirty hours of tape, while the rest of the interviews were documented with handwritten notes.

The first section of the book, "The Landscape of the *Zona da Mata* to the 1930s," contains three chapters focusing on the long term of sugar cultivation in the region and on the emergence of new production methods and visions of landscape in the period between 1870 and 1930. The first chapter takes a sweeping look at the first three and a half centuries of European occupation. Chroniclers' and travelers' accounts, geographic studies, and environmental histories help me narrate the environmental changes of the colonial and imperial periods. I suggest that through the middle of the nineteenth century the environmental impact of the sugar industry was relatively modest. The arrival of the railroads and the beginning of sugar mill consolidation intensified deforestation and river pollution—a transition toward greater human impact. As the abolition of slavery neared and coffee planters from the Brazilian south consolidated their economic primacy in the country, northern sugar planters began to see themselves as a coherent group for the first time. This growing self-consciousness was accompanied by the development of a discourse about land, labor, and power, a discourse I examine through the proceedings of an agricultural congress held in Pernambuco's capital in 1878.

The second chapter explores an extraordinarily rich vein of intellectual reflection on the landscape across several generations of the Pernambucan elite, an important cultural body of work that was a byproduct of longer planter traditions. Around the same time as the agricultural congress, the statesman and abolitionist Joaquim Nabuco explicitly linked the evils of slavery and environmental destruction.[49] Gilberto Freyre gave Nabuco's ideas new life when he reworked them in his own highly influential essays and monographs. The period from Nabuco to Freyre brought crisis to the sugar industry and a shift to a new consolidated production model. Out of this conjuncture and the anxiety it produced crystallized a discourse about the sugar region's landscape. Though these authors critiqued slavery, they also inherited and perpetuated a planter *habitus* that included assumptions about their class prerogative to command and the racialization of social relations.[50] Making a contribution to intellectual history, the chapter teases out the thematic of landscape while demonstrating how the concept of a *laboring landscape* captures a key aspect of the elite *habitus*. As articulated in Freyre, Lins do Rego, and planter memoirs, this discourse invested the landscape with personal power and embedded workers organically in the environment. While many scholars have noted the nostalgia that marks this generation's work, they have not identified the peculiarities of the landscape discourse that flourished in the 1930s, which could be seen as the most creative product of the cane industry's crisis from the 1880s to the 1930s.[51] This discourse reinforced and expressed the nostalgic themes of Freyre's generation.

In the third chapter I situate workers' lives in the context of their tenantry on *engenhos* (the term for sugar plantations) in the decades following the abolition of slavery in 1888. While discussing the cyclical agricultural rhythm of the sugar zone, I address workers' sense of subjugation to the landscape, what I call a landscape of captivity. The metaphor of captivity appears throughout the sparse biographical accounts, oral histories, folklore, and visitors' stories—the heterogeneous sources, thin by comparison with elite production, that allow us to capture the landscape discourses workers have created. Though the agro-environment they worked in lay outside their control, it provided a rich setting for their lives, with intimate ties between their homes and the fields. I also describe how workers' prospects for being heard increased as new actors began to reach out to the *zona da mata* between the 1940s and 1960s, from social scientists and reformers to politicians.

The next two chapters comprise the second section of the book, "Opening up the *Zona da Mata*, 1930–1963," which offers a labor history informed by the material realities of the changing agro-environment within which workers toiled. The agricultural modernization efforts of this period, along with new political formations, continued to open up the formerly closed spaces of power in the *zona da mata*. These changes fueled the well-known burst of worker mobilization that started in the late 1950s. Important elements of the landscape discourses discussed in the first section manifested themselves in the language of these struggles.

The fourth chapter brings workers and planters together into the same narrative, describing the growing importance of agricultural technicians and the planters' ambivalent relationship with modernization as well as workers' responses to changed conditions. The chapter provides a history of successive turnovers in sugarcane varieties, including an important case from the 1950s when a fast-growing but low-sucrose variety known as 3x was rapidly introduced in Pernambuco. Its swift spread took place against the advice of agronomists, reflecting the uneven relationship of the planter class with the modernization of agricultural science. I then examine the impact of this modernization on workers, connecting agro-environmental change to labor and politics. Besides losing access to land, workers faced new modes of measuring labor and allocating compensation, which drove an increased pace of work. Rapid organization among the workers, first into Peasant Leagues and then into unions, has generally been seen as a function of the political work of the Communist Party and the Catholic Church. While these institutions obviously played important roles, I place workers' decisions in the context of their work experiences. The geographical pattern of labor mobilization bears out the importance of specificities in work experience around different parts of the region.

The Northeast burst onto the international scene in the early 1960s with the election of a progressive governor in Pernambuco—a former cane technician—and a strike by two hundred thousand cane workers. In the fifth chapter I approach the unprecedented 1963 strike with the aim of illuminating agro-environmental connections to labor rather than providing a full narrative of the now well-studied mobilization. I provide the political background of the region, grounding it in agro-environmental experience, and turn to the issue of fire. Treated as a complex, multifaceted phenomenon, fire offers glimpses into the materiality of life and social relations at the same that it helps us understand the period's turbulent politics. In the years following the rural tumult, fire became adopted as a standard tool for preparing the cane harvest; a new agricultural labor practice emerged out of the politically charged period of transformation. Together, chapters 4 and 5 show the mutual influences of agricultural, social, and political processes.

Led by progressive politicians in the early 1960s and followed decisively by the military dictatorship, the state became an important force in the cane zone, one to which workers and planters were both subservient. The last section, "The Dictatorship Commands the *Zona da Mata*, 1964–1979," addresses this powerful actor in the political and agricultural development of the region, covering the years between 1964 and 1979. The sixth chapter details the period following the 1964 coup, when the military seized command of the *zona da mata* and reconfigured the relationships between planters and workers. The state's intervention privileged technocracy and fostered a new, quantitative, regulatory perspective toward the agro-environment and toward labor. This new perspective, combined with the restricted political space the regime allowed, pushed negotiations between workers and planters into the labor courts, where core questions about the new shape of work in the cane fields were fought out. This process, in turn, "deprivatized" the realm of the *engenhos*, removing them from submission to the private power of planters and exposing them to state view. Tenantry relationships were further changed by state enforcement of labor obligations, displacing more workers from the plantations. Workers increasingly saw themselves as wage earners, and efforts to secure access to land diminished.

In the seventh chapter I track an agricultural boom wrought by the large-scale efforts of a modernizing dictatorship that expanded the sugar and cane alcohol industry. The ambitious regime did not control as much as it thought, however, as the boom exacerbated the agro-industry's environmental impact, including river and stream degradation, producing destructive floods in the 1970s. Workers adapted to the institutional boundaries of their new roles under the interventionist state. After honing their skills in the legal arenas of the labor courts, union leaders seized the opportunity when the military regime began to loosen in 1979. A new strike, organized with discipline and precision, brought the movement for redemocratization into rural areas and increased pressure on the government. The political geography of the strike (distinct from that of the 1963 strike) suggests connections between agricultural impacts on labor and political decisions by workers. The strict, legalistic approach taken by the unions in 1979 and routinized in yearly strikes afterward responded to the political climate but also resulted from the new roles workers had assumed as a result of the increased regulation and oversight of the state.

In the conclusion I reflect on the book's lessons and briefly examine changes in cane agriculture from the 1980s onward. The end of the military regime and the rise of the neoliberal era relaxed government control of the agro-industry. But rising interest in clean energy such as cane ethanol insured that the *zona da mata* would remain bound to cane. I also discuss the new politics of land reform and ongoing challenges for workers.

HOW DO WE make sense of a history characterized as the infliction of wounds? In Freyre's metaphor, society and landscape suffer injuries unleashed by a sugarcane monoculture. Implicit in his construction is a dialectic between society and environment. The wounds opened over the course of time in a society that imported and expanded a single crop and established a system of bonded labor to cultivate it. Freyre's idea persisted in Pernambucan thought. Francisco Julião, a leader of the 1950s Peasant Leagues and himself from a cane-growing family, described a "tragic symbiosis between the poor and a dying soil" in the *zona da mata*.[52] The ties between environmental change and social processes have hovered over debates on and in the region for generations. This book makes them the center of analysis, retrieving them from the realm of metaphor and taking them seriously as a driving force in the sugar region's past.

History lies embedded in the cane fields and forested hills of the *zona da mata*. Those fields have expanded and retracted, creeping up and down the slopes in obeisance to national and international sugar prices, the command of planters, and the sweat of workers. Forests have fallen, new growth has risen to be cut again, rivers have slowed as they filled with silt, canals have been cut, hills have eroded. And people have wielded hoes, machetes, axes, torches, and whips; people have bent to their work, ridden plodding donkeys, bathed in streams, and arrived at the end of their lives. Men have worked jobs once reserved for women; women have made men's work their own; stories have been told about the fields, the rivers, and the forests; legends have been built. This work and these lives, tied to the cultivation of sugarcane, shaped the landscape and drove history. Though Freyre saw wounds in the character of the people, the workers who toiled away in the sugar fields managed to mobilize in common cause at two critical points in the twentieth century. This book tells their story, acknowledging how closely that story intertwines with the history of the fields themselves.

PART I

The Landscape of the *Zona da Mata* to the 1930s

ONE

An Eternal Verdure

The *Longue Durée* of the *Zona da Mata*

Gazing around the forest on an excursion to a sugar plantation in the 1810s, the French cotton buyer L. F. Tollenare described the surrounding forest as "an eternal verdure, an active vegetation that knows no rest, fruits and flowers one on top of the other without end, adorning the hills to their peaks." The trees appeared to him so massive, and the undergrowth so thick, that he found Pernambuco's "sublime and virgin nature" to be almost mystical.[1] By the time of Tollenare's visit, the area of Pernambuco's coastal lowlands he saw had been a site of sugarcane cultivation for more than 250 years. His impression speaks to the richness of that region and to the limits of sugar's reach despite its long tenure. Already subjected to long-term human impact, the landscape Tollenare saw would continue to be transformed over the next two centuries. In order to grasp the dynamics of social and environmental change in the twentieth century, it is important to look first from this long-term perspective.

This chapter describes a space, a terrestrial zone in the tropical latitudes lying along the eastern coast of South America. From the earliest arrival of humans migrating down from the land bridge connecting the continent with North America to the spread of European colonists and African slaves to the impact of comparatively dense Brazilian population today, this area was molded into a landscape. As people shaped the landscape physically, they also added contours of meaning and understanding. The physical processes take precedence in this chapter, which describes colonization, the beginning of sugar production and its rapid growth, the later spread of cotton farming, and finally the expansion and renewed emphasis on sugarcane at the end of the nineteenth century.

In the first section of the chapter, I examine the natural history of the coastal strip of Pernambuco that came to be called the *zona da mata*, or forest zone. In the second, I show the changes the *zona da mata* underwent in the first three centuries of European colonization. In the third, I address a moment of transition when sugar retreated and cotton briefly took root. This episode was followed by a resurgence of cane fueled by new mills and the railroad, and the addition of these technologies tipped the balance between "nature" and a society of humans bent on agricultural exploitation decisively toward the latter.[2]

From Forest to "Forest Zone" (*Zona da Mata*)

A wooded promontory greeted the first visitors to the stretch of northeastern Brazil that would become Pernambuco. After crossing the equator, European sailors expected to cover the eight degrees of latitude to Pernambuco in a couple of days, at which point they began looking for the cape "known throughout the world as Santo Agostinho" that would guide them toward the port of Recife, twenty kilometers back north along the coast. A reliable point of reference because of its elevation, unmatched by any hills along the nearby coast, Cabo de Santo Agostinho may well have been seen by a Spanish sailor several months before Pedro Cabral's "discovery" of Brazil in 1500. With the cape guiding boats in toward the natural harbors of the Pernambucan coast, those on board could begin to make out the rest of the coastline, also wooded for the most part and interspersed with low swampy areas.[3]

Forests meet in Pernambuco. The forest covering Cabo de Santo Agostinho formed part of the northern extreme of the great Atlantic Forest (Mata Atlântica), which at the moment of European contact covered some 1 million square kilometers of the east coast of South America stretching from Cabo to the twenty-eighth southern parallel. The Atlantic Forest reached about one hundred kilometers inland from Cabo and was a mostly broad-leaved, rain-loving, semideciduous forest. North of Cabo, the forest took on a somewhat different character, with deciduous trees becoming more abundant and clustering more closely together than the larger, faster-growing species to the south. At several points deep in its history the Atlantic Forest reached far enough to connect to its massive northern sibling—the Amazonian forest— as the latter stretched southward from its shallow drainage basin. The forests north of Cabo are equatorial formations, remnants left by the Amazonian forest as it receded thousands of years ago. From Cabo south, the eastern or Atlantic formations dominate—a wet and exuberant forest, dominated by large trees with dark green foliage and filled with vines, thick undergrowth, and epiphytes.[4] This broad ecological distinction between the wet south area and the drier north would have social consequences in the twentieth century.

The forests filled early European visitors with wonder. The trees appeared so tall that one hyperbolic text from the early 1600s claimed that even an arrow launched from a sturdy bow by a strong arm would fail to reach their crests, and so broad that they must date from the time of the biblical flood.[5] The colors dazzled—the brazilwood's bright red flowers; the hulking blue and green macaw; and the red, white, and black *igbigboboca*, a snake longer than a man is tall.[6] Comparing Pernambuco to alpine European landscapes, Tollenare observed that "if the nature here is calmer and more silent, its decoration is also more brilliant; the opulence of the vegetable luxury compensates for the lack of topographical extremes." The steamy forest edges hosted countless birds, many with gaudy plumage and of impressive size. In the comparatively cooler, humid interior of the forest, birdsong dwindled and nocturnal species were more common. Reaching the forest interior, visitors reported feeling wonder and dread equally. In many places, crisscrossing vines and creepers created networks so dense they impeded human passage.[7]

The Atlantic Forest is typically rich for a tropical forest, and the profusion of species confounded Europeans seeking to make sense of the whole. Indigenous names unwieldy to the ear had to suffice for European chroniclers' bulging catalogs of trees, bushes, herbs, birds, snakes, mammals, fish, and sea monsters. The range of trees alone was impressive: *araticú, riacho das serras, mamajuba, acicapugá, visgueiro, sopocerana, iribica, camassar, páu d'arco, genipapeiro, pau ferro, tatajuba*. Tollenare recorded each of these (and many more) because of their specific utilities—particular strength being useful for mill axles, resistance to water being convenient for sugar boxes, or an organic tint being useful for yellow dye.[8] Europeans found *ombú*, a fruit that tasted like figs and "makes one lose one's teeth"; *acajú* (cashew), a nut so tasty and rich it was "better than any in Portugal"; *gibóia*, a snake that could eat a deer whole and travel through the treetops "as if it were swimming in water," faster than a man could run. The Jesuit Fernão Cardim filled many pages with these sorts of observations, a valuable record of the environment he explored in the late sixteenth century.[9]

The Portuguese encountered peoples from the broad Tupi-Guarani language group, which had spread outward toward the coast from the continental interior around the end of the first millennium.[10] These societies practiced a shifting slash-and-burn agriculture that centered around the forest edges and along the watercourses leading inland.[11] The Portuguese soon called the inhabitants of this wooded land the Caeté, which means true or virgin forest.[12] With this eponymic connection between the people and the trees, the identification of the forest with the region was also complete: the Portuguese called it the *zona da mata*, or forest zone. This name remained apt for hundreds of years, as the lush forest continued to be the region's dominant feature. Only much later, and certainly from our current vantage, does the name seem a glaring anachronism, applied to an agricultural expanse that has been almost completely deforested.

The abundant and diverse trees took their cues from the soil and water available. From Cabo southward regular hills break up the landscape, divided from one another by shallow rivers that empty into the Atlantic at regular intervals. This is the geological formation known as the "sea of hills," characterized by evenly spaced, uniform hills known appropriately as "half-oranges."[13] Largely sedimentary in character, the hills were formed by water finding courses through the sediment during the slow drainage of the area in the Quaternary period. Formed in three stages, the hills are most heavily eroded in the *zona da mata* itself, while less eroded, less evenly formed, and higher hills march into the interior. The hills closer to the coast range from one hundred to six hundred meters above sea level, tending in general toward the lower figure.[14] On the sides and tops of the hills the soils tend to be deep though not particularly fertile, sandier than the clay-rich valley floors, and well drained. Between the hills, in the areas called *várzeas*, soils are deep and enriched by the organic matter deposited by numerous streams. Like the rest of the region, they tend toward slight acidity.[15] The most highly prized soil was that of the rich, wet *várzeas* called *massapê*—"dark soils, and strong," according to one agricultural manual of the late seventeenth century, "which are the most excellent for cane." Into the nineteenth century, cane was almost exclusively a product of valleys.[16] The sweetness and fertility of the lands of *massapê* would be canonized in regional mythology, and planters sought out land most favored with this fruitful soil.[17]

Not far west of Recife, the higher hills leading to the interior begin to break up the half-oranges of the coast. These far-reaching foothills precede a north-south range roughly paralleling the coast, though reaching in a shallow V-shape toward Recife. They recede inland dramatically toward the south of the region and somewhat less so to the north. These hills are the eastern edge of the Borborema Plateau, which is itself on the eastern rim of the massive interior plateau that dominates northeastern and central Brazil. This semiarid expanse to the west and the Atlantic Ocean to the east form the boundaries of what came to be known as the *zona da mata*, though the western boundary was largely imaginary and irrelevant for the purposes of the colonists who clustered along the coast for several hundred years, content to "scuttle along the seashore like so many crabs" in the words of the seventeenth-century friar Vicente do Salvador.[18]

Recife marks the rough midpoint of the region, the boundary between the northern and southern portions of the *zona da mata*, which are known respectively as the *mata seca* and the *mata úmida* (the dry and wet forests). The dry *mata* experienced less erosion in the years following the Tertiary period, with tablelands and broad, flat ridges characterizing the landscape rather than the half-oranges more typical in the wet *mata*.[19] The rivers of the north are smaller and less regular, with a higher incidence of seasonal flow. And the trees lose their leaves collectively, while in the semideciduous forests of the south tree species lose leaves individually during the course of the year.[20] As is obvious from their respective names, rainfall varies considerably between the two regions. While the south maintains averages well above two thousand millimeters annually, some parts of the north receive as few as seven hundred millimeters in a year.[21]

Despite this variability, the *zona da mata* as a whole receives abundant rainfall, with the regional average reaching almost two thousand millimeters a year. Some areas soak up more than twenty-five hundred millimeters.[22] The rainy season between May and August results from the most northerly advance of the prevailing southeast wind that pushes up Brazil's southern coast year-round. As the northeasterly winds lose power during this period, the polar winds from the south make more progress up the coast and meet eventually with the tropical air. The resulting instability produces the predictable winter showers that increase in frequency to an average of twenty-five days of rain in July, before tapering off in August. The coming of the rainy season is signaled by a switch in the prevailing winds, easily noticeable on the breezy littoral. As planters looked out to sea during the dry season, the breeze would brush their left cheeks. When they felt the wind on their right, the towering cumulous clouds and heavy rains were not far behind. And they could expect the months of the rainy season to bring almost 80 percent of the year's precipitation. Much of the rest comes in the "first rains" or "cashew rains" that last for several weeks in January, after which the region is typically dry again for months.[23]

The abundance of water on Brazil's coast fascinated the Portuguese from their first encounters with the new land. Among the men who arrived with Cabral in 1500 along the coast of what is now southern Bahia, Pero Vaz de Caminha left a written record of the event. "So pleasing" is the land, he wrote to his king in what has become an oft-cited claim, "that if one cares to profit by it, everything will grow in it because of its waters." Padre Cardim gushed in his sixteenth-century account of Pernambuco that the land was "blessed with many waters, from rivers and the sky alike, as it rains very much."[24] The sailors who spotted the Brazilian coast on their way to India came from a dry and comparatively deforested country.[25] Early Portuguese experiments with sugar cultivation in the Atlantic islands had revealed only too quickly the ecological limitations of St. Helena, Madeira, and the Azores, none of which are hydrologically well endowed. Experts in long-distance navigation, the Portuguese were also well aware of the challenges of storing fresh water at sea and finding more when necessary.[26]

The sea breezes that bring the rains in the winter and push dry air down from the north in the summer do much to mitigate the unrelenting heat of this stretch of land lying between the Equator and the Tropic of Capricorn. In the seventeenth century, the Dutchman Joan Nieuhof gratefully acknowledged the relief the wind granted from the scorching and sultry weather.[27] The "perfectly tremendous . . . heat of the sun" took its toll on colonists and visitors alike but it was also credited with great generative power.[28] The excess of sun seemed somehow appropriate to a region with a commensurate superfluity of water and forests. Nature appeared profligate in a way that struck the Europeans as unfamiliar. Nieuhof warned planters that crops foreign to the region risked growing too fast for their own good. "The Brazilian soil is extremely fertile, pleasant, and irrigated by many rivers and lakes," he wrote. Straining a contemporary reader's credulity, he added, "To avoid excessive growth [of crops], sand is put in the soil instead of manure."[29]

The elements that captured the attention of early visitors to the *zona da mata*—forests, streams, rain, and sun—would have enduring relevance for later residents. Vaz de Caminha's enthusiasm for the new land, based on the conviction of its fertility, would typify views of the *zona da mata* for hundreds of years. His proclamation that "everything will grow" echoed down through the centuries, reiterated time and time again by those who were either genuinely impressed or had some reason to trumpet the glory of Brazil's bounty (such as the need to impress the king).[30] A colonial booster from the early seventeenth century claimed that Pernambuco would support "all the agricultures in the world, because of its great fertility, excellent climate, good skies . . . and a thousand other attributes."[31] A generation later, during the generation-long Dutch occupation of the colony, Nieuhof swore of the *zona da mata* that "it would be difficult to find one piece of ground, be it in the valleys or in the hills, that does not provide something useful and in great abundance."[32] And an English doctor visiting in the nineteenth century agreed that the region "is extremely fertile and pleasant, being watered by many large rivers and lakes . . . [producing] various sorts of fruit and vegetables in great abundance."[33] In spite of all of these testaments to the diversity and range of its fertility, the province came to be closely associated with just one crop: sugarcane. Over time praise of the region's abundance became tightly bound to ideas about the region's specific suitability for sugar. When in 1963 a Recife newspaper proclaimed Pernambuco the "ideal habitat for the cultivation of cane," it simply reiterated an idea that had been prevalent for several hundred years.[34]

Sugar's Rise: The *Zona da Mata* from the Sixteenth to the Eighteenth Century

In the early years of colonization, the area under settlement and eventually cultivation was limited to Olinda and Igarrasú.[35] Olinda lies north along the coastline from Cabo some thirty kilometers, with Igarrasú another fifteen kilometers still. At the point where the Beberibe and Capibaribe rivers meet and empty together into the Atlantic, a large reef formation stands off the coast and forms a natural harbor.[36] A couple of kilometers from the sheltered harbor, Olinda sits on a hill that offered a commanding view of the sea. Duarte Coelho, who in 1534 received dominion over the coastline between the mouth of the São Francisco River and the island of Itamaracá in exchange for his service to the Crown in India, found the hill an auspicious site for the capital of the new captaincy.[37] To the south, where the port was sheltered by reefs, and to the west, the land was flat and low, heavily watered by the two rivers. A small town grew around the port, called Recife after the reefs. During their occupation of the region between 1630 and 1654, the Dutch expanded the town into a larger settlement than Olinda, extending it to the meeting of the two rivers and making it their capital.

In the mid-sixteenth century the townspeople of Olinda and Igarassú hung tenuously to their settlements. A ship arriving in 1549 was greeted with great relief by Duarte Coelho, who implored the visitors to make haste for Igarrasú and lend aid to the townspeople there, under attack from the Caeté. The rescuers, among them a sailor named Hans Staden who wrote an account of the episode, barely made it; taking small boats up an inlet toward Igarrasú, they nearly fell prey to a trap set by the Caeté. The Indians had cut large trees, securing them with thick, sturdy vines in such a way that they could be dropped on passing boats.[38] But the attack was repelled and the Portuguese continued to scratch out their living. As with other areas along the coast of the Terra de Santa Cruz, colonists in what was then called Nova Lusitânia gathered parrot feathers and dreamed of finding gold or silver, but their primary economic activity was harvesting brazilwood, which produced a red dye. Into the 1530s, the Portuguese used a barter system to convince Indians to find the trees and drag them to the coast.[39]

Pernambuco never produced precious metals, but it did soon produce the "spice"—sugar—that garnered such high prices in Europe. The endless forests and abundant water sealed Pernambuco's superiority to the small, dry Atlantic islands for sugar cultivation. The Portuguese had produced sugar for several generations, building their first offshore mill in 1433 on Madeira, an island whose name means "wood".[40] The name *Madeira* now seems as ironic as *zona da mata*, since the Portuguese had largely deforested the island within two generations of first planting sugar. The ominous results of sugar cultivation there went unheeded in the new "forest zone" of the Brazilian littoral, soon to become the heart of Portuguese sugar production. In Pernambuco sugarcane's environmental footprint quickly eclipsed the impact of brazilwood harvesting. These valuable trees do not tend to grow in dense groves, forcing loggers to cruise the forests and identify the trees individually. Certainly many trees were sacrificed to make way for these expeditions but not on a large scale. Sugar, in contrast, demanded cleared fields for planting and an enormous reserve of fuel for heating the syrup produced from ground cane. Even as production increased, though, the forests seemed endlessly capable of sustaining the planters' pace.

Pernambucan sugar made its first appearance in the customhouse of Lisbon in 1526, but news of the first mill reached the metropole in 1542.[41] Those who see Pernambuco as the cradle of Brazilian civilization, the heart of the noble sugar industry, and a land divinely blessed and tailored to the production of cane take pleasure in noting the arrival in Lisbon of sugar six years before the "official" transfer of sugarcane to the new colony by Martin Afonso de Souza.[42] The Portuguese transplanted the cane that they had cultivated on the Atlantic islands, a sturdy variety that a thousand years earlier wended its way from India up to the Mediterranean and to Iberia, a migratory pattern that would be repeated in the twentieth century when new Indian and Javanese varieties would be introduced to the *zona da mata*. This original variety came to be known as "creole cane" when new varieties were imported.[43] From very early in the colonial period, it became "impossible to divorce the civilization of the Northeast from its sugar economy."[44] With sugar's introduction, planters began to aggrandize land for its cultivation, and its role in the culture grew ever larger.

The Portuguese king distributed land in grants called *sesmarias*, given to loyal or useful men in the amount of a thousand or more hectares. The lushness of the forests, so captivating to the Portuguese, became an index of land's value. Those areas still covered in the dense Atlantic Forest were taken to be the richest. Grantees would promptly destroy this mark of fertility, burning the forests and planting their cane in the rich ash. "Not long after," Warren Dean writes, they could be found "begging for another, as a 'remedy for my poverty' on the grounds that their lands were 'tired.'"[45] In the 1580s Gabriel Soares de Souza observed that "in regard to the fertility of the land, I say that many times the newness of a farm is worth more than the property, so that [planters] maintain themselves with little patrimony."[46] With the introduction of sugarcane, new grantees invariably planted cane and built a mill to grind it. This apparatus—a simple affair of wooden rollers, with wooden or metal teeth, powered by people, animals, or water—gave the whole concern its name. In Brazil as in some other sugar-producing lands, the word for the engine (*engenho* in Portuguese), entered the common vocabulary as a metonym for the plantation as a whole, and the *engenho* would be the basic unit of sugar production for nearly four hundred years.[47]

Planters in Pernambuco borrowed *engenho* design from the Atlantic islands, along with other aspects of sugar production, including the exploitation of chattel slaves for labor. Initially, the Portuguese bartered with Indians for labor, but as their labor demands increased with sugar's spread they began enslaving Indians. They also brought Africans who had experience in São Tomé or Madeira, putting them to work in the boiling house where their skills were needed. From the 1570s, the depredations of disease in the indigenous populations and the increased efficiency of the African slave trade led to a wholesale shift toward African slaves that would be completed by the first decades of the seventeenth century. Over the next century and a half, Brazil imported four thousand to five thousand slaves a year, the vast majority of whom ended up on the sugar plantations. The average *engenho* had between sixty and one hundred slaves, and the larger ones had more than two hundred. Demand for slaves among sugar planters continued to rise even after sugar was displaced in importance by activities elsewhere in Brazil, first gold and diamond mining in Minas Gerais and later coffee in Rio de Janeiro and São Paulo. During the middle of the eighteenth century, Pernambuco imported more than fifteen hundred slaves each year. Over the course of the trade, from 1550 till 1800, Brazil as a whole brought in as many as 2.5 million slaves.[48]

Slaves generally began planting new cane in May, continuing to plant and then weed the new cane until September, shortly before the start of the harvest. They began cutting in October and stayed harnessed to the grueling task roughly through April, depending on the fruitfulness of the growing season.[49] Cane takes at least fourteen months to mature and sometimes as long as eighteen. Once cut, it regrows spontaneously from its roots, though with decreasing vigor as the years pass (these regrowths are called "ratoons" or, in Brazil, *socas* or *folhas*). Though cane can be grown from seed, this type of generation was not mastered until the end of the nineteenth century, and it remains standard practice to replant with pieces of cut cane (called "seed cane"). For planting, cane pieces were placed in long, parallel trenches until well into the eighteenth century, when holes were dug with banks all around to prevent erosion. Brazilian planters did not generally use plows even into the nineteenth century, leaving to slaves the difficult work of turning fields with hoes.[50]

With their rudimentary techniques of raising cane and rendering sugar, colonists spread their crop. Sugarcane rolled down the hills of Olinda and crept into the forest lands, preceded by fire. The littoral had been home for millennia to people who used swidden agriculture, but the Portuguese set the torch to the forests on a larger scale. At first, the Portuguese probably mimicked the cultivation patterns of their predecessor societies, who had likely confined their burns to the forest edges and to watercourses.[51] The cane followed the streams' course between hills and through forests. These streams, which became famous as the Northeast's "rivers of sugar," formed the highways for transporting sugar to the coast and powered the mills that ground the cane and extracted syrup. Boats zipped in and out of the small river mouths and beat up or down the coast to Recife.[52]

Near the beginning of the seventeenth century, forty years after Staden helped the colonists combat the Caeté, cane fields extended all the way to Cabo de Santo Agostinho. In many of the valley floors close to the coast, from

Cabo all the way north beyond Itamaracá, sugar thrived in the *massapê*. So much sugarcane grew, in fact, that the sixty-six *engenhos* in the province could not grind it all; Cardim reported excess cane rotting in the fields in the 1580s. Two thousand African slaves toiled on the *engenhos*, though they still probably made up only about a third of the labor force. The *engenhos* produced two hundred thousand arrobas of sugar a year, or almost three thousand tons—as much as could be accommodated by the forty ships a year that came from Lisbon for it. Duarte Coelho's family made a fortune every year from the taxes on this sugar alone. A class of cane growers (*lavradores*) developed who were not mill owners, with agreements for profit-sharing (and risk-sharing) with particular *senhores*.[53]

Cardim described the landscape succinctly: "I won't speak of the freshness of the wooded groves, nor of the many large rivers," he wrote, for "these are ordinary and common things in Brazil." The region, he said, "is all *cha*."[54] Given the time Cardim was writing, we might assume that by *cha* he meant flat or level land, and indeed the area around the province's capital Olinda was largely low-lying and flat. But the term evolved after Cardim, eventually describing any area planted in cane. This discursive slippage, between the landscape of Pernambuco and the presence of its most lucrative crop, has its foundations in the early period of cane culture Cardim observed, and it eventually rooted itself in almost all description of the region and colored the local culture.

By the beginning of the seventeenth century, sugar had grown to unrivaled importance in the empire, and other European powers coveted Portugal's huge profits. The lucre from Brazil's sugar production exceeded that of the entire India trade, proof that this was sugar's golden age in Brazil. The colony had thirty thousand slaves, mostly working on sugar *engenhos*, all of which together produced seventy-five hundred tons a year in the early 1600s, rendering substantial payments to the Portuguese Crown.[55] Pernambuco's sugar fields now reached up to thirty miles inland, and dyewood loggers had to travel even farther to find their trees. Many people still lived from the sale of dyewood but "[could] not find [the trees] closer because they are so sought after."[56] In 1607, a Crown inquiry into harvesting wood from the *zona da mata* for the manufacture of wooden pikes ended in disappointment—the *engenhos*' consumption of forest had been so profligate that royal cutters would be forced to penetrate the interior for appropriate wood. One observer at the time worried about the move toward a sugar monoculture, fretting over missed opportunities with other products: Would the colonists fail to fully realize the new colony's promise of abundance?[57]

Monoculture already presented a more basic threat than crowding out other export crops. In the land that inspired such superlative descriptions of abundance, the colonists faced a genuine risk of having too little food. The focus on sugar was so complete that it was planted to the exclusion of food crops, a problem that revealed itself in deadly fashion during a 1648 famine.[58] This was also a problem on English and French sugar plantations in the Caribbean, though planters in these colonies customarily granted provision grounds to slaves for producing staple foods. Indeed, though Caribbean planters called this practice the "Brazil system," they appear to have been more likely to allow slaves to market surplus produce than were Brazilian planters.[59]

Only three years before the crisis in Pernambuco, the Dutch administration had attempted to address the looming threat of food scarcity. Responding to soaring prices for staple foods in the early 1640s, the governor, Maurice of Nassau, ordered all planters to devote an established percentage of their land to food crops.[60] Though the 1648 famine testified to the measure's failure and to the Dutch's inability to achieve agricultural balance, they proved expert at extending sugar's domination. By the 1640s they had imposed a new level of order on the land, establishing separate administrative districts, each with its own set of *engenhos*. They drained and filled land to build up their capital, Recife, extending this technology eventually to the sugar fields. Floodplains that had been too wet for cultivation were brought into the planting regime through the use of diversionary ditches. Individual *engenhos'* production was on the rise and the number of *engenhos* had swelled to over one hundred. Joan Nieuhof, who was in Pernambuco at the time, reported that the sugar shipments of the early 1640s were the largest ever seen in Brazil.[61] The concentration of *engenhos* in some areas became so intense that downstream neighbors began to complain of *engenhos* throwing cane waste into the shallow rivers.[62]

In addition to the shaky food supply and some dirtied streams, the environmental toll of the *engenhos* began to manifest itself in the forests during the seventeenth century. Each *engenho* exerted enough of a demand on forest reserves that even committed optimists took notice. A day's production of sugar required about eight cartloads of wood to supply the fires, or a quantity measuring seven palms by eight palms.[63] So for each kilogram of sugar produced, about fifteen kilograms of wood were burned, leading to the logging of 210,000 tons of wood each year.[64] Studying Bahia's sugar region about a hundred years later with 180 *engenhos* (comparable to Pernambuco in the sixteenth century), Shawn Miller estimates that one harvest season demanded 750,000 cubic meters of wood.[65] During the harvest, boilers were kept fired every hour of the day, generating a constant demand for new fuel. Transporting sugar required an average of 12,300 wooden boxes every year in the late seventeenth century, putting further pressure on the forest.[66]

The *engenhos* burned their way through the forests of the *zona da mata*, producing for each cartload of harvested wood a small amount of sugar and huge piles of cane pulp and wood ash. A later observer wrote, "The entire Brazilian agricultural system is based on the destruction of forests. Where there are no woods there is no agriculture."[67] Logging also continued, especially for dyewood. In the first half of the seventeenth century, various observers noted that dyewood loggers had to make longer and longer trips inland. Nieuhof noted that aggressive logging of these trees earned the condemnation of the Grand Council of Dutch Brazil in 1646.[68]

Both planters' and loggers' destruction of the Atlantic Forest found an odd incentive in policy. Planters practiced a form of slash-and-burn agriculture and believed the wood ash increased fertility. But Miller also argues persuasively that colonial forest policy encouraged planters to destroy the forests on their lands. Since the Crown claimed a monopoly on all valuable timber (*madeira de lei*), and reserved the right to intrude on land to retrieve that wood, "timber trees remained obstacles to the pursuit of prosperity in Brazil, rather than serving as vehicles to that end."[69] Because the Crown made the forests worthless to planters by prohibiting them from selling the timber, the planters "lost nothing and gained much by the destruction of the forest."[70]

Despite these early indications of the coastal Eden's limitations, the diversity and extent of the Atlantic Forest remained impressive and the region's abundance inspired faith.[71] "Though it has been populated for so long," a colonist wrote in the early seventeenth century, "producing in all of those years such a great amount of lumber for boxes, the forests don't fail to offer wood for many more and they never will."[72] In the middle of the century, Nieuhof admired the Portuguese for carving plantations out of the dense, unending forest—"opening a path through the trees, overcoming tremendous difficulties."[73] And at the end of the seventeenth century, a Jesuit named João Antonil claimed that no other part of the world was so rich in valuable and strong trees, and he described in detail the qualities and importance of numerous different species. One measure of the forest's embarrassment of riches was the degree of specialization associated with ox cart, sugar box, and *engenho* construction: *massarandûba* for the *engenho* superstructure, *sapucaia* for the grinder axles, *sapupira* for the gears of the grinder, *páu de arco* for the waterwheel, *sapupira-merim* for the ox carts, and on and on.[74]

Only a very diverse forest could provide the timber variety for such exact matching of wood type to construction purpose.

Antonil wrote a manual of techniques and advice about Brazil's primary agricultural and mining industries, and his careful observations provide some sense of the degree to which a planter in the late seventeenth century needed to be concerned with conservation.[75] Antonil reminded prospective property buyers to find lands with forests as close as possible. Once they had purchased the land, buyers should attend to the conservation and improvement of their holdings, both in terms of the *engenho* and the land itself. Abundant water, accessible forest reserves, and good soil formed a trinity as important to the health of an *engenho* as the well-known "three Ps"—*pão, páo,* and *panno,* or bread, stick, and cloth (the fundamentals of slave management). Antonil emphasized the level of demand for firewood, saying that boats would need to be sent out constantly, "one after the other," to gather wood. He even recommended that every slave be required to gather a cartload of wood a day.[76]

Again, the degree of specificity in Antonil's manual demonstrates how far planters were from exhausting the resources available to them. Prospective sugar planters ran the risk of purchasing worn-out land, he warned, but a smart buyer stood every chance of success in a land of Brazil's bounty. Every separate step in the process of sugar manufacture required its own resource, and Antonil ticked off the demands: food plots for the slaves, rich soil for the cane, pasture for the cattle, clay for purifying sugar, diverse forests for various woods, and mangroves for seafood. It was precisely in a situation of ecological wealth that this range of environmental demands could be met. "Only Brazil, with the immense forests it has," Antonil wrote, "could supply, as it has supplied for so many years and will supply for years to come, so many furnaces, as many as there are *engenhos* in Bahia, Pernambuco, and Rio de Janeiro, that commonly run night and day, six, seven, eight, or nine months of the year."[77]

The picture that emerges from Antonil's book contrasts with the experience of the newer sugar colonies in the Caribbean, where environmental consequences quickly forced planters to modify their practices. Pernambuco had served as the model for English and French plantation development from the 1630s onward.[78] But the process of deforestation and soil depletion on islands such as Barbados was so swift that between about 1665 and 1720 planters adopted a string of measures to conserve resources. They introduced manuring, used cane pulp (bagasse) for fuel instead of wood, designed furnaces that conserved heat and cut down on fuel needs, and developed new antierosion cane-planting techniques. Pernambuco did not adopt some of these methods until the nineteenth century, in part because planters there did not face the urgent environmental crisis experienced by the Caribbean producers.[79]

Renewed concerns about forest loss did emerge in Pernambuco in the eighteenth century. In the 1750s *senhores de engenho* petitioned for indemnities to defray losses incurred by the devastation of nearby forests. The "indiscriminate cutting" of woods, beyond those decreed "wood of the law" and therefore reserved for the Crown, had eaten into the fuel and construction reserves of the *engenhos*. This decade also saw huge increases in the mother country's demands on Brazil's forest resources; Brazilian wood flowed across the Atlantic to rebuild Lisbon after the 1755 earthquake. Tannery owners complained about the scarcity of certain barks they used in their work, accusing brazilwood loggers of overzealous cutting. In 1784 the city council of Olinda sent a message of distress to the queen, noting that *engenhos* were shutting down for lack of wood. The Crown, disturbed by the implied threat to its own wood requirements, sent word in 1789 reiterating that all logging of construction-grade wood was prohibited, because the best trees "should be preserved exclusively for royal services." Eight years later, the governor complained of the "indiscriminate cutting of the precious construction trees to send them to the Crown, without the required care for replanting."[80] The repetition of the injunctions indicates their failure to restrain the colonists from cutting. By the end of the eighteenth century, agricultural reformers in Portugal repeatedly criticized Brazil's inefficient sugar planters, comparing them unfavorably to West Indian producers.[81]

From the middle of the sixteenth century to the beginning of the nineteenth, sugar expanded through the *zona da mata*. After several centuries of cultivation, cane began creeping up out of the floodplains like a rising tide, lapping up the more gradual slopes of the region's hills.[82] Planters pushed inland and to the north and south of Recife, felling trees to build their mills and burning forests to plant cane. The boundaries of cane fields expanded and contracted, in loose rhythm with the flux of sugar prices. Nearly alone in world production at the beginning of the period, Pernambuco quickly expanded its sugar business. Exporting 10,000 tons of sugar in 1600, Brazil sent out 28,500 metric tons in 1650, staying roughly at that level for thirty years. By the 1680s, though, Europe's Caribbean sugar colonies had begun to exceed Brazil's export levels. From that time through the middle of the eighteenth century, the country's exports remained around 20,000 tons a year.[83] The number of *engenhos* in Pernambuco had risen from sixty-six in the 1580s to more than three hundred in the 1640s and upward of five hundred by the beginning of the eighteenth century.[84] Impressively, even as colonial administrators turned their attention to the gold and diamond mines of Minas Gerais in the eighteenth century, export proceeds from sugar consistently exceeded those of competing products.[85]

Dreams of Modernity in the Backward Northeast: Sugar's Ebb and the Planter Response

For all of the vibrancy and growth of Pernambuco's sugarcane agriculture in its first two and a half centuries, the human impact on the *zona da mata* should not be exaggerated. Warren Dean estimates that the first 150 years of the crop's tenure in all of Brazil claimed twenty-two hundred square kilometers of Atlantic forestland. More recently, John Richards suggested that the expansion of cane cultivation between 1700 and 1850 would have demanded an additional seventy-five hundred square kilometers of cleared land. These numbers account for all of the cane growing regions, including the significant plantations of Bahia and the smaller but still numerous concerns in Rio de Janeiro. We can assume that at least two-thirds of the Pernambucan *zona da mata*'s fifteen thousand square kilometers had eluded the reach of cane agriculture by the second half of the nineteenth century.[86] The scale of this impact contrasts sharply with the thorough deforestation of the smaller sugar islands in the Caribbean, such as Barbados. Cuba was more similar to Brazil, with substantial forestland remaining in 1815 notwithstanding its small size in comparison to the Atlantic Forest region of Brazil. In that year, though, Reinaldo Funes Monzote reports, the Spanish Navy finally relinquished control over the forests to private owners.[87]

During his 1816 stay in Pernambuco, Tollenare calculated that one of the *engenhos* he visited encompassed three square leagues of territory (thirteen thousand acres). Of the total, he estimated that over 60 percent was cultivable, yet the planter was using less than one-twentieth of that. And to Tollenare's great surprise, this *engenho* actually seemed to use more of its land than any other he saw. He concluded that "in an area reputed to be very cultivated, in the *capitania* of Pernambuco," the area planted was one-thirtieth of the total available.[88] Tollenare underlined his disbelief, writing that by far the greater part of the area within twenty leagues of Recife, Brazil's third-largest city, sat completely unutilized.[89] Some decades later, the eminent Pernambucan historian Oliveira Lima agreed, estimating that a mere 3 to 4 percent of all *engenho* lands were being planted in cane.[90] Even with this uncaptured potential, the concentration of landowning was so great and planters' focus so narrowly oriented toward cash crops that food production was still threatened. The Englishman Henry Koster reported rampant hunger, and even death, in the regional town of Goiana in 1815, while a few years later his countryman James Henderson observed, "The desire everywhere of the gain which [cotton and sugar] afford, unwisely prevents the cultivation of provisions of the first necessity in sufficient quantity for the subsistence of the population."[91]

Considering the profligate exploitation of land associated with sugar, these estimates of land use appear strange indeed. But Pernambuco's ponderous pace of production also limited its environmental footprint. If approaches to industrial and agricultural techniques were slow to change during the late nineteenth and early twentieth centuries, as Peter Eisenberg has shown, then the preceding eras were glacial. The industry copied old Mediterranean technology, such as horizontal rollers augmented by presses for squeezing the juice from cane. Vertical, three-roller mills did not spread until the beginning of the 1600s, and planters still did not manure their fields at that point. For much of the nineteenth century, most Pernambucan planters still had not adopted the vacuum pan in their mills, a technology that enabled a much more efficient extraction of sugar from syrup.[92]

The modest size of the population and obstacles presented by transportation also limited the environmental impact of extractive agriculture over time. According to Gabriel Soares de Souza's observations of Pernambuco in 1587, the population was then around three thousand (an impressive number by his lights). By the nineteenth century that number had expanded to a quarter million.[93] Although the U.S. missionary Daniel Kidder remarked of Pernambuco in the 1840s that "there is no portion of the Brazilian coast more populous or better cultivated," this probably speaks more to the backward state of the empire's agriculture than to the prosperity or density of Pernambuco.[94] And within a few short years, the southern coffee boom would reconfigure the country's agricultural profile.

Deep into the nineteenth century, the only alternatives to water transport were the unwieldy ox cart and mules or donkeys. These were not nearly as efficient as the small boats that plied the rivers and coast of the region. For several hundred years, then, the Portuguese cleaved to the settlement pattern of the indigenous peoples before them, confining the bulk of their activities to river courses and the coast. Tollenare remarked on the strict limits placed on cane cultivation in the districts of Ipojuca and Serinhaém—south of Recife—by transportation difficulties associated with the lack of access to streams.[95] With the arrival of railroads in the late nineteenth century, planters' possibilities expanded. (The end of the railroads would similarly be a turning point in sugar culture, as trucks, and therefore roads, became the dominant transportation method.) Large areas of the *zona da mata* offered themselves for cultivation as a result of the new transportation technology, but even regions like the headwaters and middle course of the Una River in the southeastern part of the zone would not be developed until the 1930s.

Just because large areas were not in cultivation does not mean that they had not once been planted. Without advanced fertilizers and irrigation, planters adopted extensive rather than intensive approaches. This set them apart from their peers in the Caribbean, where limited space and forest resources forced intensive methods. The English naturalist George Gardner visited Itamaracá, north of the capital, in 1849 and wrote, "Instead of the almost uniformly level character of the country in the vicinity of Pernambuco [referring to Recife], here there is a gentle undulation of hill and dale. There is not much large timber, the wooded portions generally consist of small trees and shrubs, which give to many parts of the island an aspect more like that of an English orchard, than an uncultivated equatorial region."[96] This type of landscape, described variously (and according to density of regrowth) as *capoeira, capoeirão,* and *mato,* probably characterized large portions of the *zona da mata.* Those more accessible locations close to Recife or other towns, along rivers or the coast, had over the preceding three hundred years grown cane at some point, if not continuously.[97]

At the beginning of the nineteenth century, sugarcane occupied less territory in the *zona da mata* than in earlier years (in this it diverged sharply from the Caribbean producers racing to fill the void in supply left after the Haitian Revolution).[98] Many visitors to the region during this period were surprised by the paucity of cultivation, and if we believe Oliveira Lima's estimate that only 3 to 4 percent of land was under cultivation, their reaction makes sense. An English traveler, Andrew Grant, noted Pernambuco's fertility at the beginning of the century but spoke of its productive grandeur in the past tense. The province "formerly contained above a hundred sugar-plantations, extensive forests, well cultivated fields, and a great profusion of the most delicious fruits," he wrote, and though it once produced fifteen thousand chests of sugar a year, "at present it scarcely furnishes four thousand." Pernambuco labored under a heavy load of debt, he observed, and had recently weathered a burst of emigration.[99]

Grant's observations were acute; sugar was mired in an economic funk. The early-century downturn was mitigated by the reshuffling of global production following the Haitian Revolution, but for several decades lands once planted in cane reverted to secondary forest. Brazil was doing its own political reshuffling at the time. In 1808 the Portuguese Crown, fleeing Napoleon, had arrived in the colony. When the king returned home in 1822, his son Pedro remained in Brazil, declaring the colony an independent empire. In 1831 he too returned to Portugal and succeeded his father on the Portuguese throne. His son, Pedro II, became emperor in 1841 and reestablished stability after the political uncertainty of his regency.

From Grant's visit in the early 1800s into the midcentury, the dynamism in Pernambuco's agricultural sector could be found on the *zona da mata*'s periphery. Drier hills and tablelands—the foothills of the Borborema—felt the gouge of the hoe for the first time. The new cultivar was cotton. Growth in this commodity, which had long coexisted with sugar on a small scale, was spurred initially by the gradual expansion into the interior of the state and later by the crisis in world supply precipitated by the U.S. Civil War. Koster, resident in Pernambuco in the 1810s, described cotton as the province's "chief article of exportation," and it was largely on the strength of this good that Pernambuco led Brazil in trade with England.[100] Sugar production did rise in the nineteenth century—reportedly sevenfold between 1821 and 1853—but cotton grew faster. In the middle of the 1860s alone Kidder, the missionary, wrote (probably with some exaggeration) that Pernambuco's cotton exports had increased tenfold.[101]

Like ranching in the *sertão* (the Northeast's semiarid interior), farmers' settlement throughout the state had expanded by the nineteenth century,[102] extending westward from Recife around eighty or ninety miles and northward more than three hundred. As Tollenare noted in 1816, "the cultivation of cotton each day extends itself further toward the *sertão*."[103] By the 1820s, cotton was the primary crop in Goiana, the *zona da mata*'s most important northern city, and in Vitória de Santo Antão, the largest town on the region's western boundary.[104] Until that point, these places had been less densely settled because they lie in areas where rainfall tails off from the regional norm. Cotton's spread drove the expansion of land clearance in the *zona da mata* during the nineteenth century. When sugar went through periods of growth later in the century—and even at the dawn of the twentieth century, well after the decline of cotton's importance—it frequently followed a path cleared by the cotton industry. The cotton boom ended in the 1870s, with U.S. cotton's return to the world market (though cotton would surge in São Paulo around World War I around the coffee areas).[105]

Despite these new engines of expansion, there remained widespread areas of Atlantic Forest. As the naturalist George Gardner reported in the early

1840s, the island of Itamaracá had lost almost all of its forest after two hundred years of concentrated sugar cultivation, but other parts of the region, even some close to Recife, he still described as "richly wooded."[106] Even at midcentury, Henderson noted that there was little agricultural production in some areas around Recife, the land "being very generally uncleared . . . and a great portion remaining in its primeval condition."[107] Traveling through the province Gardner found what he presumed to be virgin forest a mere twenty miles west of the capital.[108]

Cotton's moment of economic centrality was fleeting, and the commodity's brief tenure at the pinnacle of Pernambuco's economy draws little attention. One *senhor de engenho*, relating the story of his great-grandfather's purchasing an *engenho* with the proceeds from cotton sales, was shocked to find that I was familiar with cotton's nineteenth-century importance.[109] Because of the widespread notion that the "civilization of the Northeast" was bound exclusively to sugarcane, the *zona da mata*'s experiences with cotton have little purchase in popular and official memory.[110] This forgetting can be indexed to the relative importance of the two commodities on the world market. From French efforts to grow cotton in Maranhão in the early 1600s through the eighteenth century, the crop largely remained in local markets.[111] But cotton's elision also relates to the fact that cane colonized the understanding and perception of the *zona da mata*, as much as it did the region's environment.

Pernambuco's planters directly addressed their agricultural and political identity and the future of their sugar industry when national politics forced them to in the 1870s. In September 1877, the imperial minister of agriculture circulated a questionnaire about the state of the empire's agriculture. The emperor was still Pedro II, son of the Pedro who had declared Brazilian independence and later returned to Portugal. Pedro II was scientifically minded and interested in agriculture.[112] Pernambuco's provincial president forwarded the request for information to the head of the province's Agricultural Auxiliary Society, who formulated a response under the assumption that a congress would be held the following year to discuss the pressing issues facing planters throughout the empire. It came as a surprise, then, when in July 1878 a ministerial proclamation announced the convening of a meeting in Rio de Janeiro about *grande lavoura*, or large-scale agriculture, with participants from the provinces of Rio de Janeiro, São Paulo, Minas Gerais, and Espírito Santo—the sites of coffee cultivation. Though the proclamation gave lip service to the rest of the country's planters, suggesting that it would be desirable to address the problems facing large and small agriculturalists from around the empire, it was clear that coffee's primacy to the national treasury drove the assembly.[113] Outraged, planters from Pernambuco immediately began organizing a conference of their own.

The snub to northeastern sugar planters came at a moment of acute crisis for the region. Between 1877 and 1879 the so-called Great Drought, one of the worst in recorded history, scorched the *sertão*, killing an estimated two hundred thousand people.[114] The Northeast was also watching the decline of its national economic relevance. One historian has argued that the quality of life and economic development of the Northeast equaled or even exceeded that of the Southeast around 1870, but that by the first decade of the twentieth century the Northeast had seen its exports plummet in value and importance, to less than 5 percent of the country's total receipts.[115] At the same time that northeastern planters saw their cotton boom cut short in the 1870s, Brazilian coffee production exploded, accounting for almost 60 percent of export income during the decade. Then, early in the 1880s, world sugar prices collapsed, adding to northeastern planters' woes.[116]

The economic and climatic hardships combined to deepen a sense of anxiety about the Northeast's status and contributed to a regional cohesion that had been developing slowly for generations. Various state elites felt a growing kinship in reaction to the southern provinces' actions and achievements and from a fear of backwardness and the entrenchment of regional disparities.[117] Just as important a bond, perhaps, was the northeastern provinces' shared devotion to sugar. In announcing its dissident 1878 congress, the Sociedade Auxiliadora called a meeting of Pernambucan planters and others from "those provinces whose export goods [i.e., sugar] are commonly traded in the Recife market—Ceará, Rio Grande do Norte, Parahyba, Alagôas, and Sergipe."[118] They called for the delegates to stand together in the face of the immediate crises facing their region and their crop. The "illustrious agriculturalists of this unhappy zone of our great empire, the same zone of the *quebra-kilos* [a revolt of 1874] and of the [drought] refugees of Brazil," were completely forgotten or dismissed by the Court.[119]

According to the Rio conference's announcement, the central questions to be discussed were the availability to planters of labor and capital. After the 1850 abolition of the slave trade the "free womb law" had passed in 1871, and planters and politicians had groaned about an impending labor crisis. The southern coffee planters had steadily imported northern slaves since the end of overseas trade. Between 1860 and 1879, an average of more than thirteen hundred slaves left Pernambuco each year through this internal trade. In midcentury Pernambuco slaves had outnumbered free workers by three to one, but by 1872 that ratio had more than reversed, with free agricultural workers outnumbering slaves by five to one.[120] The Rio congress set out to discuss the threat of a labor crisis, the issue of newly freed slaves and children of slaves forming a free labor force, and the means by which a stronger credit system could be developed. The head of the Agricultural Auxiliary Society responded to this outline of the southern conference with irony. The South was the same "happy region," he claimed, that had its principal railroad elevated to the category of a political program and that had its credit establishments supported by the imperial government while those of other regions were "coldly thrown to their own fate."[121]

If part of their fate was agricultural backwardness, some planters admitted that they themselves deserved some blame. A reliance on tradition and resistance to change had characterized the sugar sector for a long time. The author of an 1839 agricultural manual wrote that field techniques in the sugar industry were deplorably backward throughout the empire. Cane agriculture was so primitive, he sniffed, that "any field hand has sufficient knowledge of the techniques" to manage an *engenho* as capably as any *senhor*.[122] At the northern congress in 1878, one speaker complained of his peers' "love of routine"—their unwillingness to accept, much less embrace change.[123] And the French consul to Pernambuco wrote a decade later that the cultivation of cane was practiced "as it was hundreds of years ago; there is not the minimum of progress to record, neither in the preparation of soil nor in the use of better agricultural instruments." He characterized the Pernambucan planters as "generally very intelligent but apathetic and wedded to routine." Ten years further on, another observer found the same situation, writing that planters ignored fertilizers and thoughtlessly abandoned "worn out" lands to "move further on."[124] This much-remarked stodginess captured the attitude of Pernambuco's planters as the century ended, but it was a characterization some sought to overturn.

Conclusion

Scolding each other for their lack of innovation, the planters at the 1878 Recife congress vowed to regain their economic footing after their brief lag behind cotton and their near-total eclipse by coffee. They touted their region's vigor and independence of spirit, but they also began to come to terms with the basic shortcomings facing their favored industry. Hundreds of old, small, inefficient *engenhos* dotted the landscape of the *zona da mata*. The province's production was handily outstripped by competitors such as Cuba, which had grown rapidly from the 1830s onward and by 1878 produced nearly five times as much sugar as Pernambuco. Even the small British colony of Demerara (now a county of Guyana) nearly equaled Pernambuco in tons of sugar produced per year.[125]

The costs of transporting sugar remained prohibitively high in Pernambuco, and the province had only just begun to take measures encouraging economies of scale through a set of underfunded central mills. Recognizing the importance of capital and infrastructure to their industry, Pernambuco sugar planters called on the imperial government to provide the same access to credit enjoyed by southern agriculturalists and to aid in the construction of railroads.[126] Pernambuco actually had come early to railroad construction; the second line built in the country ran south from Recife and reached the regional sugar town of Palmares in 1862. But that track had gone no farther by the time of the agricultural congresses, and in the meantime the South had experienced a railroad-building boom.[127] Exultant as ever over the quality of their land (one delegate to the 1878 conference declared that for no other country "had nature been so prodigal with the elements of prosperity and wealth than with our soil"), sugar planters sought a distinctly modern means of overcoming the limitations presented by their environment.[128]

Despite constant complaints of neglect, Pernambuco planters succeeded in fomenting railroad construction over the next twenty years. As the twentieth century began, well over six hundred kilometers of track connected Recife to the sugar-growing regions in the southern, western, and northern parts of the *zona da mata*, and 80 percent of the state's sugar arrived in the port by rail.[129] The faster and cheaper transportation allowed planters to make ever deeper incursions into the forests and hills. The experiment with central mills that ground cane supplied by growers failed during the late 1880s and 1890s, but out of the wreckage a compromise system emerged, combining the scale and efficiency of a central mill with the hybrid agricultural and industrial aspect of the old *engenho* system.[130] The new institutions, *usinas*, concentrated the ownership and control over growing lands and increasingly pushed smaller-scale *engenhos* to the side. The combination of railroads and *usina* productivity explains the doubling of sugar production between the late 1840s and the late 1880s. A fleeting dip in production accompanied the abolition of slavery in 1888, but the state's planters reached a new productive high early in the 1890s.[131]

By the first decades of the twentieth century the impact on the environment would become more acute as a result of the intensified pressures brought by the railroads and *usinas*. As *usinas* proliferated, so did the waste they dumped into rivers. In 1910 the government implored producers to curb the dumping of contaminants but to no avail.[132] Cane's expansion also accelerated the rate of deforestation; a study from 1912 found that of the Atlantic Forest that once covered almost the entirety of what became the cane zone, only 34 percent remained.[133] As forest cover dwindled, streams filled with sediment flowing from the denuded hills, and the waterways' flows became increasingly irregular. The government frowned on this, since some cane still reached mills by boat, and in 1915 it attempted to prevent erosion by restricting planters from cutting forests on hilltops. Predictably, the measure was disregarded.[134]

The iron tentacles of the Recife and San Francisco, the Great Western, and the Central Railways reached the boundaries of the *zona da mata*, and across thousands of hectares the bright green of cane replaced the dark green of the Atlantic Forest.[135] Manoel do Ó, who began working the cane fields in the 1880s and found a sort of freedom from the oppressive waves of cane by securing a job with the railroad, described in an oral history "the completely green horizon that encircled and enclosed the life of the people."[136] His observation is a melancholy echo of Tollenare's early-nineteenth-century description of the *zona da mata*'s "eternal verdure," invoking as it does both the exploitative transformation of the environment and the subjugation of workers.

TWO

A Laboring Landscape
The Environmental Discourse of the Northeast's
Sugar Elite, from Nabuco to Freyre

"The greatness of the land was his greatness.... There went the cattle to pasture and they were his; there went the ox carts groaning under the weight of the sacks of wool or sacks of sugar, and it was all his; there were the negresses of the kitchen, the little urchins of the stable, the field workers, and everything was his. The sun rose, the waters fell from the sky to the earth, the river ran, and everything was his."[1] So José Lins do Rego described his sugar planter grandfather, with whom the writer grew up in the early years of the twentieth century. In this passage from his memoir, Lins do Rego expressed the core elements of the northeastern sugar elite's understanding of power in the early twentieth century. When they gazed at their *engenhos* the planter class saw a laboring landscape—encompassing both human and nonhuman elements, it was a space that they commanded into material and symbolic productivity.

What Lins do Rego described on a personal level in his autobiography, the famous intellectual Gilberto Freyre took to a systemic level in his 1937 book *Northeast,* arguing that sugarcane agriculture had spawned a broad hierarchy. Freyre described "cane's lone dominance over . . . this raw land" and argued that with the ascent of sugarcane, "the nature of the Northeast" changed; "it developed relations of extreme or exaggerated subordination: of some people to others, of some plants to others, of some animals to others; the entire mass of vegetation to imperial and all-powerful cane; the entire variety of human and animal life to the small group of white (or officially white) men—lords of the cane, of the rich lands, of beautiful women, of thoroughbred horses."[2] Freyre articulates the close relationship between the culture of hierarchy and the perception of landscape. The power relationships born of extensive cane agriculture affected society and nature at the same time. In this chapter I argue that long-term historical patterns of land use, legal norms regarding proprietorship, and the weight of more than three hundred years of African slavery structured the planter worldview and led to the emergence of a landscape discourse in the period from the abolitionist politician Joaquim Nabuco to Freyre. In works such as *Northeast* and Lins do Rego's memoir, these writers both expressed and contributed to the construction of this discourse.

This chapter examines work by Nabuco and the longtime planter Júlio Bello, along with Lins do Rego and Freyre, to analyze the landscape discourse that crystallized among the elite in written, elaborated form through multiple mediums in the late nineteenth and early twentieth centuries, a crucial period of change for the region that established a pattern for much of the twentieth century. During this period some members of the traditional planter class were forced out of their positions as a new production model emerged; this crisis not only changed the enterprise of agriculture, it also transmuted itself into this distinct intellectual formation. This discourse came to define how Pernambuco thought about itself, putting in written form an understanding of the world. Most of the texts I analyze here—memoirs, autobiography, novels, sociological commentary—share the themes of memory and nostalgia, linking them to a shared past. By attending closely to the elite discourse of landscape, we can outline the planter *habitus* from which it derived, with its emphasis on the roles of land, power, and the idea of command.

Land, Labor, and Power through the Nineteenth Century

In March 1879, one year after the agricultural conferences in Rio de Janeiro and Pernambuco, Senator Joaquim Nabuco from Pernambuco stood in the imperial parliament to speak about what he called "the problem of the epoch"

—the transition from slave to free labor. As the most visible leader of the abolitionist movement in Brazil, he organized a national abolitionist political party and in 1883 published his famous tract *Abolitionism*. In his 1879 speech, he argued that the labor transition could only take place when workers attained a measure of independence. This, in turn, was only possible with the establishment of an effective land tax, which would help break up the massive holdings of the elite. Referring to the empire's lack of a cadastral survey, he claimed that "Brazil is still more unknown than equatorial Africa. . . . There is no country more in need of geographic studies than ours."[3] Near the end of the nineteenth century, large landowners still presided over domains whose boundaries were defined as much by the reach of their power as by geographic surveys. Ruled as it had been since 1822 by an emperor with broad "moderating power" over all branches of government, Brazil was an overwhelmingly rural society.

Nabuco's colleagues in parliament that day, and their friends and patrons around the country, knew that the empire would not soon survey landholding and begin taxing large owners accurately; the landed families held far too much power to allow that. Almost thirty years before, the 1850 land law had decreed the establishment of a land tax, along with a surveying requirement and a land registry.[4] But despite the interest in modernizing the civil code's provisions regarding land, the empire was still bound by colonial legacies, including a marked deference to the landed elite. So while the land law may have established a land market, it also protected rather than threatened those who already owned land.[5] Land not in active, productive use was supposed to be reported as such, to be placed under the control of the empire for future sale. However, despite enormously extensive latifundia with huge unused areas, very little land reverted to the empire and most informal (nontitled) claims by landowners were legitimized de facto. The degree of large landowner control in Pernambuco was such that, following the land law, only a couple of square miles of land were reported as unused.[6]

The opposition to Nabuco's proposal had deep roots in colonial and imperial policy and culture. In his masterful "memoir" of Brazil's Atlantic Forest, Warren Dean argued that colonial Brazilians never assigned inherent value to land, for three good reasons. First, all land legally belonged to the Crown. The Crown distributed use rights through grants, which expired on the owner's death. This guaranteed an "alienated" owner and gave the family no incentive to improve property. Second, treating property with care offered few rewards even during the lifetime of use, since more land could be requested when the original grant became "tired." Third, the extreme concentration of ownership left all others, whether squatters or tenants, without a vested interest in the long-term health of the land they occupied.[7] These considerations at once produced and nurtured the extreme concentration of ownership, and they had long-term consequences. Land both connoted and conferred power, while the aggrandizement of property required power.

Nabuco recognized this reality and wove a critique of it into his opposition to slavery. In *Abolitionism*, he argued that slavery ought to be understood "in the broadest sense," as not "merely the relationship of the slave to the master . . . [but also] the sum of the power, influence, capital, and patronage of all the masters; the feudalism established in the interior" and the subservience of commerce and the state to the "massed powers of the aristocratic minority."[8] From his perspective, lordship over land and slaves both occupied the same category of pernicious authority and control.

The connection between these elements has a clear history. It was no accident, for instance, that the land law promulgated by the imperial parliament in 1850 transformed land into a commodity at precisely the same moment that the ability to buy humans from overseas as commodities was made illegal.[9] The land law was passed only fourteen days after the law abolishing the slave trade; the elite's awareness of the urgency of responding to the impending abolition of slavery drove debate over the land law.[10] They knew that a greater portion of their wealth derived from the capital represented by slaves than from the value of land.[11] The legislators self-consciously followed the example of Australia, where bureaucrats advocated artificially inflated land prices that would force immigrants to work for a certain period before earning enough to buy their own land, as a means of maintaining a ready labor source in a situation of abundant land.[12]

The creation of a land market, Donald Worster points out, results in "all the complex forces and interactions, beings and processes that we term 'nature' . . . [being] compressed into the simplified abstraction, 'land.'"[13] In northeastern Brazil, however, the land question retained an intimate relationship to the labor question, and the abstraction that resolved itself out of the prolonged debate over land involved far more than "nature." The abstraction represented by ownership was important, but instead of being simplified to the notion of "land," ownership signified "command." Labor remained threaded into the same conceptual fabric as land—as a resource that could be commanded along with the trees or soil that were visibly bound to the earth. Nabuco was right to implicate slavery in this process, since the institution put land and labor in the common category of property. And slaveholding forged specific understandings about the roles of authority and command in the treatment of property.

Nabuco portrayed the effects of this authority—of the lordship over people and places—as inherently destructive. In *Abolitionism* he forcefully argued that slavery harmed human beings *and* the environment surrounding them. All of Brazil's development, from the "conquest of the soil" to the construction of railroads and telegraphs—in short, all accumulation of wealth—was, Nabuco wrote, "none other than a gratuitous donation from the race that toils to the race that forces it to toil." He pointed out that the source of this forced subsidy, the institution of slavery, held the country in a state of backwardness. All Brazilians are "condemned by slavery to form . . . a nation of impoverished men and women."[14]

The link Nabuco drew between slavery's brutal treatment of human beings and its brutal treatment of the environment went beyond the dogma of the time that slavery was simply an inheritance of the past. As José Augusto Pádua has shown, Nabuco shared this intellectual innovation with a few others among the last generation of abolitionists, for whom the environmental destruction wrought by slavery was key to their indictment of the institution.[15] Progress, they felt, meant an end to the anachronism of slavery and a triumph over environmental degradation.[16] Nabuco claimed that under slavery "there is no alliance of man with the earth," so wherever slavery has held sway land is "returned to nature, wrecked and exhausted." Starkly opposing it to the wholesome and free tilling of the soil, he characterized slavery as the "struggle of men *against* the land." Of Pernambuco specifically, Nabuco offered a grim narrative: "More than three hundred years ago the soil was first violated, the forests humbled, the sugar fields set to seed. . . . Slavery everywhere has moved over the land and over the people who accepted it like a wind of destruction."[17]

As Nabuco's parsing of the word *slavery* makes clear, planters put a premium on the *effect* as opposed to the *fact* of property ownership. Planters placed value in controlling land and commanding resources, not necessarily in executing the ceremony of ownership administered by surveyors, notaries, and registrars. In such a world, the power of property could make itself felt without a title inscribed in an official's ledger.[18] As Dean observes, this pattern evolved over the colonial period when planters requested grant after grant from the king in their pursuit of the resources to make sugar. This approach to landholding produced a paradox: "Despite the ease with which pretendants were awarded multiple [grants] and the consequent low price of land, it was axiomatic that 'landholding is the greatest wealth,' because only land bestowed prestige."[19] Dean correctly implies that prestige was distinct from wealth per se; in truth power and prestige accrued only to those who commanded the labor necessary to make land produce.

The result was an elasticity in the understanding of what fell under the effect of ownership, and labor became conceptually embedded in the natural terrain over which planters ruled. In Brazilian society during slavery, the "duty to obey" extended far beyond the master-slave relationship.[20] But the continuity of slavery was not necessary to the maintenance of the link between workers and land. Planters saw it as their right to command labor just as surely as soil, since the latter held little value without the former. This view both contributed to and resulted from the tight linkage between labor and land in the minds of nineteenth-century planters and politicians and had important consequences for continuities in the social formation after the abolition of slavery.[21] The perspective gained inertia over the generations of experience with a world in which power, wealth, labor, race, gender, and the environment were bound together. Those who wielded power did not simply own but rather controlled land and labor, or better put they *commanded the laboring landscape*.

The End of the "Patriarchal Tribe": The Sugar Zone in Transition

In Gilberto Freyre's admiring view, Nabuco's attacks on slavery and the seigniorial privileges surrounding the institution made him "a deserter from his caste, from his class, from his race, whose privileges he combated with a vigor and frankness that . . . left the Parliament of the time astounded."[22] Despite the fiery rhetoric of his abolitionist career, however, Nabuco's autobiography from 1900 reveals a man for whom the memory of the rolling hills of cane and the valorization of the *zona da mata* express his class and racial identity. The recollections from *My Formation* reveal a warm, affectionate view of the region's landscape. He uses lush detail to describe his childhood *engenho* where he was born in 1849, remembering "cane fields spreading themselves out over the landscape, interrupted by winding lines of old palm trees, covered in vines and moss, which hang over either side of the small Ipojuca River." Although he departed the countryside in his ninth year and never again lived there full-time, he captured it with memorable prose and deep nostalgia. For Nabuco, his years in the sugar zone "were those of [his] formation, instinctive or moral, definitive."[23]

Nabuco may also have responded to economic and political changes operating at both the national level and in his home region. In the time between the abolitionist movement and the publication of his autobiography, Brazil let go of slavery and shrugged off its monarchy to become a republic. These two developments were tied to one another, as the monarchy lost legitimacy following abolition, and they took place in sequential years: 1888 and 1889. The coffee industry flexed its muscles in Rio de Janeiro, Minas Gerais, and above all São Paulo, self-consciously driving the nation's economy and spawning development in the center-south. These changes did not favor northeastern planters. The Northeast's two primary exports, sugar and cotton, had provided nearly half of the country's income in 1822, but by 1913 accounted for only 3 percent.[24] Among Pernambucan planters there was a pervasive and deep-seated sense of crisis, of a golden age lost and a grand future deferred or stolen. Though agriculture was seen as the only appropriate enterprise for a Pernambucan man of substance, the national trends of rapid industrialization and urbanization from the 1880s into the twentieth century conflicted with this view. In just five years, between 1890 and 1895, the capital of Brazilian industrial firms grew by more than 50 percent.[25]

The sugar industry also began a major change in its productive organization with the establishment of large mills (*usinas*), which absorbed many smaller *engenhos* and brought downward mobility for many in the traditional planter class. It has often been remarked that *usina* owners frequently came from traditional planter families, but there were many fewer of the mills, and the symbolic fall of other planters from their roost at the summit of patriarchal power may have been bitterer than the material realities of the transition. From the 1880s to 1910, more than sixty *usinas* were established in Pernambuco's sugar zone. Smaller establishments continued to proliferate—in 1914 Pernambuco still had a staggering 2,788 *engenhos*—but the vast majority of these were tiny, inefficient operations, and some of them were defunct.[26]

Nabuco's tone clearly changed between the angry rhetoric of *Abolitionism* and the reflective ease of *My Formation*. Surprisingly, the distaste for slavery disappears in the passages about his childhood, as he sank into great nostalgia for the social system into which he was born. He admitted that slavery was preserved in his memory "as a gentle servitude." "I fear that this particular species of slavery has existed only on very old properties," he continued, "administered generation upon generation with the same spirit of humanity, and where a long heritage of fixed relations between the master and the slaves had made of them a type of patriarchal tribe isolated in the world."[27] Nabuco here distinguished between good and bad masters, rather than taking a monolithic stand against the system, a view that contrasts with his unequivocal assertion in *Abolitionism* that slavery is the same everywhere and evil at the core.[28] The warm "approximation between situations so unequal before the law," Nabuco lamented in his autobiography, "would be impossible in the new and rich farms of the south."[29] By implication, the new mill owners also lacked the appropriate ethic of firm order and paternalism.

Júlio Bello, the long-time owner of an *engenho* named Queimadas and author of a valuable autobiography, agreed. Bello was born in 1873 in the southern sugar-zone municipality of Barreiros and grew up on the banks of the Perşinunga River, which forms Pernambuco's southern border with Alagoas.[30] After completing his studies, he made his way back to "the patriarchal home" and donned the mantle of his father as *senhor de engenho*.[31] Bello's *Memories of a Senhor de Engenho* originally appeared in 1935 and included prefaces by both Freyre and Lins do Rego. In the book, Bello declared that "the time we're living is obviously an era of transition: a new world prepares itself and will emerge in this incessant turmoil."[32] He regretted that a real planter in the old mold was an anachronism, "no more than a shadow that the industrialization of the lands is rapidly obscuring."[33]

Bello referred repeatedly to the "industrialization" of the land, using the term to describe all of the transformations of his world, including the growth and increased dominance of the *usinas*, the increased exploitation of workers, and the degradation of the environment. Bello argued that the old Pernambuco sugar families were naturally suited to power. Workers, he claimed, did not respond as respectfully to newer planters, "obey[ing] much more willingly the orders of *senhores de engenho* from families tied to cane agriculture." He went on to say that "the very lands seem to protest" against the changes, seeming "to cover [themselves] and to refuse the new owners [*usineiros*] like honest women do conquistadors." Bello noted with sorrow that the *engenho* was "no longer a kingdom in miniature, but a miniature of its own past."[34]

Nabuco's patriarchal tribe and Bello's bygone kingdom both revolved around male power and status, but their idyllic extended family was a manufactured relic, a throwback to a mythical time. The images did, though, reflect some aspect of the reality of power holding. We can imagine Nabuco, for instance, reclining into his memory, not of slavery per se, but of the sense of well-being associated with the comfort of superiority and unquestioned dominance.[35] He likened the slaves' behavior to "the dedication of an animal that never changes" and suggested that the Northeast's sugar *engenhos* were not about generating great wealth but rather "existed only to conserve the status of the planter, whose importance and position was judged by the number of his slaves."[36] Bello remembered a similar milieu, acknowledging that "few things were so appealing to my sensibility as seeing an old slave." "Seeing them I remember with intimate nostalgia those I knew as a child."[37] The two men thought the world operated properly when the power to command secured and perpetuated social status.

This equilibrium of a stable hierarchy was challenged at the end of Nabuco's life by the establishment of the republic and the translation of coffee's economic might into the political dominance of São Paulo and Minas Gerais. Thirty years later, as he in turn grew old, Bello witnessed the liberal revolution of 1930, which toppled Nabuco's Old Republic, bringing federal intervention to Pernambuco's politics (as well as to those of other states). The revolution established Getúlio Vargas in power, and the forces that supported him and his modernizing instincts fostered a specter of rising popular power.

As the two men wrote their autobiographies and struggled with the respective changes they had witnessed, both were drawn to landscape metaphors to describe their feelings. Nabuco had addressed the environment in *Abolitionism*, but in *My Formation* he struck a different tone in linking slavery to place.[38] It becomes clear that his fondness for the sugar landscape was linked powerfully to patriarchal social relations, and he described both in strange and revealing ways. "Slavery will remain for many years a national characteristic of Brazil," he wrote, "it spread over our vast solitudes a great smoothness. . . . Its contact first gave form to the virgin nature of the country, and it was this that it watched over; it peopled the country as if it were a natural and living religion, with its myths, its legends, its enchantments." Nabuco thus evoked what was for him an ineffable, but far from malign, aura around the institution of slavery. Over the course of this particular passage, Nabuco's use of the word *slavery* shifted subtly. At first, it signified an aspect of the society as a whole, a "national characteristic." But then it became an active force, giving form to or shaping the country.[39]

Nabuco's writings range from full-throated abolitionism to nostalgic memoir. But consistent in his writing is the idea that slave labor shaped and left its imprint on formless nature. It specifically molded the Northeast that he knew and loved, producing a discrete landscape. Moreover, Nabuco differentiated little between the elements of that landscape. His description of slavery melded into his description of slaves, implying a tendency to recognize no distinction between the African laborers and their condition of bondage. Black workers were placed at the same level as the diligent donkeys and oxen. Indeed, they were also the same as the land itself, which had been "violated" by the institution of slavery. Nabuco might at turns deplore slavery and mourn its loss (and in doing both he would be no more contradictory than many a nineteenth-century Brazilian liberal), but he recognized it as an active and integral component of the sugar landscape. The landscape of the Northeast, for Nabuco, was a tragic terrain of African labor—a laboring landscape.

"God Had Wanted Us to Be Born White and to Rule": The Sugar Landscape in the Literature of José Lins do Rego

The novelist José Lins do Rego's importance to articulating and perpetuating the crystallized elite discourse about the sugar zone is unquestionable. His novels distill most elements of the discourse in their presentation of the environment, labor, and the changes brought by the last decades of the nineteenth century and the opening of the twentieth. Starting with the 1932 publication of *Menino de engenho* (*Plantation Boy*), Lins do Rego produced a series of novels, essays, and memoirs that have penetrated the region's mental world remarkably, a feat realized in part through the books' ubiquitous incorporation into school curricula. Lins do Rego acknowledged that he was strongly influenced by Freyre, but in his own way the novelist questioned some of his peer's ideas about patriarchal hierarchy.[40]

The processes Bello described with dismay as the "industrialization of the lands" ground along during Lins do Rego's childhood on his grandfather's *engenho* in the state of Paraíba, along its border with Pernambuco—"José Lins do Rego's region of southern Paraíba that is almost Pernambuco," as Freyre put it.[41] The expansion in production volume brought by more *usinas* meant that Pernambuco increased its sugar production 50 percent between the beginning of the century and 1930 (Lins do Rego was born in 1901).[42] Lins do Rego had a keen eye for the changes going on around him and self-consciously observed that his class lacked a literary voice. The planters' sons among his classmates were "crude like their fathers," he said, and "indifferent to any and all distractions with books"; he singled out Bello as an exception. The marginality of intellectual and literary production that Lins do Rego identified makes his body of work all the more important as a register and repository of planter discourse.[43]

Plantation Boy was the first in a series of five novels that came to be called the "Sugar Cycle," the last of which appeared in 1936. Lins do Rego published *Fogo morto*, his most critically acclaimed novel, in 1943, after which he wrote mostly short pieces, fictional and autobiographical, until his death in 1957.[44]

About the Sugar Cycle, Lins do Rego wrote that he wanted "merely to write some memories that would be those of all children reared in the big houses of the northeastern *engenhos*." But the books took on a life of their own, he claimed, and brought out things "hidden in his interior."[45] Indeed, when we compare his memoirs to *Plantation Boy* it becomes immediately clear that the latter is patently, transparently autobiographical. His memoirs focus exclusively on what Lins do Rego called "my green years"—the early years of childhood. He sought to regain those early experiences, he wrote, "which fade with time but stay embedded in the writer's memory and provide so much material."[46] That material proved extremely compelling to a large readership, especially since it was woven into a discourse that was familiar to at least his northeastern readers.

Lins do Rego ended the Sugar Cycle with the novel *Usina*; appropriately, since he felt that the advent of the *usinas* marked the end of an important epoch in the sugar zone. Throughout the cycle, *usinas* embody a set of moralistic ills as they set about destroying a way of life: they devour land insatiably, destroy the balance of *engenho* life, displace common workers, and rule the lives of their owners.[47] Like Bello, Lins do Rego framed the transformations of his time through the effects on the land, linking changes in the social structure to changes in the environment. One of his protagonists ruminates on the impending changes after he sells his *engenho* to an *usina*: "Tomorrow a factory chimney would dominate the cashew trees. . . . The land would learn what it was to work for a factory, and the urchins would come to know what it meant to go hungry."[48] Chronicling his present and predicting the future, Lins do Rego expressed environmental and social changes as one multifaceted complex. When he referenced the conservation efforts of his protagonist's grandfather (reminiscent of measures Bello took to protect the remaining forest on his *engenho*), it appears as a truly *conservative* attempt to save a landscape rich in meaning.[49]

Many of the meanings Lins do Rego associated with this landscape carried assumptions about race, gender, and power. His meditations on changes in the environment were not separate from his perspective on a social structure that included assumptions about race and gender. In *Fogo morto*, for instance, Lins do Rego describes the once powerful and vital *senhor de engenho* Captain Tomás, who becomes sick with despair over an ill daughter. A slave's escape from Tomás's *engenho* revives him, as he engages in vigorous and single-minded pursuit. Tipped off that the slave has sought refuge on another man's *engenho*, Tomás has a humiliating interaction with the defiant *senhor*. After the slave returns, he vents his frustration by ordering a harsh punish-

Engenho Araticuns, northern *zona da mata*, 1940s–50s. (Acervo Fundação Joaquim Nabuco, Recife)

ment.⁵⁰ The episode demonstrates the links between masculinity and the planter worldview in its portrayal of the revitalization Tomás experiences when he springs into action following his slave's escape and in the battle of seigniorial will between Tomás and his recalcitrant peer. At stake in the confrontation is the exercise of the masculine capacity to command.

In his memoirs, Lins do Rego describes a friendship with his grandfather's stable boy, a young black man who brought him to the realization that "life was great for whites because there were blacks."⁵¹ But Lins do Rego viewed the young man as a curiosity more than anything else, an "abandoned Negro" full of sage observations of the world but never an equal.⁵² Even this modest reflexivity about the racial politics of the sugar zone is remarkable, and this quality sets Lins do Rego apart from progenitors such as Nabuco. He questioned his own place in the hierarchy of power that placed landowning white men above black workers. Lins do Rego even wrote an entire novel from the perspective of a poor, African-descended boy (*Moleque Ricardo*), indicating his willingness to address social disparities.⁵³ Even his more frequently recurring (and autobiographical) character Carlos reflects that "books had begun to teach me to pity such poor people" as the workers on his grandfather's *engenho*.[54] A later book brusquely dismissed these fancies, when Lins do Rego's narrator acknowledges that "it is very easy for anyone who has all the comforts and all the advantages to say that he envies the happiness of the poor, who live like animals. It is just hypocrisy and meanness."[55]

Lins do Rego's alter ego, Carlos, further clarifies the natural order of things. He observes of his grandfather's workers: "I got so used to seeing these people every day in their degradation and misery that I accepted their state as something natural. . . . I thought it quite natural to see them living and sleeping in a pigsty, having nothing to eat, and working like beasts of burden."[56] The repeated use of the word *natural* underlines the point that planters viewed their power as running straight through workers and into the natural world. The novelist makes this even clearer in the rest of the passage. "My understanding of life at that time made me accept all this as part of God's plan. They had been born like this because God had wanted it that way just like God had wanted us to be born white and to rule [*mandar*] over them. Did we not also rule over the cattle, the donkeys, the fields, and the forests?"[57] Lins do Rego could hardly have been more explicit in describing the planters' view of the laboring landscape over which they ruled. His words recall Bello's contented description of his family's old slaves, "bound to the service of the *engenho* like oxen and donkeys."[58] The passage from Lins do Rego also underlines the crucial salience of color, demonstrating continuities from race-based slavery. God mandated that the narrator and his family be born white, to rule over "them"—the laborers—assumed to be African-descended.[59]

As the power of old *senhores* relative to the new *usineiros* eroded, Bello and Lins do Rego both feared, so would the "natural" equilibrium of the sugar region. The old planters' control over their lands and workers held all of those elements together in a rough balance. But workers and land alike were degraded by the new modern and efficient industry. For Bello, the sight of an old slave reminded him of the days when the social order, including the bondage of African slaves, was intact. But the "industrialization" of the land rendered some familiar agricultural practices obsolete, along with the social relations they structured. The many references to the destruction wrought by a modern, industrial-scale sugar industry and to the uplifting qualities of nature contribute to what Luciano Trigo calls Lins do Rego's "ecological vision."[60] Lins do Rego's and Bello's true interest lay in the resurrection of a disintegrating social structure. They loved the land because, as Bello put it, "the land conferred privileges of nobility."[61]

Lins do Rego's contemporaries and planters who came later recognized themselves in his books, perhaps the most compelling evidence of his relevance to our understanding of the elite worldview. A cynical reading of his landscape discourse sees it mobilized toward the "idealization of the oligarchic order, camouflaged by praise of its integration with nature."[62] But it was more open than that and more reflexive. Even Francisco Julião, who was born to a landowning family but fought against and tried to bring down that order as a peasant leader and perceived radical in the late 1950s, remarked in the 1980s, "I have the memories of a plantation boy, because it was where I lived, was born, and grew up. . . . It was a childhood that I find in the books of José Lins do Rego, who wrote about the Northeast, sugarcane, and the cane field."[63]

Lins do Rego's work distilled nostalgia and conjured the shared memories of rural power for an entire class. His good friend Gilberto Freyre integrated these same themes into his formidable and influential corpus, and the two maintained a close dialogue. Freyre took seriously the environmental themes prevalent in the work of Nabuco, Bello, and Lins do Rego, explicitly analyzing the sugar-producing Northeast as a landscape. Along with Lins do Rego, Freyre helped shape the way several generations have viewed the Northeast, and indeed Brazil as a whole.[64] Born in 1900, Freyre responded to the same factors that so preoccupied others of his generation. When he returned to Recife in the 1920s after several years of teaching, study, and travel in the United States and Europe, Freyre eventually found a patron in Governor Estácio Coimbra, serving as chief of staff to this politically minded scion of a sugar family. So Freyre's response to Getúlio Vargas's revolution in 1930, which he suspected for its antitraditional cast, also carried a personal edge as Coimbra fell from power.[65] Vargas broke the traditional power bloc of the economically important center-south states—Rio de Janeiro, São Paulo, and Minas Gerais—but he also threatened the power of large landowners and other traditional oligarchs in other states.

With the publication in 1933 of *Casa-grande e senzala* (translated in 1946 as *The Masters and the Slaves*), Freyre reoriented the prevailing discussions about the meanings of the nation, the significance of race, and the nature of Brazil's cultural heritage. The book, as a major critic put it, represented a "rupture" with the traditional historiography and confronted "the intellectual, theoretical, and methodological backwardness that characterized social and historical studies in Brazil."[66] The book "turned on its head" one of the elite's central preoccupations: whether miscegenation had doomed the country to backwardness and degeneracy.[67] These contributions brought him fame, but Freyre also forced intellectuals to confront the issue of human society's interactions with the environment.

In *The Masters and the Slaves* and his second book, *The Mansions and the Shanties,* Freyre took up the two sides of Nabuco's discourse about slavery and the sugar zone's past. *The Masters and the Slaves* shows Freyre exultant, describing the rich, hybridized creation of authentic Brazilian culture, while in *The Mansions and the Shanties,* where he records the beginning of this society's deterioration, his tone shifts to one of loss and bitterness. So in his writing Freyre made explicit a chronology implicit in Nabuco's work—the old time of the "patriarchal tribe" giving way to the soulless capitalism that revealed slavery's rotten core. A close look at Freyre's work reveals a significant debt to the earlier Pernambucan statesman; the two certainly shared the conviction of the superiority of the traditional patriarchal social relations on the *engenhos*. With the growth of factories, the obsolescence of the old plantation system, the efflorescence of class politics, and increasing entrepreneurialism among recent immigrants, Freyre "confronted a new style of life and politics and [was] not quite sure [he] liked it."[68] He glorified the northeastern sugar fields as a response to this anxiety even though, like Nabuco, he himself spent his life in the city. It was as an older man that he retreated to genteel repose at Apipucos, a Recife suburb that had long before been an *engenho*. Before then, his experiences in the sugar zone were limited to visits with family members and friends. His own father was a businessman from Recife a generation removed from any family history of sugar production.[69]

It was precisely as a city dweller that Freyre developed an enthusiasm for the cane countryside. Lins do Rego observed Freyre's eagerness when the two traveled together to Paraíba as young men. The novelist "show[ed] my people and my land" to Freyre on a trip to family *engenhos* in Paraíba's stretch of the sugar zone. "The fields of cane, the old *engenhos*, the uncles, the aunts, and everything," Lins do Rego wrote, "appeared to him better than I thought it could possibly be. I took him afraid that it would disappoint him and, on the contrary, he liked it all very much."[70] Calling it the "most aristocratic" of the country's regions, Freyre treated the Northeast as a metonym for the country as a whole and as the true cradle of Brazilian civilization and culture. In his work, the sugar zone's history, heritage, and legacies blurred into the nation's culture with little mediation. He argued that the constituent parts of the northeastern sugar society had been, and in some important way continued to be, "the most typical elements of the Brazilian landscape, physical as well as social."[71] Brazil's basic pattern of social organization, Freyre implied, mirrored the sugar zone's.[72]

A Brazil "Sweetened by Sugar and the Negro": Gilberto Freyre's *Northeast*

In *Northeast*, from 1937, Freyre explicitly addressed the region's environment and, more subtly, the labor by which it had been transformed. Freyre described the book as narrating the "drama of monoculture," but it reads like an extended metaphor for the social and cultural history of the Northeast. Indeed, he made clear his position that the natural history of the region was at the same time its social history. It is as if Freyre rewrote *The Masters and the Slaves*, with the environment playing the central role instead of the Portuguese, Africans, and Indians who were the protagonists of his first major book. The same core elements remain—violence, sex, accommodation, and the forging of a new culture (in this case, a monoculture). "Never more violent in its beginnings," he wrote, "was the drama of monoculture than in the Northeast of Brazil." The rivers of the Northeast were "prostituted by sugar," while the cane field "deflowered this virgin forest" of the sugar zone. The "virgin vegetation" he paired with the "native life" of the region (i.e., the indigenous peoples), furthering his sexualized and racialized conflation of people and land.[73]

Northeast is in many ways a breathtaking essay, presented with force and passion. But it is as contradictory as it is fervent. Even as Freyre mourned the effects of sugar monoculture in general, certain passages seem to indict only the most recent agricultural changes for the crop's more grievous offenses. He felt that the new *usinas* made the most brutal assaults on "a nature degraded by sugarcane to the ultimate extremes" and aggravated the "enormous deformation of man and the landscape by monoculture."[74] Once again, Freyre's conceptualization of the consequences of sugar agriculture owes its shape to Nabuco. This particular quotation recalls Nabuco's statement that sugar and slavery "scarred the land and the society with every symptom of premature decline."[75] Also like Nabuco, Freyre took special issue with the more recent, industrial-scale developments in the region, which both men feared would disrupt the landscape. Freyre wrote that the "harmony" between a society and a landscape is a "cultural expression" more than the "imposition of nature." This balance was achieved over the long term and could sometimes be disrupted, as with the development of *usinas*.[76]

Freyre's utopia of a society integrated seamlessly into its surroundings, which he described as "landscape" (*paisagem*) or occasionally "region," was borrowed from the geographer Carl Ortwin Sauer, who taught at the University of California. Freyre acknowledged this debt and praised Sauer as a scholar of the highest achievement. "The regional environment tends . . . to make the man, the group, the human culture in its image," Freyre writes, "but, for its part . . . the human culture acts on the regional environment, altering it in some cases profoundly." He described his own work in *Northeast* not as geography but as regional or ecological sociology, the best of which he said "never discounts the 'qualities' of a landscape."[77]

Freyre's celebration of the Northeast ran against the grain of national intellectual culture, since he achieved prominence during the middle of a period that would come to define the region as backward. As Cristovam Buarque recently observed, "toward the end of the nineteenth century, Brazil discovered the Northeast—as a problem."[78] Some of the very characteristics that conspired to create this image Freyre proudly touted as the ones that made the region so important to the nation's history—the prevalence, for instance, of self-styled aristocratic traditions. Belying his championing of a Northeast that was the core of the nation, Freyre acknowledged that "there are more than two Northeasts, not just one, much less the monolithic North of which they speak so much in the South with the exaggeration of simplification."[79] He recognized the separation between "the agrarian and the pastoral," the wet coastal region and the dry *sertão*.[80] His Northeast was the sugar zone or, as he said of Nabuco's passages about home, the "Brazil more sweetened by sugar and the Negro."[81] At the same time that an image was developing of the region as being most out of step with the country's march to modernity, Freyre constructed a defiant paean to tradition and attempted to redirect the nation's gaze toward the sugar zone and the past.

Freyre also refused to adopt prevalent ideas about the physical characteristics of his tropical region. When Freyre was born, "Brazilians lived in the shadow of two determinisms: race and climate," as Thomas Skidmore writes. While separable in theory, the two were closely linked in people's thought.[82] Freyre referred to these determinisms repeatedly and was explicit in his intention to challenge the work of "those who, oblivious of [the influence of] diet, would attribute everything to the factors of *race* and *climate*."[83] He emphasized the importance of diet to both Portugal's and Brazil's history, arguing that the "physical inferiority of the Brazilian" frequently attributed to race or climate is actually "due to the bad management of our natural resources in the matter of nutrition."[84] The culprits to be blamed for an attenuated empire and a flawed colonialism were monoculture and land monopoly; these had produced Brazil's "deepest wounds." Freyre thought that a comparison with slavocratic societies in other geographical areas, such as the Antilles and the United States, revealed the primacy of social and economic factors over the influences of race and climate in social development.[85]

Despite his conviction on these points, Freyre still argued that climate and environment retained a central role in the development of Brazilian colonialism and by extension contemporary Brazilian society. Before Freyre, one of the most influential Brazilian voices on the subject was Euclides da Cunha, who declared in his widely read book *Os sertões* (1902) that "the man of the backlands stands in a functional relation to the earth." Da Cunha found in the *sertão* the "bedrock race of our nation," a race that was forged in the heat and dust of the rugged interior, a hardy mix of different bloodlines that represented something new and uniquely Brazilian.[86] From Freyre's perspective, the shaping influence of the environment represented a democratizing force—the different races of Brazil, even as they mixed promiscuously in the bedroom, were also converging due to their shared environment.[87] Freyre also elaborated his position in relation to those scholars who had made more contemporary statements on the issue, negotiating a careful path between Ellen Churchill Semple, Ellsworth Huntington, Carl Ortwin Sauer, Griffith Taylor, Franz Boas, and others.[88] He rejected Huntington, whose *Civilization and Climate* (1915) carried great weight from the 1910s through the 1940s, as a "fanatic" on the subject of climatic influence.[89]

Freyre also did not fully agree with Semple, the pathbreaking geographer from the University of California whose *Influences of Geographic Environment* (1911) challenged deterministic links between environment and race.[90] In a discussion of her work, Freyre granted that "no one any longer looks upon climate as the almighty being of old," but he also signaled his continued belief in the transformative influence of the environment. He wrote, "it is impossible to deny the influence that climate exerts upon the formation and development of societies: if not a direct influence, through its immediate effects upon man, an indirect one through its relation to the productivity of the earth, the sources of nutrition, and the means of economic exploitation available to the settler." From this cautious position Freyre made occasional forays, staking more ambitious claims about the impact of "the tropics" on the Portuguese and their cofounders, as he put it, of Brazilian society—the African and the Indian. He romanticized, for instance, the connection between Indians and the land, and lamented that the European intrusion disrupted "the balance in the relations of man to his physical environment." "It was sugar that killed the Indian," Freyre concluded, and he thought that an expanded cane monoculture spelled grave consequences for the northeastern environment.[91]

The Portuguese, in contrast to Indians, had no tradition (or not for several hundred years) of closeness with the environment, something Freyre attributed to "a bourgeois and Semitic mercantilism on the one hand and, on the other, Moorish slavery followed by Negro slavery." He detailed the disjuncture between the colonizers and their surroundings, a gulf that he argued still existed: "We are still in large part a people incompletely integrated in the tropical or American habitat." He pointed ironically at his fellow Brazilians' tendency to refer to any creature as, simply, a *bicho* (animal). "Even in the rural districts," he concluded, "every June bug is a beetle and nothing more."[92]

Freyre wrote in a style driven by environmental images and metaphors, and he frequently conflated cultural and geographic characteristics. Waxing verbose about the fertility and sensuality of mulatto women, or "women the color of fertile soil," Freyre compared their virtues to "the best farming land in the colony." Elsewhere he wrote that the Portuguese "took from Africa the topsoil of a black people that was to fertilize its cane fields and coffee groves, assuage its parched lands, and round out the wealth afforded by its patches of fertile soil."[93] This passage shows the mark of Nabuco's influence, recalling the abolitionist's description of African slaves giving "form to the virgin nature of the country."[94] Even in the book *Northeast*, Freyre's "ecological study" of the region, he is charmed by aspects of the "civilization of sugar." He loved the aristocracy, the ladies bathing in rivers, the endless lists of sugary desserts prepared in the noble big houses of the cane fields.[95]

Freyre felt that the sugar civilization took the only form that would bring success to the colonial venture, but its particular development determined understandings of race and labor. In *The Masters and the Slaves*, he exhorted his readers to have the honesty to acknowledge that only a society characterized by latifundia and slavery could have surmounted the "enormous obstacles in the way of the European civilization of Brazil." Only "the rich plantation-owner and the Negro capable of agricultural exertion and compelled to it by a system of slave labor" could have accomplished the "civilization" of Brazil. This civilizing process transformed the environment as well as the agents of that transformation. The combination of "nature and the system of labor" resulted in the "Brazilianization" of the African slaves. These forces also racialized the very nature of labor. Freyre referred several times to the linguistic development that produced *mourejar* (from *Moor*) as a verb for work, similar to the common phrase "to work like a moor."[96] One can understand, he wrote, why those "who founded the culture of the sugar-cane in the American tropics, under physical conditions so adverse, should have become imbued with the belief that 'work is for the black man.'" The corollary of this steadfast belief, Freyre pointed out, was a "horror of manual toil" among slave owners.[97] The planter class accommodated itself to command, not work.

Senhores in Command: The Foundations of the Laboring Landscape

The texts I have analyzed in this chapter articulated a class-based discourse that was expressed in terms of a landscape—a thing that exists in the mind as much as the world. A landscape is both an expression of a society's impact on its environment and the environment's record of influence over a society; the concept implies the shared product of the dialectical relationship between humans and their surroundings. Critically, the term *landscape* carries within it a reference to the ideas about and perspectives on the environment prevailing in a given place. For members of the sugar elite, the landscape they created was deeply influenced by experience with and assumption of the right to command. They expressed command as an ethic, or a guiding code of conduct on the *engenho*. This understanding helped shape a particular planter *habitus*, which in turn provided the foundations for the elite's discourse of the laboring landscape.

The laboring landscape was a portable, communal, and durable discourse; it was shared by more than one person, was persistent over time, and was seen throughout the sugar zone.[98] This consistency was owed to the discourse's basis in a planter *habitus*, by definition widespread and stable. Their *habitus* centered on authority and the will to command, notions captured by the Portuguese verb *mandar*. The will to command was reinforced through violence, but it was also assumed by planters as a class prerogative. It may be that further research will demonstrate the concept's applicability in the Caribbean and other sugar-growing areas or places of dense plantation cultivation generally. My point here is not to argue for the distinctiveness of this tradition of seigniorial authority but to emphasize its relationship to the crystallization of a particular discourse of landscape during a period of crisis and transformation in the cane zone.

Nabuco and Bello, Freyre, and Lins do Rego tracked social and economic changes that they regarded with discomfort. They described the decline of a benevolent rural patriarchy and the rise of impersonal capitalism.[99] Nabuco observed the early experimentation with large, "central mills," while the other writers witnessed the consolidation of *usina* domination. Unease and foreboding marked the earlier experiences, while profound regret and nostalgia marked the second. But the period as a whole generated a common discourse, one that linked land and labor into one production- and prestige-oriented whole—the natural and human fuel that drove the manufacturing of sugar.[100] These men shared a belief in the nobility of Pernambuco's sugary past and the philistinism of Brazil's new, southern, industrial future. This regional antagonism resulted in part from conflicting, or increasingly divergent, notions of the meanings of land and labor. The northeasterners, though, did share with their southern peers a common foundation in African slavery.

Nabuco described how a planter worldview revolving around the will to command spread outward from its institutional home in slavery. "Our character, our temperament, our whole physical, intellectual, and moral organization are profoundly afflicted by influences which . . . were infused into Brazilian society [by the institution of slavery]," he wrote in *Abolitionism*, and "the task of annulling those influences is certainly beyond the ability of one generation." Society would have to address racism, he went on, as well as eliminate "the gradual stratification of three hundred years of slavery, of despotism, superstition, and ignorance."[101] Nabuco insisted that the word *slavery* signified "the spirit, the living principle, which animates the entire institution," which he saw as unbridled power and control. He saw this power to command, which extended from masters and slaves to patron-client relations, as the malign root of the country's problem, not just the foundation of the legal institution of slavery. "There is no denying the truth of the formula 'Power is Power,'" he wrote. "It is our entire history."[102]

That history did not end with the end of slavery, of course. The formula "Power is Power" continued to hold true, as did the norms of coercion, discrimination, and prepotency that accompanied it. Despite slavery's abolition, workers labored under what Denise Soares de Moura describes as "the paternalistic Brazilian cultural heritage, associated with violence and with personalized relations of power."[103] These ongoing norms of power and exploitation parallel the emancipation experience in Cuba where, as Rebecca Scott has shown, an incremental process of abolition "did not require planters to break their habit of direct coercion," whether physical or economic.[104] Moura writes, in her study of freemen during the waning days of slavery in São Paulo, "the latent will to command [*mandonismo*] in the sentiments of the *senhores* spread into the relations of work" at times, and well beyond it, one might add.¹⁰⁵ These sentiments were equally if not more applicable to Pernambucan attitudes about power and control.

Freyre captured the subtlety and centrality of command in an essay on changes in Pernambuco during the century leading up to 1925. He cited a mid-nineteenth-century planter who asserted that *engenho* owners were tied to the land "like oysters to the rocks." The tie that bound them, Freyre argued, was "the imperative to command [*a tarefa de mandar*]." Unlike the easy rule enjoyed by "today's millionaires," the sugar planter needed to apply constant attention and action. Freyre imagined planters directing their slaves in the fields "speaking harshly to enormous Negros armed with knives and machetes—and them without any weapon, certain of the influence of their voice and gesture."¹⁰⁶ They were not truly linked to the land itself "like oysters to rocks"; they were sustained by the power land provided. They inhabited and embodied that power, assuming that their voices and gestures would transmit their authority. The centrality of command also appears repeatedly in the writing of scholars and observers of Pernambuco's sugar zone. Carlos Mota describes the contradiction between the "modernizing perspective" of capitalist development early in the twentieth century and "the *senhor de engenho*'s will to command [*mandonismo*]."¹⁰⁷ The literary critic Sérgio Milliet, writing about a planter in a Lins do Rego novel, described "the peaceful satisfactions of authority [*mandonismo*] and wealth."¹⁰⁸

Gregório Bezerra, who would later become an important Communist Party organizer in the sugar zone, recounted childhood experiences on an *engenho* shortly after the twentieth century began. Almost exactly the same age as Freyre and Lins do Rego, Bezerra presents rural worker memories that confirm the shape of the patriarchal, patronage-driven society evoked by those upper-class members of his generation. In the young Bezerra's world, the planter's rule went unquestioned, and to it everything was subjugated. He was struck by the warlike appearance of the *engenho*'s strongman, Pelegrino, who carried on his rounds a knife, a dagger, a club, and a double-barreled pistol. "He had been the planter's slave," Bezerra wrote, "but despite being formally free, he remained in slavery." Pelegrino "could seize, disarm, and beat the workers, if he so wished." His prerogative was sanctioned by the "authority of the *engenho*"; that is, the word of the planter. When the owner rented the *engenho* to a different planter, Bezerra wrote that "the new occupant was the lord of the big house, of the mill and of everything, including the inhabitants."¹⁰⁹ This matter-of-fact statement from the perspective of an impoverished worker exactly parallels the quotation from Lins do Rego's memoirs with which I began this chapter; both evoke the sweeping reach of planter power.¹¹⁰ Both Bezerra and Lins do Rego explicitly classify human and nonhuman nature under the planters' domination.¹¹¹

Bezerra's reference to the former slave Pelegrino's continued subjugation also points to continuities in racial understandings in the sugar zone. These were not separable from ideas about labor, and land as well. As slippery as the categories of race and class were in this area, there was a presumed overlap between workers and blackness. This comes through clearly in a short 1934 essay by an Afro-Brazilian named Jovino da Raiz, who compared the "black worker" under the old *engenho* regime and under the new, *usina*-dominated regime. Throughout his four-page text, which he presented at an Afro-Brazilian Congress organized by Freyre, Raiz assumed perfect synonymy between the terms *black* and *worker*.¹¹² In Freyre's argument from *Northeast*, race, like all other hierarchies in his society, derived from the placing of cane above everything in the "natural" world. With that claim, he implied that black workers' inferior position reflected an inexorable ordering of the world, in which the power flowing through social and environmental processes affected them all in the same way because it was the same power, originating from planters.

The sense among the dominant class of how it experienced and wielded power originated in the practices of social interaction forged under slavery. In the everyday practice of their lives, they exercised the prerogative of command that their power granted them. Pierre Bourdieu offers no shorthand definition, but he refers to *habitus* as "a socially constituted system of cognitive and motivating structures." He further describes it as "schemes of perception and appreciation" that form an organizing principle revealing itself only in actions.¹¹³

Bourdieu's concept of *habitus* refers to the lived expression of beliefs and understandings that accrete over generations.¹¹⁴ At his most Delphic, Bourdieu calls *habitus* "history turned into nature." That is, individuals absorb the dispositions, understandings, systems of meaning, and habits from their particular social group and the larger social structure, all of these elements having constituted themselves over time. "Individuals carry with them, at all times and in all places," Bourdieu writes, "their present *and past* positions in the social structure." For this reason, he insists, "'interpersonal' relations are never, except in appearance, *individual-to-individual* relationships and . . . the truth of the interaction is never entirely contained in the interaction" because all individuals carry with them accumulated cultural dispositions. *Habitus* not only encompasses ways of thinking about the world, it embeds itself in individuals' bodies, guiding the way they stand, walk, and crack a whip.[115] With nearly four centuries of slavery structuring their *habitus*, the planters' worldview and behavior dominated social relations and individual perceptions even after abolition.

As their access to total control slowly began to erode with economic transformation, some members of the elite reacted with acute nostalgia. This explains a powerful attraction to childhood evinced by the authors studied here. Nabuco wrote in his autobiography that "the entire line of life is for many an outline from childhood forgotten by the man, but to which he will always be unconsciously bound."[116] In Freyre's introduction to an edition of Nabuco's autobiography, the younger scholar wished the old politician had included even more from his childhood, since Freyre felt that "nostalgia of the past" was so important.[117] For his part, Lins do Rego wrote down the memories of his "green years" in order to "regain all that remained of that 'aurora,' which . . . was nostalgia."[118] This absorption with childhood reflected more than a constant yearning for a personal past; it also represented nostalgia for a collective past.[119] The emphasis on nostalgia in analyses of these authors should not obscure the insights their work offers into the nature of authority in this region.

The roots of the elite's landscape discourse become clearer when we reverse one of Bourdieu's formulas. The process at work in the sugar zone can be perceived not only as history turning into nature but also as nature turning into history. The ways the members of the elite took the "nature" around them and incorporated it into a particular history helped structure their memories of the environment and their disposition toward the elements of their personal stories. They embedded in the sugar fields around them a history of their own naturalized power and the subjugation of African-descended workers and the environment. Power operates through "nature" just as it operates through race, class, and gender, since environmental change and the effects of environmental constraints on human beings have consequences that are levied on societies in close accordance with power differentials.[120] The extension of elite power into the environment took place through the bodies of workers and thereby bound the environment and workers together in elite memory. In naturalizing this state of things, the elite also projected its views on its workers. Bello, for instance, wrote that former slaves remained with their former masters, performing the same work as ever. Since they stayed at work on the *engenhos*, Bello assumed, those former slaves must have recognized their natural positions.[121]

Landscape became the repository for shared understandings of authority. "Nature," or the space of the sugar *engenho*, became the carrier of shared meanings, the spatial embodiment of the collective past. Those meanings include the assumption of planter power, as well as the enforcement of that power through violence. The elite expressed its nostalgia for the bygone world as nostalgia for place because what it mourned was the loss of a *landscape*—an environment overlaid by a set of assumptions and understandings—that characterized the world of the *engenhos*.[122] This was, as Simon Schama has written of landscape generally, "a work of the mind. Its scenery is built up as much from the strata of memory as from layers of rock."[123] The fact that the elite was not nostalgic about a particular *physical* environment is clear from the curious ways the old *senhores* were actually distant from that environment.[124] The former planter Henrique Milet wrote in 1878 that all of his book-learning about sugar agriculture was inferior to the local workers' knowledge, since they "knew conditions better by eye, from the nature of the spontaneous products of the soil."[125] There was a definite territorial aspect to planter families' lives and to their mourning for traditions gone by, but there was no delicacy in their relationship with the environment. What they proclaimed as telluric love was not passion for nature but fondness for the dynamics of racial, class, and environmental exploitation that constituted the laboring landscape of the sugar zone.

THREE

A Landscape of Captivity
Power and the Definition of Work and Space

Manoel do Ó, born in Pernambuco's cane-growing region in 1869, began working in the fields at the age of twelve. Interviewed almost a hundred years later for a book about his life, he said that "the green horizons of the Salgado Mill lands were the limits of the world." He desperately sought to break free of those limits, working thirty-six different jobs in fifteen years before finally achieving a kind of freedom with a railroad job that he secured at the end of the 1890s.[1] Ó's memories poignantly evoke a widespread sense among workers of planters' suffocating power, visibly represented by the monotonous green cane fields. Workers' urge to flee the oppression of the cane fields reflected their feelings of captivity, a part of their apprehension of landscape strongly linked to the legacy of slavery in the region.

Manoel made his escape a few years after the abolition of slavery. Though most cane workers were already free by the time of abolition in 1888, the period did bring changes to working conditions and relationships between workers and bosses. Over the decades after Manoel's flight, new norms emerged governing the provision of work and its compensation. Cane work dictated the rhythm of workers' lives. The details and materiality of labor, especially agricultural labor, shaped workers' understanding of and ideas about their surroundings. "Humans have known nature by digging in the earth, planting seeds, and harvesting plants," Richard White has written. Through these and all the other tasks of work in "nature," "they have achieved a bodily knowledge of the natural world."[2] The work that produced that "bodily knowledge" was performed in the context of specific power relations.

Labor and the power relations governing its execution molded cane workers' experiences of the landscape. Having depicted the planters' discourse in the previous chapter, arguing that they saw their domains as "laboring landscapes"—productive wholes in which human elements (workers) and natural ones were equally subject to patriarchal command—I now turn to the lives of workers and their central metaphor for describing the landscape. The laboring landscape reduced workers to natural elements, alongside oxen. The elite view necessarily dominated narratives of the region, since so few voices emerged from among workers (their voicelessness being an index of planter power). Despite the paucity of sources for resolving their perspective, however, we find traces of workers' struggle to distinguish themselves from the natural world even as they consistently characterize their conditions using the metaphor of captivity. We must reach forward in time to find evidence of these discourses, which nevertheless rest on longer-term traditions (as with folklore, for instance). After describing workers' living arrangements and working conditions, I discuss efforts by social scientists in the 1930s and 1940s to study labor conditions. These efforts helped add to the number of worker voices registered in durable form, at first among those rural workers who had "escaped" to the city, and it led eventually to investigations of the rural world. I end the chapter with an analysis of workers' engagements with their environment.

From Slavery to Tenantry: The Continuity of Captivity

Joaquim Manuel dos Santos, called "Paulo" after a grandfather, was born in the northern *mata* town of Nazaré da Mata in 1922. Without land of their own to cultivate, Paulo and his family lived on an *engenho*, his father working the cane fields. After his parents died when he was eight years old, Paulo began working on small farms, clearing brush and picking cotton. There was less cane then in the northern *mata* and more cotton and vegetable farms, he said in a 2003 interview. Like Manoel do Ó, Paulo moved from one job to the next, but when he was fourteen, after several years on small farms, he began cane work. Work on the *engenhos* was "such a cruel system," he said, that even cane workers today do not understand how hard it was. He continued to perform cane work into the 1980s, when he entered union administration.

Between Manoel do Ó's first years in the fields and Paulo's, an increasing percentage of Pernambuco's sugar came from *usinas*, or large mills, and less came from traditional *engenhos*. In characterizing a sluggish pace of change in social norms over the course of his long life, Paulo believed that he saw deep continuities with periods long before his birth. He suggested that the transition represented by abolition was illusory, simply a change to a new system "on the same terms." The social system he was born into, he thought, owed its shape to "habits that the bosses had before, [from] four hundred years and more of slavery."[3] Paulo was right, in the sense that power relations on *engenhos* were structured under slavery, but they also owed their shape in the early twentieth century to dynamics surrounding slavery's end. A look back to earlier in the nineteenth century, and through Manoel do Ó's lifetime, will situate Paulo's experiences.

The French cotton buyer L. F. Tollenare's satirical description of Pernambuco's cane zone in the 1810s has become a standard reference for the region's social organization under slavery: "With bare legs, wearing only a shirt and drawers or a calico dressing gown, the *senhor de engenho*, armed with a whip and visiting the dependencies of his estate, is a king who finds only animals about him: his black slaves whom he mistreats; his squatters or *moradores*; and some hostile vassals who are his tenants."[4] His dress changed over time, but the planter continued to wield unquestioned and extensive power over his workers. Tollenare estimated that the tenant category, consisting of "mestizos, mulattos, free blacks, and Indians," constituted 95 percent of the nonslave population.[5] As the slave population dwindled over the decades following Tollenare's visit, the tenant category absorbed many of these workers. The brisk internal slave trade to the southern coffee plantations drained part of the slave population and drove their prices higher, compelling planters to turn to their squatters and "hostile vassals" for labor. While squatters already provided some level of "service" to landowners, planters began to demand obligatory work from their tenants instead of merely a share of their yearly harvests or earnings.[6] Gradually, the boundaries between categories blurred, while the provision of labor became increasingly important to all relationships between planters and residents on *engenhos*. Crucially for planters, the system kept workers bound to *engenhos* even without the legal tie of slavery. In other areas, such as the Caribbean, planters resorted to other legal solutions to the problem of fixing workers on plantations. But in Pernambuco poll taxes and vagrancy laws were not crucial tools as in the Caribbean; nor were indentured laborers imported.[7]

Despite the longing by workers such as Manoel do Ó for a terrain beyond the *engenho* and away from the planter's sweeping power, the vast majority of workers remained on plantations. Júlio Bello commented that the end of slavery inspired no movement back to Africa, or even a large-scale departure from the *engenhos*. The former slaves "stayed in the work gangs, respecting the white man and serving him almost with that humility that habit and atavism had taught them."[8] Lins do Rego had a similar recollection: "My grandfather's former slaves remained on the plantation even after Abolition. . . . There they stayed until they grew old. Their children and grandchildren succeeded them as servants, with the same loyalty toward the plantation and the passivity of well-trained domestic animals."[9]

As Bello and Lins do Rego testify, workers remained on *engenhos* after the abolition of slavery in 1888. In fact, the formal end of slavery is a deceptive temporal benchmark for examining labor relationships in the cane zone. The transition to free labor took a protracted course in Pernambuco, in contrast to the core of coffee production in Brazil's center-south and far different from the Caribbean, where apprenticeship meant a four-year transition to freedom for British colonies but emancipation was otherwise abrupt. Already by the 1870s free laborers made up 80 percent of Pernambuco's sugar plantation labor force; about forty thousand slaves worked in the cane, while the region's total population numbered about half a million.[10] In a certain sense, the legal transition to postslavery simply formalized a process that extended over many decades of the nineteenth century. The final end of slavery, then, did not badly hinder planter access to labor, since "abolition represented a financial, political and emotional problem, but not a labor problem." Indeed, with the impact of railroads and the first experiments with central mills, the sugar industry became more productive than before. Though the slave population fell by almost 70 percent between the late 1840s and the late 1880s, sugar production doubled.[11]

The continued presence of some slaves, and the simultaneous employment of enslaved and free workers on *engenhos*, likely had an influence on collective assumptions about labor. A rich historiography treats the transition away from slavery in Brazil's south, addressing among other themes the evolving definitions of freedom, but far less work has appeared for Pernambuco.[12] Writing about urban nineteenth-century Pernambuco, Marcus Carvalho suggests that the "opposition between being free or not being free should be analyzed within a concrete historical situation," and notes that the line was frequently unclear in Recife.[13] And about a different northeastern city—mid-nineteenth-century Salvador—João José Reis notes that "the fact that slaves and freedmen worked side by side gave a sense of liberty to the former, and also a sense of enslavement to the latter."[14] It seems fair to suggest that free workers toiling alongside slaves in Pernambuco's cane fields would likewise have felt a sense of captivity by association, especially since planter treatment of workers clearly reflected the tradition of bonded labor. These were the conditions against which Manoel do Ó chafed, the reason that "his greatest ambition as a child was always to free himself of cane and the injustice that it symbolized in his life." Observing the continuities on *engenhos*, he commented that "the enslavement of the blacks ended and the enslavement of the whites began." The observation's meaning turned on the colloquial usage of *black* as a synonym for *slave*; after formal slavery, he meant, workers in general suffered exploitation.[15]

External economic pressures facing the industry also influenced the development of tenantry as a system of labor exploitation. Despite the growth in production in previous decades, crisis came to the sugar industry around the time of abolition. The problem had little to do with the end of slavery—a worldwide depression in sugar prices began in 1884 and worsened in the 1890s. Pernambuco led the country in sugar exports during this period and, as Peter Eisenberg writes, the province "exemplified well the problems of the national industry." The value of sugar as a percentage of total Brazilian exports plummeted, from 10 percent in the 1880s to 6 percent in the 1890s and less than 2 percent in the 1900s. This was the period of the first republic, when coffee dominated and government was oriented toward and run by coffee interests. Sugar exports by weight in 1910 dropped to just over a fifth of 1880s levels. Since more than 80 percent of Pernambuco's sugar was destined for export at that time (and production had been expanding), the collapse in world prices was particularly damaging to the state. Between 1903 and 1905, planters sold less than 10 percent of their product abroad and by late in the first decade of the twentieth century the state's share of the world market had slipped to only 0.3 percent (down from 4 percent a few decades earlier). Not coincidentally, this trajectory was the opposite of Cuba's. The island had drastically increased production from the beginning of the nineteenth century and built on its gains as it began a new century with an influx of U.S. capital. Puerto Rico also competed heavily; even though its yield was far smaller than Cuba's, it nearly matched the production of all of Brazil in 1900.[16]

This same period brought the *usina*-based production model. Initially, these large mills were intended to operate like the *centrales* of Cuba, buying their cane from growers and concentrating on the industrial side of the business.[17] But after being unreliably supplied by growers, mills ended up buying some primary material and growing the rest on their own. As the mills grew, many traditional *senhores* sold their land and left the industry while others became cane suppliers. By the turn of the century, there were forty-one large mills; hundreds of *engenhos* stopped milling their own cane and sold to the mills.[18] The system became more stable in the 1930s, when relationships between *usinas* and *engenhos* were clarified by a new agency President Getúlio Vargas created called the Instituto do Açúcar e do Álcool (IAA, or Institute of Sugar and Alcohol); some in the state credited Vargas with having "saved Pernambuco from ruin."[19] The transition in production methods affected the culture of the industry in ways that bothered Júlio Bello and others, as we have seen. Some workers, too, felt that the *usinas* negatively impacted their quality of life. Jovino da Raiz, the black worker who delivered a short paper at the Afro-Brazilian Congress organized by Gilberto Freyre in 1934, said, "After the *usina* appeared the black field worker found himself deprived of community, nutrition, and leisure."[20]

The increased global competition and the new mill-based production clearly affected tenant workers. Their average pay at the end of the nineteenth century was only three-quarters what it had been in the 1870s, and, according to Raiz, workers lost their formerly easy access to land during the first few decades of the twentieth century.[21] Bello also complained that during the early twentieth century mill owners "invaded every corner of the *engenhos* with cane cultivation," absorbing some areas traditionally given over to subsistence plots.[22] But cutting workers off from land for subsistence took much longer than a few decades. What Raiz and Bello perceived was the crucial place of land in planters' maintenance of power over workers. Very broadly, the same was true of the coffee plantations that held such crucial economic significance in southern Brazil. The immigrant workers who increasingly took the place of African-descended workers following abolition enjoyed as part of their contracts the use of land around the coffee groves. However, they also had a great deal of control over the cultivation and care of the coffee trees and their pay eventually became linked to harvest size.[23] Comparatively, cane workers enjoyed less freedom in their use of land and much less control over the provision of their labor with the cash crop.

In the context of the final abolition of slavery and the emergence of *usinas*, cane workers' living and work arrangements settled into a system that revolved around the concession or withholding of access to land. Defined most simply, the system that became known as *morada* involved the planter's granting a house and garden in exchange for labor from the *morador*, or worker. According to some, the growth of *morada* was an obvious development—the planter's intuitive response to the end of slavery. The planters simply "exchanged something they valued lightly in cash terms—land—for labor."[24] This arrangement of productive relations was common in many plantation contexts; though less research has been done on Pernambuco, it appears that like most Caribbean societies, slaves were given garden plots to work as a means of relieving planters of the burden of providing all food.[25] However, though the process in Pernambuco appears seductively straightforward, this version simplifies what was actually a complex relationship. First, as has been shown, the transition away from slavery was a long-term process in Pernambuco, so the emergence of this form of tenantry likely had as much to do with contemporary economic exigencies as it did with the move to a wholly free labor force. Most important among these was the transition toward mills. Second, *morada* was not a straightforward exchange; it was a labor relationship, but through the mechanics of patronage it also mediated workers' access to living space and land. The quality of the relationship, not necessarily just the amount of labor performed, determined the nature of the reward.

Tenants fell into several categories, including those who simply paid an annual rent for land and those (called "yoke carriers") who gave a certain number of days of work per year in exchange for their living quarters.[26] But the greatest number fell into the category of conditional tenantry, which most people imply when they speak of *morada*. The Portuguese verb *morar*—to dwell, or to inhabit—provides the root for both *morada* and *morador*, the word for tenant.[27] *Moradores* were expected to work on a daily basis for pay, usually also providing free work days for their rudimentary houses and access to land for a subsistence plot.[28] Paulo entered this sort of agreement when, as a fourteen-year-old orphan, he began life on his own on an *engenho*. His boss held all the power in relationships with workers; there was "no one to complain to," Paulo said. He owed the *senhor* free days for the privilege of living in a house on the *engenho* and, working six full days a week and planting food in a garden, he made just enough to survive.[29] If a relationship of trust developed, the *morador* might earn access to a larger piece of land (a *sítio*) and a house removed from the rest of the workers. The pervasiveness of *morada* in the early twentieth century is demonstrated by the straightforward definition provided in a midcentury study of the Northeast's sugar industry: "In general, the tenant [*morador*] is simply the rural worker."[30]

As one of those rural workers in the first few years of the twentieth century, Gregório Bezerra experienced the complex interplay between captivity and freedom implicit in tenantry. In 1907 the *senhor* of the Engenho Brejinho granted Bezerra, his siblings, and his mother admission to residency on the *engenho*, which gave them a home and placed them at the planter's disposal. If a job appeared, they were expected to work and were paid some small sum. If no work was available, the family members gathered what food they could and waited anxiously for the call from the *senhor*. Bezerra recounted the desperation of the period between cane planting and the harvest, when his family had to scrabble for survival and depended on the availability of fish in nearby streams for food. The young Bezerra had feared the *engenho* enforcer Pelegrino, though he had also recognized Pelegrino's continuing subordination to the *senhor*. Yet when Bezerra noted that the renting of the *engenho* to a different planter meant that the new occupant was "lord of . . . everything, including the inhabitants," he also acknowledged the contingency of his own freedom.[31]

Relationships began and developed under the terms of the *engenho*'s hierarchy, so the work and living relationship of *morada* consisted of a worker's negotiation of a position in the space controlled by the *senhor*. The dynamics of his relationship with the *senhor* determined the location of a worker's residence on the *engenho* and the terms of his access to land. The anthropologist Lygia Sigaud writes, "in the act of giving an established *sítio*, or authorizing a tenant to make one on a determined plot of land, the patron also said symbolically that he appreciated [the tenant], that he wished him well and hoped that he would stay."[32] Workers exerted a modest influence over the terms of their agreements, since they could decide simply to move from one *engenho* to another in search of a more favorable arrangement (and as Manoel do Ó's thirty-six jobs in fifteen years makes clear, workers did exercise this freedom). But since a good arrangement depended in part on a worker's stability, tenants stood little chance of receiving access to much land unless they had proved some degree of "loyalty" to a planter. And for planters loyalty generally had to do with work.

The Work of Cane

Severino Moura began cutting cane in the 1930s, around the same time as Paulo, eventually earning promotion on the *engenhos* belonging to the Usina Pedrosa. In an autobiography published in the 1970s—a valuable document of worker life—Moura includes a memorable tableau of a busy harvest when the last of the season's cane was being furiously cut and brought to the mill. The winter rains had already begun, though the temperature had yet to drop. Drenched workers made their way through the fields as drivers struggled to guide their laden oxen and donkeys along paths and roads that had turned to sticky mud. The drivers swore at their donkeys: "You mangy dog, you wrathful, syphilitic, rat-diseased, yellow feverish animal, I'll kill you!"[33] The donkeys' "perfume" of sweat and fetid mud mixed with their drivers' smell of cheap cane liquor and tobacco. As they struggled with the soupy ground, tough canes, and recalcitrant animals, the workers complained about their salaries; it took two and a half days of work to earn enough for just a kilogram of meat. "If the President of the Republic . . . knew how the poor peasants live," Moura concluded, "he would have two paths to choose: change the system of government or resign."[34]

Cane work is notoriously tough, especially during the chaotic urgency of the harvest. The harvest takes place fourteen to eighteen months after planting and generally runs from September to March or April.[35] Workers cut cane manually with a sickle, which they must sharpen three or four times a day since the cane is tough and resistant. They bend to slash the cane off as close to the ground as possible, then trim the leaves with several rapid passes of the blade before cutting the cane in half and proceeding to the next one.[36] Skilled cutters make the sequence look like one extended motion—a fluid coordination of back, shoulders, and arms. Even as the two halves of a cane fall to either side of the cutter, she or he bends to grasp the next. Moura indicates his respect for the work as he remembers "watching the cutters wielding their sickles in the harvest with an admirable mastery." He describes precisely how they used their hands to cut, trim, and halve, leaving the cane tops in a neat row for wrapping the cane pieces. Eventually, they stopped cutting and spent the end of the day tying the top leaves around bundles of ten to twenty canes.[37]

During Paulo's early years in the fields, he cut several hundred bundles of cane every harvest day. Another worker named Severino had a similar experience, saying he was expected to cut four hundred bundles a day. Born in 1936, Severino was younger than Paulo. He began work in the cane fields as an eight-year-old (the same age at which Paulo began hoeing cotton fields), when there were still sixty *engenhos* in his home municipality of Vicência. Over the course of his lifetime he has seen all but a handful of the *engenhos* brought under the control of the three *usinas* in the area. Once workers had bundled their pieces of cut cane, donkeys and oxen carried the bundles out of the fields, often to a waiting train that would supply the mill. *Engenhos*

Donkeys carrying cut cane, 1940s–50s. (Acervo Fundação Joaquim Nabuco, Recife)

would have as many donkeys as they did workers, Severino said, shuffling them in and out of the fields to haul cane. The large Catende sugar mill in the southern cane zone, for example, had two thousand mules during the time Paulo and Severino were beginning their working lives.[38]

Compounding the cyclical pressure of the harvest, the common subsistence crops of corn, brown beans, and fava beans needed to be planted in March, with the early rains that preceded the rainy season proper of May through August.[39] Then, during the "dead" time between harvests, many in the cane zone faced something worse than the grueling work: no work at all. For others, life still revolved around cane: planting, cultivating, and weeding. Fields had to be prepared before planting cane; when young forest or brush covered the field, the foliage was cleared and burned.[40] Then a plow turned the field, loosening the soil and blending in the brush's ashes. Workers began planting cane around May and continued for several months, their work calibrated to the maturation periods of the different varieties being planted. Traditionally, one of two methods was used for setting the "plant cane," the short cane pieces that will germinate new plants. In drier or exposed soil, workers opened holes in horizontal lines at intervals of a foot and a half up to three feet, placing plant cane in each hole. In wet, low-lying soils, large beds were slightly raised and workers dropped plant cane into deeper holes.[41]

As the cane grew, the fields had to be "cleaned" of weeds, a task done a month after planting and perhaps two or three more times before harvest.[42] Gregório Bezerra described the brute drudgery of cleaning cane fields as a child: "The work was hard and the foreman didn't let us lift our heads. Not even to breathe better. No talking, no lagging, and drinking water only when we got to the end of the cane field. We stopped work at midday to eat a little something. If there was anything. If there wasn't, we only stopped when the sun hid under the horizon and it began to get dark."[43] In addition to the sun, the rain, the dust, and the insects, workers contended with slopes. Half of Pernambuco's cane grew on slopes with a 20 percent grade or steeper, and planting and then weeding around and finally cutting the cane on the inclines challenged workers' endurance.[44]

Workers were intimately familiar with the many varieties of cane they worked with, though they did not refer to the alphanumerical codes agronomists used to denote varieties. Workers referred to black cane and purple cane, crystalline, ashen, red, bronzed, and "little Cayenne" cane. They liked "hairless" cane because it lacked the spiky edges that abrade the skin, and they referred to some varieties as "butter" canes because as they matured they slowly turned from light green to a creamy yellow. Some canes, banded purple and green, were called "striped" by the workers.[45] The most common cane color, though, was a bright yellowish green called simply "cane-green." Numerous shades of green are recognized and differentiated in the cane region, each denoted through reference to a particular plant.[46]

Just as it had a palette, work in the cane had a soundtrack; mixing with the percussion of hoes heaving soil or sickles hacking cane was the "song" of ox carts creaking along dirt tracks. Before he was old enough to join the regular work gang, Bezerra minded oxen. He rubbed the ox cart with soap and oil so that the cart would "creak and sing beautifully." Drivers took pride in the song of their carts, claiming that the oxen worked harder and more dutifully when they heard the loud squeaks of the axles grinding under the cart bed—two distinct notes produced by the side-to-side sway of the animals' stride. In order to feed the animals well, the young Bezerra decided to provide them with corn from the *senhor*'s special storehouse. Gregório recognized that he had no right to eat the corn himself, but he figured the oxen could because they belonged to the *senhor* and therefore needed to be well fed. When he moved on to field work, Bezerra missed the sound and smell of his former charges and grudgingly faced the difficult conditions "at the foot of the cane."[47]

Workers dealt with the difficulties of their jobs in various ways, includ-

Workers plowing with oxen, 1940s–50s. (Acervo Fundação Joaquim Nabuco, Recife)

ing by internalizing the demands they faced. The famous education theorist and former Harvard professor Paulo Freire carried out his doctoral research studying families in the sugar zone, where he found parents using punishments that "ranged from tying a child to a tree . . . [to] thrashing [children] with leather straps." They justified these measures with phrases such as, "Hard punishments make hard people, who are up to the cruelty of life."[48] Yet as Bezerra's apparently fond memories of his work with oxen indicate, people did sometimes find pleasure in their work. Bezerra's feelings appear to follow a long-term pattern of workers' sharing pleasant moments with draft animals, perhaps because they felt a sense of kinship with the working beasts. A traditional folk play with origins in the eighteenth century called the *bumba-meu-boi* dramatizes the death and revival of an ox, with various comedic plot twists. According to a Brazilian folklorist, the ox plays the central role because "the Negro slave was the ox's brother in suffering and in work."[49] The *bumba-meu-boi* was compellingly echoed in a song by the popular 1950s performer Jackson do Pandeiro, which referenced the shared experience of animals and workers. Jackson, a native of the cane zone, memorializes a "stubborn ox," singing, "This ox is a reminder of the time of slavery," an inheritance from a grandfather (implicitly both the ox and the memory were inherited). "In that time the Negro was [treated] the same as the ox," the song continues, "the reminder of this friend hasn't left my memory; he died and I remained to tell our story."[50] Besides touching on the relationship to draft animals, this evocation of a legacy of hard labor recalls Paulo's description of cane work as a cruel system.

Manuel do Ó's stories of the time of abolition, Gregório Bezerra's memories of the early twentieth century, and Paulo's and Severino Moura's descriptions of work in the 1930s reveal a considerable continuity of experience over that crucial half-century in terms of work performed in the fields. Indeed, the persistence of basic cane agriculture techniques arguably reaches much farther into history. Júlio Bello, writing his memoirs at the same time that Paulo and Moura began cutting cane, claimed that he recognized from his life as a *senhor de engenho* all of the techniques described in the Jesuit Antonil's early-eighteenth-century agricultural manual.[51] This observation also fits the pattern I discussed at the end of chapter 1 of stubbornly traditional cane farmers. In any event, the consistency in the "folk" view of work was paralleled by a steady vision of labor rhythms among the planters.

With the apparent stability of routinized drudgery, occasionally punctuated by more attractive tasks, it comes as little surprise that scholars have claimed over the years that cane workers detested their work. A wide-ranging study of 1920s Pernambuco politics and society includes the assertion that cane workers' migration to Recife during that decade was "born of a veritable repugnance for field work and the economic and social conditions of the agricultural life that repels the inhabitants."[52] Around that same period, a new cultural form was emerging in the sugar zone—a rural version of a Carnival celebration with deeper urban roots called *maracatu*. Rural *maracatu* musters an ensemble of characters and references from African and indigenous traditions, including characters of mixed indigenous descent called *caboclos* dressed in colorful costumes carrying spears or wearing elaborate feather headdresses. In their annual *maracatu* performances, sugar workers communicated a message of defiant resistance, seen especially in the fierce, spear-wielding *caboclos*. An anthropologist studying the form in the 1960s heard from every participant she interviewed expressions of abhorrence for sugar work.[53]

Maracatu emerged at the same time that *usina* production was being consolidated—the target of Bello's criticism and the context for Paulo and Severino Moura's first years in the fields. An agronomist of the time affirmed that "the man who abandons the rural zone, moving himself to the urban zone, will never go back. The worker of the city will [never] voluntarily return to the land."[54] Cognizant of ideas like these, Sigaud expanded on the basic idea of sugar workers' distaste for their work, arguing that they also placed little value on proficiency at field tasks. This, she suggested, distinguished them from urban workers "who pride themselves on the *métier* they perform." Deriving no satisfaction from their work, according to Sigaud, cane workers privileged their relationship with the planter over their own skill in labor, per se. She pointed to the fact that workers rarely spoke of their jobs or of tasks in the fields. "Whenever asked about work or jobs," she wrote, "workers speak . . . above all about the relationship with proprietors." One can see the logic in this, since the demands placed on workers and the remuneration they hope for depend on that relationship.[55]

However, Sigaud's theory contrasts with a joke that Severino told during an interview, about a tense standoff with a mill owner during a union dispute. With his adversary persistently condescending to him, Severino declared to the man, "I have a degree in the question of land!" "You have a degree!?" the man asked incredulously. "Yes," Severino affirmed. "I was educated in hard ground and thick brush." He presented this response as a punch line and claims that the mill owner always laughingly repeats the line when they see each other.[56] In addition to catching the man off guard, Severino claimed that his detailed knowledge about the regulations and demands of field tasks—based on a work life that began when he was eight—impressed the mill owner and helped bring the pay negotiations to a successful conclusion. The self-regard Severino expressed comes forward from other workers as well. "We cut our teeth here," a foreman said in an interview. "Many don't have schooling but have a clear mind that knows the norms of work in the fields; it's experience."[57] Workers' capacity for and knowledge about work clearly served as the hinge for their broader relationship with planters, but it was also a source of pride. This oversimplifies, though, the complexity of workers' feelings toward labor and the place of these feelings in their worldview.

Given this evident pride, perhaps what Sigaud interpreted as workers' devaluation of their labor was not a reaction against the work itself but a response to the coercion they so frequently faced. An analysis of workers' everyday language, analyzed by one of Sigaud's collaborators in the 1970s, helps explain the distinction. Some tenants referred to themselves as "know[ing] how to do every job [*sabe fazer todo serviço*]". Not only did workers take satisfaction in this thorough competence, but it earned them a privileged tenantry arrangement. Knowing how to do every job was different from actually *doing* every job. Those who did every job (*faz todo serviço*) did so because they lacked a privileged relationship with the planter. And these slightly disadvantaged workers were different still from those who simply worked with the hoe (*trabalhar na enxada*). This last group merely did what *everyone* knows how to do and therefore occupied the lowest rungs of prestige and access to the benefits of a good relationship with the planter.[58] This rough breakdown was verified in more contemporary interviews, as workers showed obvious pride in claiming, "I can do every type [of task]. I can prepare for planting, load wagons, cut cane, clear brush, cut brush, everything. I know how to do just about every job in the field."[59]

Feelings toward work were influenced by the ties linking the performance and provision of work to living conditions in the tiers of tenant privilege. Work in the cane fields was demanding, and workers needed to master a range of skills (as in the case of banana work elsewhere: "Tasks were specialized, but workers were not"). They took pride in having a range of knowledge and the capacity to apply it to their work. But workers resented the subjugation of *having* to perform every task, the menial as well as the complex, in no small part because such subservience implied inferior living arrangements. The willing application of a broad set of skills earned confidence, trust, and a better relationship, but being made to perform the full gamut of tasks exacerbated resentment and feelings of captivity. Again, this parallels the experiences of workers in other plantation contexts, such as Ecuador's south coast. Workers there resented the "extreme paternalism and daily indignities" of the cacao boom during the early 1900s and referred to the period as the "time of slavery."[60]

The Urban and Rural Connection: Social Science Studies Workers

In *Northeast* Gilberto Freyre claimed that he and several collaborators embarked on a detailed study of workers' conditions in the *zona da mata* in 1935 but that their efforts were thwarted by the antagonism of the planters. Unwilling to allow access to their domains, landowners undermined the team and forced it to abandon the project. One of Freyre's partners, his uncle Ulisses Pernambucano, settled for a study of urban workers, published in 1937.[61] Their experience illustrates the challenges social scientists faced at the time working in rural environments. Until the late 1940s, research, reform, and advocacy, in their limited forms, were essentially restricted to the city.

As earlier references to rural-to-urban migration during the 1920s and 1930s imply, there were important connections between the two spheres during this period. During these decades fully half of the working class in Recife had migrated from the interior of the state. This movement preceded the migration of large numbers of northeasterners to the center-south cities of São Paulo and Rio de Janeiro in the ensuing decades, a process that helped transform Brazil into the urban country that it would become decades later.[62] The movement of people played out against a background of political movements and upheaval. A restless, dissatisfied generation of mid-level officers came into the armed forces in the 1910s and 1920s and agitated to break the vise grip on power held by a privileged oligarchy. These *tenentes* (lieutenants) led small, unsuccessful barracks revolts in 1922 and 1924, but their frustration with what they saw as a pattern of failed governance reflected a widespread dissatisfaction in society. The 1930 revolution owed its success to this simmering discontent.[63]

Most of the workers moving to Recife likely came from the cane zone, considering that the area held one-half of the state's population. The rate of urban migration was high, averaging 13 percent over the first four decades of the century and accelerating as time went by.[64] A significant textile industry and a rapidly growing population developed in Recife under the increasingly watchful eye of the state. After the revolution of 1930, Estácio Coimbra was replaced as governor by Carlos de Lima Cavalcanti, whose family owned the Usina Pedrosa where Severino Moura worked. Lima Cavalcanti was ousted when Vargas transformed his regime into the corporatist dictatorship called the Estado Novo (New State) in 1937.

Vargas installed Agamenon Magalhães as his interventor, and Magalhães became the state's political boss until the Estado Novo ended in 1945. During this time Magalhães attempted to shape the direction of change along the lines established by his modernizing patron. Magalhães prosecuted a vigorous "campaign against slums [*mocambos*]" and organized public works projects aimed at reforming the city.[65] He also confronted an urban labor movement that had gained confidence and organizational capacity since the beginning of the 1920s.[66] A new group of social researchers carried out studies on a parallel track to Magalhães's efforts, and gradually their focus shifted from the urban core to the cane zone, drawn by the growing industrialization of the sugar industry and by an increased sense of connection between the problems of urban and rural labor.

At the same time that Freyre's uncle Pernambucano carried out his research on urban workers, a young, energetic doctor in Recife conducted a similar survey. Josué de Castro's study of Recife workers' living conditions and dietary standards appeared in 1936. The coincidence of these efforts points toward social science's growing interest in the working class, an impulse that expanded in the next decade and ripened further in the 1950s. The same year that Castro published his findings, Freyre published his second major book—*Sobrados e mocambos* (translated into English as *The Mansions and the Shanties*)—which took Recife as its main subject. Freyre tied the growth of the city and its sprawling slums to the rise of merchants and financial capital in the nineteenth century, a process that he thought sapped the power of the rural elite.[67] This transition also presaged the further loss of traditional prestige associated with the growth of *usinas*, which was taking place as Freyre wrote.

These developments also drew the attention of Lins do Rego, who addressed the rural-to-urban experience and the urban labor movement with his one literary foray into the lives of the working class. The novel *Moleque Ricardo* tells the story of a poor young black worker and covers the strikes and demonstrations of 1919 and the early 1920s. Having fled the suffocating repression of an *engenho*, Ricardo finds that workers in the city were also trapped in desperate conditions. He puts in long hours at a bakery and hears haunting tales from a friend about work in a textile factory.[68]

That factory, by far Recife's largest industrial concern during this period, was the Companhia de Tecidos Paulista (CTP, Paulista Fabrics Company). A sprawling complex on the northern fringe of the city, the CTP was owned by a German family named Lundgren and employed as many as twenty-three thousand workers. Though the Lundgrens and Magalhães were political antagonists, he greatly admired the company town they had built, considering their complex "a factory and a civilization."[69] Just as Magalhães lauded the Lundgrens, he praised some mill owners' efforts, and there were similarities between the most advanced cane operations and the urban factory. With such a huge workforce, the Lundgrens had established a tradition of allowing workers access to subsistence plots. Seeking efficiencies in the 1930s, however, they used a government directive to reforest (the factory consumed prodigious quantities of wood) as an excuse to displace workers from their gardens.[70]

Considering such parallels between the urban and the rural, it was a natural step for social scientists to extend their work from the former to the latter. It was arguably Castro, born in 1908, who helped forge a new intellectual generation concerned with the health and well-being of the population and with the development of a more equitable society. Already by the 1930s, Castro had shaped a set of interests that he would go on to pursue across the globe in the following decades, working for the U.N. Food and Agriculture Organization.[71] Advocating for the interests of the poor and focusing specifically on the issue of hunger, he developed a partisan voice, writing, "I do not believe in neutral literature, literature without tendencies, without roots, without connections with the other social aspects that define a culture." He thought he recognized around him in the 1930s a cohort of like-minded writers, whose emergence he saw as a "volcanic eruption" in the region's intellectual life, replacing "pretty conventional lies" with sincerity and a "true art" that depicted plain life without hypocrisy. Castro omitted Freyre from this young group, a statement of his distaste for the celebrated writer. Without much enthusiasm, he did include Lins do Rego, in appreciation of the novelist's "capacity to describe the human landscape of the sugar cane area."[72]

Throughout his life, Castro would concern himself with the human landscape—with the challenges facing people in the particular settings of their lives. It was "to the landscape of the Northeast," Castro claimed in an essay, that he owed "the direction of [his] intellectual life."[73] Castro's merchant father moved the family to Recife when Josué was a young boy, and they lived in a comfortable house on the bank of the Capibaribe, one of the two rivers running through the capital city. The young Josué could probably gaze out from his house to tumbling rows of shacks—called *palafitas*—perched on stilts over the river's edge. The misery of these shacks' residents, living lives tied to the river, made a strong impression on Castro. After earning a medical degree in Rio de Janeiro, he returned to teach medicine and human geography in Recife and write sociological essays (and occasional fiction) about the city's poor.[74]

In the early 1930s, Castro led his investigation "Living Conditions of the Working Classes of the Northeast." Surveying the daily eating regimens of more than twenty-five hundred working people in Recife, his team found pitiable nutritional norms. "The only way to feed oneself worse than this," Castro wrote, "is to eat nothing." His conclusions were illustrated by the experiences of CTP workers, many of whom relied for sustenance on crabs gathered from the polluted river near the factory.[75] Like Freyre, Castro presented his findings as a challenge to environmental and racial determinism. "Today no one would consciously affirm that miscegenation is the true cause of the low vitality of our people," Castro wrote, citing "modern anthropologists'" debunking of racial inferiority.[76] Among the workers that he studied, Castro argued, race was irrelevant compared to the challenges of desperation wages, horrendous nutrition, and chronic disease.

As Castro pursued his work and influenced a generation of aspiring doctors and geographers in Recife, some researchers extended his approach to the sugar region and arrived at similar conclusions. The agricultural engineer J. M. da Rosa e Silva Neto issued a stinging critique of the sugar industry based on the state of its workers. He wrote in 1948 that the typical worker, "malnourished, full of worms, syphilitic, and in difficult living conditions, falls into a feeble conformism." "Without motivation and hungry," he continued, "his production falls with each new year."[77] The last phrase clearly contained a warning for planters (and indicated the limits of Rosa e Silva's sympathies): ill-fed workers could not work hard. A year after Rosa e Silva published his work, Freyre founded the Joaquim Nabuco Foundation with the stated aim of carrying out "social research destined toward the study and valorization of the agrarian North." It was the centennial of Nabuco's birth in 1849.[78] In the ensuing decade more social scientists, as well as people with a more political bent, engaged the rural workers of the sugar zone.

The Landscape of Captivity: Articulations of the Worker Worldview

Workers' feelings about their labor played a crucial role in structuring their distinct view of the landscape. Besides giving them knowledge of and insight into natural processes, work also contributed to the accretion of meanings around the environment being worked, informing workers' sense of place. "Labor involves human beings with the world so thoroughly that they can never be disentangled," writes Richard White, and the intimacy of workers' connections to the land partially developed through the physicality of the work, so expressively rendered in Moura's and Bezerra's descriptions.[79] Workers carried on their bodies the marks of their labor as they were scratched by abrasive cane leaves, developed a stoop from bending to wrap bundles of cane, or bore the deep, sickle-inflicted leg wounds that Moura described as "characteristic of the rural worker."[80] This labor, in turn, transformed the land through harvests that turned a vast sea of green-leaved cane stalks into a brown, stubbly field or through drainage canals that made soggy bottomland arable. The means by which workers engaged their environment through labor contrasts with the planters' view of workers simply existing as part of the landscape. If a landscape is "an enduring record of—and testimony to—the lives and works of past generations who have dwelt within it, and in so doing, have left there something of themselves," then workers participated in constituting the landscape of the *zona da mata*.[81]

Analysis of the landscape must include both workers' "lives and works." Workers' living conditions put them in close contact with their surroundings. The built environment did not separate them from the nonbuilt environment as much as emphasize the connection between the two. Workers cobbled building materials together from the leaves, plant stalks, and mud available. A social researcher writing in the mid-1940s for *Brasil açucareiro*, the magazine of the IAA, said "the rural habitation of the worker shows perfectly the material conditions of the environment."[82] Around the same time the magazine documented "habitations in the sugar zones of Brazil" with photographs showing the decrepit shacks that housed northeastern sugar workers. Low mud-brick huts with overhanging roofs of unfired clay tile enclosed adjoining rooms, each housing multiple worker families. The buildings had little natural light and sluggish air circulation under the roof. One picture shows a house tucked into a hillside, with a couple of banana trees in the background and a bare dirt slope in front. The small building has six doors, and very likely an entire family lived in each room.[83]

The classic *engenho* layout revolved around the big house: the planter's home and the seat of patriarchal authority.[84] The porch proper and the space immediately in front, commanded by the big house, was also important as the space where workers resident on the *engenho* were called to see the *senhor*, or where they went to make a special request. Close by was the worker housing described in *Brasil açucareiro*—the lines of rooms in long block buildings (called an *arruado*), that descended directly from the design of slave quarters. Most *engenhos* also had a store (*barracão*) for subsistence needs that could not be met through garden plots (e.g., sugar, salt, meat, and dry goods). Structured like a company store, the *barracão* marked up prices on basic necessities, extended credit, and garnished wages, frequently driving workers deep into debt. Besides tightening the bonds between workers and the *engenho*—in the form of monetary obligation—the *barracão* served as a meeting place and a mediating site of relations on the *engenho*.[85]

Workers gleaned what pieces of wood they could from forests jealously guarded by planters who required timber for fuel to drive their mills and railroad engines. The lines of worker housing frequently nestled close to the stables and corral that held the donkeys, horses, and oxen so important to *engenhos* for cane transportation while the big house generally commanded high ground.[86] Older *engenhos*, established before cane processing was largely consolidated at large mills, located this *engenho* core close to prime cane-growing areas in order to minimize the distance cane had to be transported. This meant that cane often reached right up to workers' doorsteps, and subsistence plots were tucked far away on marginal land. The plots shrank as, between the mid-nineteenth century and 1930, the area planted in sugarcane increased almost fourfold.[87]

Among Severino Moura's many valuable observations are those regarding the importance of personal relationships to tenantry arrangements and the importance of power to the space of *engenhos*. Moura came from the transitional area between the cane zone and the semiarid *sertão*, leaving his home town in search of work in the cane. He found a job in the fields of the Cavalcanti family's Usina Pedrosa and was soon promoted to foreman.[88] His hard work came to the attention of the mill's field administrator, who promoted him again, to task manager (*apontador*).[89] In describing each promotion, Moura detailed the new benefits he received. While he remembers the exact pay levels, he seemed to place more importance on the nonmonetary but no less material perquisites legible to those familiar with relations between planters and workers. On rising to the position of task manager, Moura moved to a "tile house" wired with electricity. After a mere six months in the position, Moura received yet another promotion, to administrator of the entire *engenho* (the mill owned quite a few). With this new step up he moved into the big house, which was fully equipped with a servant. "I felt a little like my own man," he wrote. "I was already someone."[90]

Two important points emerge from Moura's treatment of his trajectory in the cane zone. First, he narrated his rise through successive strata of responsibility and power as a story of physical movements into better and better living quarters. In foregrounding places of habitation while relating his tale, he shows just how intertwined were work and living arrangements in the region, all brokered through the tenantry relationship. As I have already noted, it is instructive that the term describing what was ostensibly a labor relationship—*morada*—refers not to work but to dwelling, to a house. Indeed, only the dwellings ceded to *moradores* were called "houses," while similar structures workers occupied in towns were called *ranchos*. The houses of tenants were "something definitive"; they implied the maintenance of a set of social and labor relations that gave tenants some stability and security, in part because the houses located them within the social system.[91] Second, Moura began to feel like "someone," like his "own man," when he moved to a big house. The association of a fully rounded identity or selfhood with movement into a space associated with power recalls the planter landscape discourse, which equated workers with nonhuman resources. It also indicates that workers felt deprived of self-determination, that their ability to define themselves derived from improvements in their work and home lives, which were one and the same.

The link between identity and control also manifested itself in the norms and oversight of land use under tenantry arrangements. Usually, in the smaller garden plots tenants could plant manioc, beans, and other subsistence crops with short growing cycles. But if granted a larger field, tenants sometimes gained the right to plant bananas, mangoes, and other fruit trees or "root crops." Workers coveted this freedom; one man interviewed in the early 1980s said, "If I lose [my garden plot] I'm a captive and stop being my own man."[92] And planters, in restricting some practices and allowing others, played on the value workers associated with these privileges. A worker named Benedito said of the rule prohibiting tenants from planting "roots" in their plots, "Planting bananas is a gesture that [shows] you feel you're the owner of the land, even a mill owner." "The owners say, 'He'll plant and later he'll want to own my land.'"[93] Prohibiting workers from using the land in any way they saw fit meant keeping them from feeling any ownership over the land, a sense that would have implied that they had the power, or right to command, associated with ownership. Metaphorically, a worker's tenure was as vulnerable and fleeting as a bean plant's gestation period.

The rules of use emphasized the fragility of workers' access to land, which meant that the cultivation of slow-growing fruit trees and that activity's implication of stability could not be tolerated. These workers' descriptions remind us that the construction of meanings that constitute recognizable places occurs through social interaction and its norms and practices.[94] The operation of power through control over land speaks to the importance of land in defining the shape of the tenantry relationship. Because planters had a near monopoly on land ownership, workers' only "legitimate" access to land lay through tenantry of some sort; the path to access led through the provision of labor. Such was the case for Paulo's parents and later for Paulo himself. The role of land ownership in labor control has been recognized since long before the emergence of liberal economics, though the early 1800s brought clear expressions of the connection between the two.[95]

Turning again to worker language, we see the importance of power and specifically the control of land to social relations on the *engenhos*. Workers' labels for different types of people associated with the *engenhos* contain revealing patterns. While discussing their relationship to the wider society, workers might refer to themselves using the term *camponês*—peasant, or rural person. The term carried an implication of labor power, and workers used the label "to transcend the limits of the *engenho*." However, "the term *camponês* would never be employed in a context strictly internal to the [*engenho*] system." When discussing relations with planters, work contractors, or others within the *engenho*, workers used the terms *trabalhador* (worker) or *pobre* (poor person).[96] The one term, then, that connoted a sense of strength was never used in discussions of the *engenho*, a fact that points to the demobilizing power of the system of social relations with the *senhor* at the top.

Further, *camponês* relates linguistically to land, meaning through its root "of the land." Again, the fact that workers did not use such a term to describe themselves in relation to *engenho* power relations tells us that they felt disempowered in their relationship to both land itself and the landscape.

Alongside the operations of power on the *engenhos*, workers' sense of the regional environment manifested itself in routine daily practices and naming. Workers did not write sociological texts or novels, but they did have a set of shared discourses. The folklorist Luíz da Câmara Cascudo pointed out that workers built a rich toponymic vocabulary, naming all *engenho* landmarks from hills and valleys to paths and individual trees, using people's names or events from collective memory.[97] Moura's memoirs make clear just how important specific places in his corner of the region became to him, how substantially his surroundings impinged on his perspective. He frames many of his memories spatially, or through reference to particular places in which he lived or worked. For instance, when he was offered the job of running field operations for an entire mill, he remembers that he was standing "under a cashew tree, on the *engenho* named Polar, but known as Levas."[98] Practices of naming both reveal and contribute to the process by which spaces acquire meanings and importance over time, and they indicate how subjectivities become powerfully linked to spaces. This is how, in "places [that] are actively sensed, the physical landscape becomes wedded to the landscape of the mind."[99]

The rules of tenantry and the larger conditions of production in the cane zone, including the extension of planter power to the "natural" world around the workers, found their way into the folklore of the region. A legend about the green serpent of the cane fields, recorded by a folklorist in the 1970s, reveals these connections. As cane matures, workers said, a gigantic green snake grows with it, taking on the role of guardian. Though people like to duck between cane rows and cut a stalk to suck the sweet juice, "no one is brave enough to interfere in a world defended by the green serpent."[100] Some also believed that "cane fields guarded the living spirits of their dead workers," entities that were part of "another population, supernatural and visible" that inhabited the *engenhos* in the workers' collective imagination. Cascudo writes of workers' fears of "the dense, interminable canes, closed in the uniform green of the adjacent rows."[101]

This fear, so reminiscent of Manoel do Ó's aversion to the suffocating green horizons around his home, derived from the cane fields' lying under the explicit and demonstrated power of the *senhores*. The green snake is a fierce personification of planter power, and the exhausting interminability of the cane's green a reminder of the crop's centrality to an oppressive labor system. Workers could find themselves subject to the very real violence of planters if they stole even a small amount of cane. Severino recalled that the owner of the *engenho* where he lived in the 1930s yanked out the teeth of anyone he caught surreptitiously sucking a cane for the juice.[102] The snake legend also informs Rosa e Silva's language in his 1948 critique of the *usinas*' aggrandizement of land, likening the "disorderly expansion" of cane to "devouring serpents" that "surrounded the workers' houses, the old *engenhos*, and the stables."[103] We might also point out that in the biologically impoverished milieu of the cane fields, rodents have flourished, while snakes' predators have been eliminated; so snakes are ubiquitous.[104]

Ronivaldo, a worker I interviewed in 2003, described his relationship to the *engenho* where he lived in the following way: "The people who live here, these people are part of this land. . . . This is why people say, 'Each person in his place.'"[105] The comment combines the two basic elements of my analysis of the workers' descriptions of the landscape: the interlocking of power and quotidian practices. Ronivaldo referred to a basic sense of home and belonging but also perhaps to locations in the hierarchy of power: he knew his "place" socially as well as spatially. The strict power relationships of the region are mapped onto the physical environment, so that Ronivaldo's small mud and wood shack on the side of the meandering highway was both geographically and socially "his place" and his (lack of) power over the surroundings was clear in his comments about the landscape.

As Manoel do Ó's escape tells us, and as the rates of early-twentieth-century urban migration indicate, workers were not irredeemably captives of the cane fields. But enough evidence from linguistic patterns, folklore, oral history, and scholarly literature exists to show us that workers consistently used the metaphor of captivity to describe the landscape of the cane zone. Even researchers found it difficult to penetrate the world of the cane fields until the 1940s. But aside from flight, how did workers define freedom under the conditions of traditional *morada*? Workers felt themselves more fully realized when they had a stable residence, preferably a freestanding house, and access to land with the fewest possible restrictions on its use. This status was important as much for what it implied as for what it made possible. It meant that workers were less subject to direct, onerous control of their labor ("having to do every job"). It meant that workers could expand their self-sufficiency with personal cultivation (though within boundaries, lest they "feel [they are] the owners of the land"). And it meant they had achieved "something definitive," living in a house rather than a *rancho* or an *arruado*. But all of these achievements, it should be pointed out—a house and access to land with few restrictions—depended always on the relationship with the planter. As agriculture changed, labor needs and demands changed and so did these relationships. Along with these shifts came new definitions of freedom for workers.

An author collecting material during the 1940s for a book on the history and culture of the ox cart received the following carter's song from a Pernambucan planter.[106]

> Creaking along without cease
> Pulled by four bulls
> Here comes the ox cart
> With that wounded weeping
> Arriving here at the *engenho*
>
> But still much sadder
> Than the crying cart
> Are those four oxen
> Living to pull the cart
> Passing their lives
> In that sad labor
>
> Ox of the cart! Of the two of us
> Who suffers most?
> You go to the *engenho*
> I don't even know where I'm going
> You pull a cart of cane
> I pull a cart of pain

The song expresses important elements of workers' conditions: the difficulty of their work and its ceaseless rhythm, their sense of solidarity or kinship with work animals. The lament also returns us to the question of the values rural cane workers associated with their surroundings. "This is the best life there is, in terms of *life*," the worker Ronivaldo told me. "Now, in terms of work, no; a lot of people are forced out of here . . . because of the [hard] work."[107] Many workers cited their birth places in response to questions posed in 2003 about the best place they had ever lived. "The best place was where I was born," Ronivaldo also said. "The best spot that I've lived in was where I was born and raised."[108] This makes sense, since "through living in it, the landscape becomes a part of us, just as we are a part of it." Through these daily, routine practices workers tried to build a home for themselves in a landscape characterized as the home of sugarcane.[109]

But this was very much a process of resistance, a struggle to carve meaningful space out of conditions they described as captivity. The apparent endlessness of cane fields—the tall, unvaryingly green canes marching over hills and up valleys—metaphorically evoked the reach of planter power. Hemmed in by cane, forced by the domination of ownership into relations of domination in labor and home life, workers felt themselves captured by their surroundings. The contrast between the workers' descriptions of the landscape and the planters' elaborated landscape discourse emerged through distinct forms of engagement with space and their particular social relationships.[110] The recurring metaphor of captivity—a reflection of the planter discourse of the laboring landscape—demonstrates the workers' conflicted relations with the environment and their bosses.

PART II

Opening up the *Zona da Mata*, 1930–1963

FOUR

Modernizing the Sugar Industry
Cane Expansion and the Path toward Rationalization

In July 1963 the state's broadly circulating newspaper, the *Diário de Pernambuco*, reported on workers striking at the Usina Roçadinho, in the far southwestern corner of the cane zone. The workers demanded that the mill pay for work by the "count"—a unit measured out by a foreman that was supposed to approximate one day's work. Instead of the count, Roçadinho had begun using the *tarefa* (task), which took two or three days to complete. The conflict received statewide attention because Roçadinho was owned by Cid Sampaio, who had just stepped down as governor the year before and remained a powerful political figure.[1] But the *Diário* reporter added an interesting twist to an otherwise straightforward story; his tone dripping with sarcasm, he wrote that the workers had so little reason for complaint "that the pretext for the strike was to demand lower wages."[2]

How could the reporter have made this claim? In this chapter I suggest what the workers actually wanted. This explanation depends on an investigation of agricultural change and its reception by both planters and workers. A professionalizing agricultural science sector and federal support began opening up a region that had traditionally operated under the tight control of planters. The intervention met ambivalence, as some planters eagerly adapted, while others were less interested in change. One fruit of the new dialogue between planters and technicians was attention to cane varieties. Varietal management and other new agricultural techniques contributed to a rationalization process that impacted workers directly and sometimes radically. This chapter examines the complex and diverse forms and traditions of payment as well as how reforms of them generated conflict. The reaction of Sampaio's field workers to the changes they faced on the job provides insight into these dynamics.

Pernambucan planters' attachment to tradition was legendary, a subject of debate at the 1878 Congress and over the decades afterward. Even the movement away from *engenhos* and toward *usina*-based production did not generally bring more refined agricultural techniques, and the state lost ground to newer south-central producers. Age-old conventional wisdom about Pernambuco's innate superiority as a site for cane appeared to be wearing thin. The *senhor de engenho* Júlio Bello remarked sarcastically in 1935, "If our soil had the dumbfounding fertility of our imagination, we would truly be that fabulous country, privileged by God and nature . . . that shines in certain false books."[3] New support from the federal government in the 1930s and the professionalization of agronomy brought a more concerted process of modernization, but it also impacted social relations and the region's politics.

Sugar's Ambivalent Modernization:
New Cane Varieties and New Partners for Planters

Nothing matters more for sugar agriculture than cane itself, and over the course of its cultivation planters have produced different varietal strains. Pernambucan planters played a role in the first two major turnovers of cane varieties in the history of Brazil's cane industry. From the first cane cuttings planted in the sixteenth century until the beginning of the nineteenth century, Brazilian planters had grown one type of cane. When a new variety found its way into Brazilian fields at the beginning of the nineteenth century, the old standby became known as "creole cane."[4] Planters had complained for years about diminishing yields from their canes, and while creole cane held up remarkably well, its centuries of cultivation meant a long-term reliance on a narrow gene pool. This opened planters up to the danger of catastrophe when the variety began flagging, or became the target of attack from another organism. These problems were particularly acute with sugarcane, which was reproduced asexually by planting cuttings rather than through the germination of seeds. Indeed, planters around the world were unsure until the late nineteenth century whether cane *could* reproduce sexually.[5]

Brazil received an infusion of new cane genes in 1810 from an unlikely military engagement. The Brazilian army occupied French Guiana between 1809 and 1817, and the occupying governor, Brigadier Manuel Marques, made it his business to ship plant samples, including sugarcane, to the royal botanical gardens in Pernambuco. The garden director propagated the cane and quickly distributed it through the province.[6] The new cane's name—Cayenne cane (*cana caiana*)—revealed its provenance. Called Bourbon cane by French planters, the variety had arrived in the 1780s from Tahiti.[7] Planters used Cayenne cane almost exclusively for more than sixty years, until around 1880, when gumming disease decimated their fields.[8] At that point, a series of new varieties arrived mainly from Mauritius and Réunion. The importation of the new strains indicated a growing interest in broadening the range of available varieties.[9]

Another new variety emerged from Pernambuco's own fields, the product of experimentation by an enterprising *senhor de engenho*. In 1892, with the memory of gumming disease still fresh, a planter named Paulo Salgado announced to his colleagues that researchers in Barbados had recently demonstrated that cane could be reproduced by seed; similar results later appeared from a research station in Java. Salgado noted that experiments continued in the Caribbean, and he encouraged his fellow planters to try fertilizing seeds on their own.[10] Manoel Cavalcanti de Albuquerque took up the challenge, successfully fertilizing cane flowers and eventually producing new strains. His "butter" and "hairless" varieties became popular in Pernambuco (the first matured into a creamy yellow color, while the second lacked spiky hairs along the stalk).[11]

The auspicious (though unfortunately short-lived) botanical gardens and the personal enterprise of Cavalcanti notwithstanding, the pace of modernization was slow as the twentieth century began. Pernambucan planters had demonstrated their ambivalence about change at the 1878 Congress, where many decried any alterations to "their" routines—whether political, social, or agricultural—as an implied threat to the naturalized operations of power within them. And even as he criticized his peers' backwardness, Bello was distrustful of the modernizing mills.[12] The opposition between *engenhos* and *usinas* characterized the industry's changes during the first half of the twentieth century. The state had fifty-six active *usinas* in 1914, but there were also over twenty-five hundred *engenhos*. It was clear where the energy and innovation was located; in the 1920s, the length of private railroad tracks belonging to the *usinas* surpassed all of the lines of Pernambuco's major railroad company, the Great Western. By 1933, sixty-eight *usinas* operated in the state, before contracting and reaching a rough equilibrium at fifty-four by the 1940s. Also by that decade, the relative contribution of *engenhos* to state sugar production had dropped significantly.[13] The growing influence of large mills reflected similar changes elsewhere; in Cuba and Puerto Rico, for instance, the countryside had been largely assimilated into sugar production by the late 1930s.[14]

Just as the expanded production model began to be consolidated, a disease struck the fields in the mid-1930s. Cane leaves revealed the disease's advance, developing striations of lighter- and darker-green bands. The mosaic virus had arrived. Since the Campos region of Rio de Janeiro had faced a mosaic infestation a few years earlier, Pernambucan planters were not surprised by the virus's arrival.[15] The response across Brazil's sugar regions—the introduction of mosaic-resistant cane varieties to the fields—brought the third major varietal turnover in the history of Brazil's sugar industry. This time, careful attention to the health of the canes coincided with an increased concern for overall technical renewal in the industry. In the context of budding professionalization in agronomic work, expanding sugar production, and a government interested in arrogating power to itself, planters came into more frequent contact with actors external to their domain who tried to guide their management of the fields.

In Rio de Janeiro, mosaic forced a switch from varieties such as *bois rouge*, from Réunion, to Indian and Indonesian strains (these new canes were respectively identified by the prefixes CO and POJ). Cavalcanti's Pernambucan varieties succumbed to the virus later in the same decade, and Pernambuco followed Rio's lead.[16] Compared to the earlier renovations, this turnover was characterized by rapid replacement and an increased awareness of the need to select varieties for their appropriateness to particular conditions of soil, slope, and moisture.[17] Though the switch was primarily to CO and POJ canes, Brazilian research stations also sought to develop new national varieties.[18] Some in Pernambuco were especially concerned with matching strains to environmental conditions—such as hardier varieties for drier soils—because of the severe drought between 1933 and 1938.

The rapid response to the mosaic virus, both in terms of identifying and distributing other varieties and pursuing research and development of new strains, reflected the increasing professionalization of the agricultural science sector generally. During the early 1930s and into the 1940s scholarly specialization increased, and new universities developed courses in geography and agriculture. In 1936 the first Brazilian Congress of Agronomy convened, with participation by scientists studying sugarcane.[19] Researchers working for the São Paulo secretary of agriculture, industry, and commerce brought to the conference a study, "The Creation of New Canes in the State of São Paulo," which reported comprehensively on hybridization efforts at São Paulo's Piracicaba Research Station.[20]

Agronomists working with cane were further empowered by institutionalized federal support, most importantly the creation in 1933 of the Instituto do Açúcar e do Álcool (IAA, or Institute of Sugar and Alcohol). The IAA was a "public corporation" or autarky designed to bring central planning to the industry, just as other autarkies created by Getúlio Vargas oversaw coffee, salt, lumber, and other commodities. These institutions gave the federal government an aggressive new presence in parts of the economy, which reflected the turn the Vargas administration took following its ascendance in the 1930 revolution. Coffee, Brazil's most important commodity in the nineteenth century, had flourished with virtually no state assistance until after the abolition of slavery, and even afterward the government swooped in to aid in crises rather than rigorously controlling the coffee industry.[21] But the 1930s brought greater intervention in many sectors; comparatively, the sugar industry experienced more thoroughgoing reform under the IAA than most other commodities under their autarkies. With the institute's establishment, a coordinated network of research centers emerged, reflecting an effort to "rationalize" the industry through the application of science and modern technologies.[22] These were also hallmarks of the Vargas administration, expressed through central planning, attention to higher education, and the professionalization of technocratic sectors.

Pernambucan planters had established a sugarcane experimental station in 1910, but for a generation it produced little research, occupied scant space in discussions of improvement, and moved several times around the sugar zone in search of support. In 1934, however, with IAA financing and the institute's arrangement of consistent support from the federal Ministry of Agriculture and the Pernambuco mill owners' and cane suppliers' organizations, the laboratory settled in Curado, an industrial suburb of Recife, and took on a more active role.[23] Nationally, the agricultural research community developed closer relations with international agricultural scientists and monitored studies carried out abroad, reprinting papers from Jamaica, Trinidad, India, Hawaii, and elsewhere, primarily through the IAA journal *Brasil açucareiro*.[24] One of the leading lights of the new cane research effort,

Adrião Caminha Filho, frequently published articles in *Brasil açucareiro* on a wide range of cane-industry topics and traveled to other sugar-producing areas.[25] His research trip to Indonesia in 1930 proved crucial to Brazil's fight against mosaic as he secured the Javanese cane that renovated the nation's cane fields: POJ 2878.[26] By the mid-1950s, this adaptable variety occupied 65 percent of Pernambuco's fields; some attributed to the POJ varieties the salvation of the sugar industry from beet sugar competition.[27]

The IAA tried to strike a balance between the older centers of production in the Northeast and Rio de Janeiro and the developing areas in São Paulo through state-by-state production quotas and price supports. At the same time that the federal government attempted to finesse the regional disequilibrium associated with rationalization, it also addressed tensions within Pernambuco related to the state's uneven pattern of production with both *usinas* and *engenhos*. The IAA director, a Pernambucan named Barbosa Lima Sobrinho, wrote the Estatuto da Lavoura Canavieira (ELC, Sugarcane Farming Statute) in 1941 to organize the industry and to protect *engenhos* from the aggrandizing force of the *usinas*.[28] About five hundred extended families owned and directed Pernambuco's sugar production, according to an estimate from a scholar of 1920s Pernambuco. The rise of mill-centered production probably displaced around two thousand *senhores de engenho*, most of whom became cane suppliers for the new mills.[29] Vargas's government recognized the potential pitfalls of sweeping the entire industry into mill-based production—such a thoroughgoing shift could create a group of powerful opponents to the regime. Pernambuco's expansive and secular relationship with cane (unlike São Paulo's, for instance) increased the collective inertia of older production models, creating the conditions for greater internal tension.

The growing sophistication of agricultural science overlapped with the development of social science, reflecting the aspirations of social engineering that characterized much of Vargas's regime. Vargas had already strengthened the central government when, in 1937, he established a dictatorship he called the Estado Novo (New State). The Estado Novo went further in molding a corporatist social structure that herded people into groups linked to the state, from unions to indigenous groups to sports leagues.[30] Especially during the great economic challenges of the 1930s, governments around the world wrestled with methods of organizing and controlling social tensions. The push for democracy following the war was powerful, however, and the Estado Novo came to an end in 1945. Vargas would return in 1950, elected on the strength of powerful populist appeals.

The ethic of investigation, planning, and regulation in social policy and the economy espoused by the Vargas regime in the 1930s was expressed in the work of people such as Vasconcelos Torres, a geographer and eventually a legislator from Rio de Janeiro. In the late 1930s Torres carried out research throughout the country's sugar-producing regions, publishing his findings as articles in *Brasil açucareiro*. Compiled eventually into a book, Torres's analyses kept an eye always on the health of the agro-industry, but they simultaneously explored questions such as worker mobility, labor conditions, housing quality, and food prices. The book opened with a preface by Francisco Oliveira Vianna, the intellectual architect of much Vargas-era public policy. This personal connection underlines the real links between agricultural research and social science during this period.[31]

National-level agricultural modernization was refracted through Pernambuco's specific context. The sugarcane research station at Curado set out to support the state's planters and the state's Instituto de Pesquisas Agronômicas (IPA, or Institute of Agronomic Research) was founded in 1935 with various departments that carried out research and support functions for the cane industry.[32] Some scientists based at these institutions pursued work not only for the improvement of the sugar industry but also in accordance with nascent conservationist and ecological perspectives, which linked sound environmental management to economic development—connecting nature's economy to society's economy.[33] The emerging language of ecology could be found sprinkled through their work. Reflecting this influence, an agronomist wrote in 1946 that planters should seek to work in harmony with nature, since "the soil is the patrimony of each generation, which no man has the right to destroy."[34] The same year, a state "Permanent Commission" dedicated to water protection was founded and succeeded in pushing through state legislation explicitly forbidding the dumping of mill effluent into rivers (a law generally ignored).[35]

The modest influence of conservationism was most clearly visible in scientists' concerns about deforestation. An IPA scientist and geography professor named João de Vasconcelos Sobrinho, who was influenced by the early U.S. ecologist Frederic Clements, warned of the critical condition of Pernambuco's forests and argued that their devastation would cut into rainfall levels. Totaling up the wood consumption of mills and *engenhos*, he declared that the sugar industry burned through at least fifty-two hundred hectares of forest every year.[36] Another scientist acknowledged that the need for reforestation, too long ignored, "today . . . is an obsession for all who have their eyes attuned to the desert made of the prodigious forest by the woodcutter's ax." He proposed that all sugar mills using more than five thousand tons of wood annually be required to replant trees. Based on his calculations, the Usina Catende, the state's largest consumer, would have to plant fifty-six hectares of trees per year, modest in comparison to Vasconcelos's estimate of the amount of forest destroyed annually.[37]

Scientists and Planters: Usina Catende and the Limits of Modernization

Catende was the *usina* that most fully embodied the spirit of research and innovation that these Pernambucan scientists and the IAA technicians encouraged. During the course of his life, Antônio da Costa "Tenente" Azevedo erected the largest sugar mill in Brazil at Catende. Tenente's story captures one aspect of the trajectory of the Pernambuco sugar industry as a whole, from *engenho*-based production to the consolidation of mill-based production in the 1930s. The scion of a modest Pernambucan sugar family, Tenente owned two *engenhos* in succession before building the Catende mill. He bought out *engenho* after *engenho* to establish a huge territory of control in the southwestern corner of the *zona da mata*. At its height, Catende included nearly thirty-five thousand hectares of land and employed as many as seven thousand workers.[38]

Tenente's efforts represented the same overlap between agricultural and social improvement discernible in the work of people like Vasconcelos Torres. Besides introducing large irrigation projects, experimenting with fertilizers, and modernizing the industrial aspects of production (the mill proper), Tenente built his domain into a showcase of social engineering, founding a hospital, schools, and even a children's scout program for the families he employed. Barbosa Lima Sobrinho, the one-time director of the IAA, lauded Catende for its leadership among Pernambuco's *usinas*.[39] Tenente's agronomist Apolônio Sales eventually became another link between Catende and the ideas emanating from the center of state power. After serving as a key member of Tenente's management and scientific team, Sales went on to lead Vargas's Ministry of Agriculture.[40]

Looking back on the beginning of Catende's rise, Sales described the challenges it faced. "In 1937, the situation of cane agriculture was as precarious as could be," with production well below other major sugar-growing areas. "The varieties cultivated were in utter degeneration," he continued, noting that the appearance of the mosaic virus compounded the industry's problems, and a prolonged drought lasted from 1933 to 1938.[41] Working with IPA, Sales set a vigorous agenda for improving on the state's dismal performance.

He visited Hawaii, which at the time boasted the world's most efficient sugar plantations, and published his observations of the techniques employed there.[42] Brought to Catende by the ambitious Tenente, Sales guided cultivation on the mill's fifty-six *engenhos* and oversaw a hierarchy that included 7 general administrators, 21 agricultural service monitors, 168 auxiliaries, and 5,500 workers.[43]

Sales used mill waste as fertilizer and introduced innovative irrigation schemes in 1937—desperately trying to ease the effects of the drought. He also directed the construction of a fertilizer factory, exploiting phosphorus lodes discovered on land near Recife owned by the *usina*. While Pernambucan planters have never had to devote much energy to "correcting" soil acidity, since the region's soils generally fall within the preferred pH range for sugarcane (5.5–6.0), fertilization is another matter.[44] During the nineteenth century, some Pernambucan planters procured nitrogen from Chile but widespread and systematic use of the fertilizer began only in the 1950s.[45] In 1941, the mill planted cane on 12,000 of the 28,600 hectares it then owned but Sales, impressed by the intensive cultivation in Hawaii, dreamed of producing more cane than ever on only 3,000 hectares. Although it never achieved such intensive production, in the decade and a half following the beginning of Sales's efforts, Catende increased its sugar output 220 percent.[46]

In the midst of Catende's rise, Adrião Caminha Filho—the scientist responsible for bringing the POJ canes to Brazil—visited the Northeast. In a 1945 article in *Brasil açucareiro*, he declared that the state of the region's cane industry was "appalling" and demanded the "renovation of the culture, through rational cultivation processes and fertilization." Caminha found a substantial minority of planters still using antiquated varieties such as Demerara, Manoel Cavalcanti's "hairless," and *bois rouge*; canes that were "degenerated and receptive to mosaic disease, no longer appropriate to the contemporary national sugar industry."[47] Caminha's biting words revealed the increasing assertiveness of the agronomic community as well as the social role it envisioned for itself.

Some of the problems Caminha found could be accounted for by the divergence in resources and support available to *engenhos* as opposed to the larger, richer *usinas*. But scientists also faulted *usineiros* for their preoccupation with just the processing aspects of their operations. A couple of years after Caminha's visit, for instance, a *Brasil açucareiro* article criticized northeastern planters for purchasing advanced milling machinery while ignoring new techniques in the fields. The author asked with exasperation, "Of what value powerful and highly efficient mills if the lack of primary material persists?"[48] An *usina* not far from Catende provides an example of the limits of technology in the fields: of its more than six thousand hectares, the Usina Frei Caneca had only twenty-five under any form of irrigation.[49]

Catende's success became a powerful argument for adopting the mill's innovations, and some other planters demonstrated a heightened interest in new techniques over the course of the 1940s. In 1947 Recife hosted a conference of northeastern planters where all aspects of the industry were discussed.[50] The state's producers had rebounded from the 1933–38 drought, and by the 1943–44 harvest Pernambuco produced more sugar than ever before. The state agriculture agency released an enthusiastic bulletin in 1946 declaring that "the technical organs of the government should move ahead of private initiative, controlling the race for fertility and planning ahead for the new agriculture, on a broad scale and on a secure economic and technical foundation."[51] Much of the industry's new production, however, derived from the expansion of cane fields rather than from increased efficiency or productivity. When Tenente died in 1950, Pernambuco continued to lag behind its competition in São Paulo, where planters increased their production nearly 400 percent between 1944 and 1954.[52] *Brasil açucareiro*'s obituary for Tenente claimed that he "embodied the spirit of renovation—the antiroutine" and that his life's work had been to overhaul the "primitive processes" of Pernambucan cane agriculture.[53]

Much remained to be accomplished, and many of Tenente's peers still clung to traditional practices. The same state agriculture agency that had been so optimistic in 1946 condemned the "deplorable" backwardness of the state's cane industry two years later, indicating the mixed reception of modernizing initiatives among Pernambucan planters.[54] The importance of sugar to Pernambuco's economy meant that journalists, intellectuals, and politicians, as well as the planters themselves, devoted an enormous amount of attention to the industry. But the standard terms of debate—over quotas, pricing, and subsidies—skirted the important questions about agricultural advancement on which the agronomic community focused. It is telling that the subject most discussed at the regional sugar conference in 1947 was the fixed prices established by the IAA, rather than agriculture techniques or industrial efficiency.[55]

Caminha Filho and other technicians active at the national level struggled to drag northeastern producers into the conversation about agricultural modernization. The federal government supported their efforts, allocating large sums of money for technical support for the industry in the Northeast.[56] In addition to IPA and the state agricultural agency, the Joaquim Nabuco Foundation that Freyre had founded began to contribute publications on the agricultural sector.⁵⁷ And, following the lead of the IAA and other sources of federal support, the state government formed the Commission to Combat Sugarcane Pests in 1953. The most active scientist affiliated with the new group was Bento Dantas, an agronomist with vast experience and a career that would eventually span six decades. Dantas authored or coauthored more than half of the commission's first dozen publications. Two of his colleagues at the commission, who released a basic guide to antipest planting techniques in 1959, were realistic about the agency's reach. They acknowledged the gap in knowledge and capacity between *usinas* and cane suppliers. While *usineiros* were aware of the commission's work, the scientists wrote that many suppliers did not even know their office existed.⁵⁸

Because of these limitations, and planters' adherence to tradition, the embrace of agricultural science was modest. This pattern parallels experiences elsewhere, including in the Spanish Caribbean somewhat earlier. Stuart McCook observes of Puerto Rico that agricultural scientists' legitimacy was "based on their ability to deliver practical results to the island's sugar planters"—if immediate, tangible benefits were forthcoming, they would respond.⁵⁹ (For the 1930s specifically, production in Puerto Rico rose, though Cuba did not fare as well.)⁶⁰ There was also a divide between resource-rich *usinas* and the cane suppliers, though even *usinas* failed to take full advantage of agricultural science. In the mid-1960s, less than half of the state's *usinas* employed a professional agronomist.⁶¹ The title of the pest commission's first publication offers a clear signal of the low threshold from which its scientists began their work of disseminating modern methods: *Instructions for the Installation of Pluviometers and the Registry of Their Observations*. The fact that agricultural professionals were trying to convince planters to put simple rain gauges on their properties as late as 1954 demonstrates the scientists' marginal progress in convincing planters of the benefits of agricultural science.⁶²

Putting Science to Work in the Fields: The Mishandling of Cane Variety 3X

The disjuncture between modern agricultural science and the planters' proverbial traditionalism comes to the fore in the story of the emergence, adoption, and crisis of a particular cane variety in Pernambuco in the decade and a half following Adrião Caminha Filho's disappointing 1946 trip to the Northeast. For a generation following the major 1930s varietal turnover, the state's sugar industry relied in large part on three varieties of cane: POJ 2878, CO 290, and CP 27-139 (from Java, India, and Florida, respectively).[63] Usually, planters cut CO 290 at the beginning of the harvest and planted it on valley floors. POJ 2878 could be found on slopes, and CP 27-139 dominated in drier areas. These varieties balanced each other well, since CO 290 and CP 27-139 developed quickly and could be harvested early (they are "precocious" canes), while POJ 2878 took longer to mature and could be cut at the end of the harvest (known as a "tardy" cane). However, in the mid-1950s concern arose about the "degeneration" of POJ 2878 and the susceptibility of CO 290 to disease; enthusiasm spread about the high productivity of a newer variety—CO 331, often called 3x.[64]

Researchers at the Curado experimental station began experimenting with 3x at the end of the 1940s and had noticed its high weight yields and adaptability to poor conditions, ranging from waterlogged and clayey to dry, sandy soils.[65] The cane was a hybrid produced by geneticists at the Coimbatore Sugarcane Breeding Institute in Tamil Nadu, India—the first lab to hybridize noble and wild canes at the beginning of the twentieth century. "Noble" cane varieties, such as creole cane, are descended from the secularly cultivated *Saccharum officinarum*, but industrial canes are hybrids of *S. officinarum* and "wild" cane, *S. spontaneum*. Noble canes have thick stalks, high sugar contents, and shallow roots, while wild canes are thinner, higher in fiber, and have deeper roots.[66] The Coimbatore researchers produced 3x in the 1940s. A thin, fibrous cane like its wild ancestors, it was fast-growing and disease-resistant and seen as an improvement on successful varieties recently produced by the lab.[67] In 1950 *Brasil açucareiro* publicized the enthusiasm about 3x, and more than one federal cane expert recommended that Pernambucan planters introduce 3x into their fields.[68] Two years later, the geneticist Clóvis Coelho de Andrade Lima wrote up the results of five years of varietal research and commented on 3x's "high agricultural production" and robustness; he also recommended that Pernambuco planters adopt it immediately.[69]

Lima's report emphasized the benefits of balancing precocious and tardy varieties. The trio of POJ 2878, CO 290, and CP 27-139 had been such a potent combination because in combination they gave planters high sucrose contents throughout the harvest. Lima explained that he designed an experiment comparing weight yields after noting 3x's growth patterns, so he could distinguish between the beginning and end of the harvest. He reminded his readers at the end of his report, "We should not forget [3x's] high fiber content and late maturation, making imperative the planting of other varieties for the initial phase of the harvest."[70] He recommended that 3x play a significant role in future harvests, balanced with precocious canes such as CO 419 and IANE-C 46-177 (the latter developed at the Curado research facility).[71]

Lima found in Pernambucan planters an audience eager to follow at least some of his advice. In addition to the declining vigor of their reliable strains, planters faced a comparative disadvantage in the gross tonnage of cane they produced for every hectare of land, lagging far behind most other large sugar-producing areas, including other regions in Brazil. Their 30–40 tons of cane per hectare were dwarfed by the 100 tons per hectare in Florida and the 150 tons per hectare in Hawaii. São Paulo planters gradually increased their yields in the 1950s, and by 1964 they pulled 65 tons of cane out of each hectare of land.[72] With 3x Pernambucan planters saw yields increase as much as 40 percent. Though still not in São Paulo's neighborhood, the average production reached closer to 50 tons per hectare.

In response to the enthusiasm of Lima and the other agronomists for the variety, and pleased with the results they saw in their own fields, planters spread 3x as quickly as possible. An agronomist writing in the daily *Jornal do comércio* in 1956 praised planters' willingness to adopt new canes, especially the Javanese and Coimbatores, which he felt were particularly productive.[73] The planters' experiences confirmed that 3x could be planted virtually anywhere and did not demand increased fertilizer or irrigation. In 1954, 3x occupied almost 8 percent of Pernambuco's cane fields; by 1958, the variety's share had grown to one-half of the state's cane. In 1963, 3x accounted for nearly 80 percent of the harvest, having grown to ten times its earlier acreage in less than a decade.[74]

The increasing share of 3x in Pernambuco's harvest came in the context of an overall expansion in the industry; almost sixty thousand hectares were added in roughly the same period.[75] So at the same time that the state increased the territory planted in cane by 30 percent, it increased the gross tonnage of production by almost 50 percent. The increased yields came partly from 3x's continued productivity after the first year of growth. After being cut in the first year, the drop-off in weight for the second and third years of regrowth (ratoons) were not as sharp as that of other varieties.[76] Planters found this quality attractive, especially since they had grown accustomed to low ratoon yields. Agronomists had tried to convince planters of the importance of caring for cane fields with regular weeding and aeration, not just in the first year of growth but for ensuing ratoons as well, but the practice was rarely followed.[77] The vigorous ratoon growth of 3x further undermined efforts to extend care to ratoon fields and reinforced the argument for the variety's adoption.[78] But as they aggressively spread 3x through their fields and watched with satisfaction as per-hectare weight yields rose, planters and scientists noticed that another crucial indicator fell: the yield of sugar per ton of cane. Between the mid-1950s and 1964, sugar yields dropped by as much as 20 kilograms per ton of cane.[79]

A cane variety's "industrial quality" is measured by the concentration of sucrose, which is a function of fiber levels and by the purity rate of the juice.[80] In 1959 Bento Dantas noticed that 3x had very low sucrose levels, and therefore juice purity, until late in the growing season. Dantas echoed a caveat that the geneticist Lima included in his 1952 report: "The widespread diffusion of CO 331 in recent harvests, indiscriminately replacing 'precocious' varieties like CP 27-139 and CO 290 and tardy clones like POJ 2878, demonstrates its exceptional agricultural value in Pernambuco's conditions. If we consider, however, its very slow maturation, only ready for grinding after January . . . we can see the danger it represents to our state to maintain such a large area with this variety, which should only be collected at the end of the harvest and is completely immature when cut at the beginning."[81]

The former president of the Curado research station also reacted to planters' selective response to directives emanating from the research infrastructure so diligently installed over the previous decades. In 1960, he wrote a letter to the editor of the *Diário de Pernambuco* asking, "Who can blame the Curado technicians if the majority of sugarcane planters align themselves with CO 331, the so-called 3x, a slow-developing, hardier cane?" His frustrated rhetorical questions continued: "How many times have those technicians said and written that the problem of a rational cultivation of sugarcane cannot be solved with the exclusive utilization of an obviously tardy variety?"[82] Also in 1960, Frederico Veiga, the director of the Campos Experimental Station in Rio de Janeiro, visited Pernambuco and likewise called attention to the state's declining output. In conversations with his northeastern colleagues, he emphasized 3x's late maturation, pointing out its low sugar concentration at the beginning of the harvest. In addition, by the time the cane matures at the end of the harvest, its juice volume dips below the norm for other varieties. Veiga suggested five other varieties that Pernambucan planters might use, each with a particular advantage in productivity, hardiness, resistance to drought, early development, and high sugar content.[83]

Three years after the Curado chief's letter and Veiga's visit, with 3x accounting for almost four-fifths of the cane in Pernambuco, the industry was struggling with the effects of its headlong pursuit of increased tonnage.[84] The *Diário* reported on an IAA directive to Pernambucan planters advising the replacement of 3x with several new varieties. The predominance of 3x in the state's fields had had disastrous effects on industrial yields, and urgent changes were required. The tone of the article was almost forensic: "CO 331 has been identified as responsible for the fall in the *usinas*' industrial yield, in virtue of its late cycle of maturation and its elevated fiber content."[85]

Perhaps planters breathed a sigh of relief that the culprit had been found and their difficulties could be explained by a troublesome variety. But agricultural researchers knew that the blame did not lie with the variety itself. A former state agriculture secretary weighed in on the rampant premature harvesting of 3x in an essay from the same year, expressing bemusement but not surprise that such a shocking problem could take place "still today, when Brazil maintains an open cultural and social exchange with the largest cane-growing centers of the world, when magazines, bulletins, journals, and official publications go not only to governmental organs but to professional organizations and even the sugar companies."[86] Another agronomist defended 3x the next year, through a backhanded critique of planters. He wrote that many factors contributed to the mills' low yields and pointed specifically at outmoded techniques, warning, "Either we rejuvenate and improve (update) our work methods in the sugar agro-industry or it will not long survive." Yet another scientist wrote the same year that despite having "a sugarcane experimental station in this state for fifty years . . . the cultivation methods have scarcely been modified."[87] These sorts of complaints had been typical of agronomists from their emergence and professionalization into the mid-1960s. Their chronic frustration derived from combating the entrenched cultural values of planters, which militated against a purely technical approach to agriculture.

The IAA and state agencies continued to ship substitute varieties to Pernambuco and tried to convince planters to overhaul their fields. Adding insult to the injury of the whole episode, one of the loads of cane cuttings shipped to Pernambuco from Rio de Janeiro apparently carried spittlebug larvae, a pest known in Brazil as *cigarrinha*. By the middle of the 1960s, still struggling to address low per-hectare yields and low industrial yields, the Pernambucan planters confronted a new battle with this cane pest familiar to other sugar-producing areas around the world. Planters' immediate recourse to a new cane variety and the new problem they faced as a result were further irritants to agronomists. A scientist complained in 1967 that cane planters saw "new varieties as a sort of panacea for all the problems of the cane industry."[88] The planters' problem was not so much their focus on one aspect of agricultural science as their tendency to ignore the full prescription and think too narrowly about production. But this sort of thinking fit their discourse of the purpose and nature of the laboring landscape.

Despite the planters' uneven responses to agricultural research, one constant from 1940 to 1960 was the expansion of cane. Sugar production rose during this period, a result of the expanded cane fields rather than of increased efficiencies or yields.[89] The growth was marked enough in 1936, in fact, that Tenente went so far as to suggest that Pernambuco might run out of land for planting cane. The area of land planted in cane expanded from 146,724 to 193,270 hectares between 1948 and 1955, and then made another startling jump to 254,133 hectares by 1962.[90] This was a 30 percent increase in seven years followed by another 30 percent over the next seven years; in less than a decade and a half, the area cultivated in cane grew by nearly two-thirds. With sugar yields dropping as a result of the 3x mistake, the gross tonnage of cane had to rise in order to maintain production levels. Increasing fertilizer use in the late 1940s and into the 1950s also led some planters to abandon the practice of fallowing fields. The hasty expansion rescued the industry from what one analyst called the threat of total collapse, but it also pressured workers, who had to plant, weed, and cut ever more cane.[91]

Resistance to Rationalization:
Workers Confront "Standard Customs" and Increased Demands

Agro-environmental change such as the drive to increase production with new cane varieties impacts the daily lives of workers. In the 1950s more and more workers were caught in a vise: cane crept outward *and* increased in density as previously uncultivated areas were planted and fields were less frequently rotated out of use. *Usinas* employed 93,182 workers and suppliers an additional 42,915 at this time, and these were the people bearing the burden of the industry's growth.[92] The spread of cane impinged on workers' lives in multiple ways. It increased labor demands and planters' desire to mobilize workers at precisely the moments they chose. The need for space meant that less and less land was available for food crops. All of these effects led planters to gradually cut off worker access to land for garden plots. New payment regimes left workers less time to cultivate their own crops anyway, but tenantry arrangements also less frequently included land access. Júlio Bello acknowledged displacing a number of the *moradores* on his *engenho* (though he claimed that he tried to reserve some space for gardens).[93]

This trend was clear enough and worrisome enough to Barbosa Lima Sobrinho as early as 1941 that he included in the ELC a stipulation that workers resident on an *engenho* for more than one year receive access to plots close to their houses and of a size large enough to provide their families with food.[94] The framers of the statute feared that decreased access to land would increase worker mobility and transience. An author in *Brasil açucareiro*, explaining the logic of the ELC a few years after its passage into law, described the work conditions that made the legislation necessary: "Uprooted from the land by the prohibition from planting it, living from wages, the peasant . . . begins to move in search of a better price for his labor. No remaining tie binds him to the land."[95]

Critics of the reform, upset by what they interpreted as an attack on the prerogatives of the *usinas*, pointed to the mills' recent efforts to provide "social assistance" to their workers. Barbosa Lima Sobrinho acknowledged these programs, and singled out Catende as particularly praiseworthy, but he also distinguished the situation of the field worker from that of the mill worker. He explained that the latter had access to labor rights, but for the former help came from the mill as a voluntary measure, if at all. "What the worker might receive would be a favor," he wrote, "dependent on the generous soul of the donors." In the ELC Barbosa Lima Sobrinho thus sought to engineer that sort of support by mandating assistance to workers.[96] In the early 1950s, however, the state secretary of agriculture joined with the director of the IAA and the archbishop of Olinda in urging *usineiros* to grant their workers access to land as a means of increasing the production of subsistence crops. Though the program does not appear to have gone anywhere, its very existence revealed the inefficacy of the ELC's guidelines for making land available to workers.[97]

Barbosa Lima Sobrinho's comments made clear that even as more and more people worked in the fields and mills of the supposedly modern *usinas*, relationships were still brokered through patronage networks. Industrial workers earned legally prescribed benefits thanks to policies established by the Vargas administration, and these advantages drew workers to the mills themselves. Yet securing such jobs still had as much to do with patronage as the tenantry agreements these same workers had struck back on an *engenho*—getting the jobs depended on securing a favor from a boss.[98] Even Catende exploited its field workers, despite its reputation for enlightened management. In 1945 Catende was accused of mistreating workers, including withholding pay. Mill officials responded that "anyone who has contact with our rural areas" knew that sporadic payment was common. "The agricultural operations of the Catende Mill are orderly," they wrote, "but they do not diverge from the standard customs of our rural area."[99]

Those "standard customs" included manipulation of payment systems, and when this tradition continued in force during the expansion of cane and the increased pace of production of the 1940s and 1950s, workers suffered. This explains why one scholar characterized the expansion of cane cultivation as "the key element that marked the beginning of the irreversible process of proletarianization of the Pernambucan sugar worker."[100] Certainly the creeping tide of cane and increased production rates had severe consequences for tenantry relations and the labor rhythms of rural workers, but we will have to explore just what "proletarianization" meant in terms of the nature of these consequences and the shape of workers' responses. Besides losing access to land, workers faced new modes of measuring labor and allocating compensation, which increased the pace of work.[101]

Work performed in the fields had a secular rhythm: planting, weeding, cutting. At the most basic level of operations performed, the 1940s and 1950s did not represent much of a break; the salient changes of the period involved the pace and organization of work, and these shifts were reflected in shifting payment schemes. Referring to the apparent changelessness of field tasks, the anthropologist Moacir Palmeira explained in 1976 that the perception of stasis "is belied, at each turn, by the diversity of forms of labor organization and remuneration of workers that peppers the history of the sugar plantation and hints to the researcher, in a symmetrical and inverse mode, of an absolute discontinuity." He points analysis away from the tasks themselves and toward their compensation, and he emphasizes that workers made strong distinctions between different sorts of work and payment regimes and knew the ins and outs of many different systems.[102]

During the transition from slave to free labor, planters paid workers for a day's service (a *diária*, or *jornada*) of working in a gang (the *eito*). With the rise of mills over the following decades, payment slowly shifted away from rewarding periods of work rendered toward quanta of work performed; in other words, task work or piecework emerged.[103] This pattern, similar to other sugarcane areas, contrasts with the labor patterns for other commodities. The coffee regions of Brazil's south turned to *colonos*, families of workers who lived on plantations and were responsible for maintaining a certain number of trees over an extended contract period. This more closely resembled the sharecropping system that emerged in the cotton areas of the United States during the postemancipation period.[104] The adoption of task work rather than the daily wage spread unevenly through the sugar region, as with so many changes. Indeed, this ought to be expected, as very rarely do changes of this sort take place in one rapid transition. Other sugar areas had similarly long trajectories. The Puerto Rican cane cutter don Taso told the

Work gang, or *eito*, 1940s–50s. Note the foremen on horseback. (Acervo Fundação Joaquim Nabuco, Recife)

anthropologist Sidney Mintz in the late 1940s that over the previous two decades he had earned money through both piecework and daily wages.[105] For Pernambuco's cane zone, the overall direction of change in the mid-twentieth century was toward a rationalization of payment regimes, carried out in the interests of increasing efficiency and productivity by increasing pressure on workers.

Task assignments evolved from traditional agrarian units and the habits of field foremen. The *tarefa*, or task, was, as the geographer Affonso Várzea put it, an "old agrarian measure." In his book *Geography of Sugar* (1943), Várzea spends several pages discussing the wide variability in the *tarefa*'s size. Traveling around the sugar-producing Northeast, he found numerous different versions of the unit, ranging from 3,025 square meters to 4,400.[106] From the origins of the sugar industry, the *tarefa* was a unit equal to the amount of cane a mill could grind in one day. In the seventeenth century, this was interpreted roughly as twenty paces square.[107] It then evolved into a unit for farmers to estimate field sizes. Though linked to labor, it was not explicitly a unit for assigning work. Equally traditional, and variable, were the long rods that overseers carried as they rode the fields monitoring workers. Called *varas*, these long wooden or iron poles with hooks at the end were used for measuring jobs (or for wielding against workers).[108] Eventually and haltingly, the

Cane cutters and foremen on horseback, 1940s. (Acervo Fundação Joaquim Nabuco, Recife)

vara achieved a uniform length of 2.2 meters, or one *braça*, and became the standard unit of job measurement.[109]

The first widespread work unit measured out in square *braças* was a "count" (*conta*), which consisted of an area intended to take one day to complete, whatever the job, and therefore approximated one worker-day. For many, the count was simply a synonym for "measure" (*medida*), with count areas changing according to the nature of the job required and the remuneration depending on the established wage of a given location. The count spread in the 1920s and 1930s and by the mid-1940s had become fairly generalized throughout the cane zone, though flat daily payments had not disappeared. As it spread, the count increased in refinement. By the 1940s some workers could perform more than one count during the course of a day and therefore could sometimes earn more than even the official rural minimum wage—a rate set by the Vargas government in 1943. In the years leading up to Vargas's decree, producers had debated the idea of minimum wages and followed similar policies in other sugar-producing areas. In practice, though, the minimum wage prescription had almost no effect on workers' actual pay.[110]

A prominent agronomist alleged in 1946 that workers could earn more in less time than they had before. Though this was the opposite of the intent of the changing payment regimes (and therefore a suspect assertion), some workers probably could earn more through extreme exertion. The agronomist said that workers responded to these conditions by working less rather than by "improving themselves," and the changes in remuneration were thereby creating "opportunities for laziness" and a false labor scarcity. Increasing wages, he concluded, was against the interests of the workers, who without being compelled to work would never aspire to a better life.[111]

Just when the count stabilized in size at ten square *braças* in the early 1950s, people began referring to "tasks," likewise areas measured in *braças*.[112] Unlike a count, however, a task was not expected to be completed in one day and therefore was a means of delaying worker pay, which was withheld until completion of the full assignment. Workers from the cane zone still remember the arbitrary measurement of tasks during this period, when it emerged as a unit of work assignment. Paulo, the Nazaré da Mata worker who began cane work in the 1930s, remembered in 2003 that "the *tarefa* was twelve by thirteen [*braças*], [or] eleven by eleven; it was just a measure."[113] Paulo's colleague Severino, who cut cane as an eight-year-old, agreed: "The *tarefa* was a square, and it was measured according to whatever size the boss wanted," he said. "Twelve by thirteen, fifteen by fifteen. Anything."[114]

The fitful regularization of agricultural measures that took place in the mid-twentieth century was a process replicated all over the world across the modern era, part of what Max Weber described as rationalization. Traditional units of measurement were, as James Scott observes, sometimes quite literally human-sized, and generally specific to particular regions.[115] And so it was in the sugar-producing Northeast, as Várzea found. But the increased scale and pace of production in the 1940s, 1950s, and 1960s demanded a more regularized system. Planters argued that they needed to streamline their processes and trim a bloated workforce. A cane industry report from 1964 observed with frustration that "Pernambuco is probably the only locale in the entire sugar-growing world in which canes are cut, collected, and wrapped in small bundles during the harvest."[116] And these quirks made Pernambuco impressively inefficient compared to other areas. On average, 3.6 workers per day were required to harvest one ton of cane at late as 1965, whereas São Paulo used only 1.2 and Hawaii just 0.3.[117]

Paying workers by the day or week reflects a view of labor as an undifferentiated resource to be applied to the fields. The task, on the other hand, was a unit tailored to specific quanta of work in particular conditions. With tasks, planters began to regularize particular processes and assign tasks in such a way that the pace could be increased by adjusting measurements, or adjusting payment rates based on production. The "count" was a transitional unit,

Cane cutter with bundles of cane (*feixes*), wrapped in the leaves from the cane tops, 1940s–50s. (Acervo Fundação Joaquim Nabuco, Recife)

which still linked payment to the notion of a generic day of labor. That link was severed with the rise of the task, which outlined a designated job that generally took two or more days to complete. Attempts to increase efficiency and manage expanding and increasingly complex agro-industrial businesses translated into an increased technocratization of fieldwork.

A very similar pattern emerged in Puerto Rico, which, as don Taso explained, had complex and varied payment systems.[118] A 1951 study of the sugarcane industry in the British West Indies also tracked a similar phenomenon. Suggesting that payment systems emerged from the task measurements assigned to slaves, an analyst wrote, "Over the course of time, a complicated system of task definitions has been developed which has become progressively more refined. . . . What has developed in the British West Indian Sugarcane industry is a complicated incentive wage payment system which staggers the understanding of outsiders."[119] In the British West Indies, though, unions protected the traditional system, since workers had mastered ways of maximizing their earnings within the system. In Pernambuco, until workers found a means of exerting some control over the process in 1963, these shifts tended to erode rather than enhance worker control and opportunity.

Though workers interviewed forty years later generally remember being pushed to the limit to meet their assignments, there is some evidence of workers' developing strategies to increase the efficiency of their work. In Goiana in 1959, workers collectively demanded that the payment per bundle of cane be increased.[120] That effort failed, but planters there complained later in the year that "a new problem relating to the field workers has arisen." They told police that workers were cutting cane without removing the leaves at the tops.[121] Workers compelled to improvise timesaving measures like this were clearly confronted with the decision of either producing more or earning lower wages. Leaving the cane tops meant one less sickle-stroke for each cane—the self-imposed adoption of a Taylorist technique by workers compelled by pressure to produce. The agronomist Bento Dantas belatedly arrived at a conclusion similar to that of the workers in 1965, when he pointed out in an essay that burning the leaves off with fire could eliminate the extra sickle strokes used to trim leaves. The savings in time, he claimed, would amount to half a worker's production per day.[122] It is not surprising, therefore, that workers sometimes set fires to cane fields in an effort to expedite their work and maximize their earnings.

The switch in remuneration from time to space—from the day's wage to pay by the task—might have been a "rationalization" of the work process, but it did not eliminate planter interference in the murky interstice between work performed and pay received by workers.[123] The planter could reduce payment amounts, or eliminate them entirely, according to an arbitrarily determined decision about whether or not the work was "well done."[124] Struggles over control of units and tasks on the job mirror the battles over "worker control" and labor process in the early-twentieth-century United States so carefully studied by David Montgomery.[125] Unlike U.S. industrial workers, Pernambucan cane workers had very little control of any sort until the 1960s, but they did sometimes resist the changing payment methods and compromised access to land.

In mid-1960s, state police officials investigated a conflict on an *engenho* where workers had refused to complete their jobs. Though the planter complained to the police of leftist agitation, the investigators found a different situation. In a detailed explanation of payment systems included in their report, they stated that a worker typically received a work assignment measuring ten by ten *braças* for one day's labor (which is to say, a count). "An honest *braça* is two meters, ten centimeters [sic]," they wrote, but "dishonesty reigns" on this *engenho*—and, they speculated, most others—as workers faced tasks measured with a 2.3-meter *braça*. This amounted to an additional eighty-eight square meters of hoeing, weeding, or cutting each day. They noted that when the proprietor adopted a fair measurement unit and tripled the pay for each task completed, fifty of sixty tenants returned to work.[126] A worker I interviewed named Benedito, describing his parents' experiences in the 1940s and 1950s, said, "They always said a *braça* [*vara*] should be iron, so nobody could rob [by using a longer measure]."[127]

At this time, the application of units and payment systems still varied throughout the cane zone, a fact that comes through clearly in testimony from the first years of rural labor courts (1963 and 1964).[128] Some plaintiffs claimed to get paid by the count, others by the task. Some said that they worked "by production," another term that began to proliferate as mills sought ways to boost worker productivity and paid wages according to weight-based quota levels.[129] And some workers still earned a simple daily wage. Since the labels had yet to settle into a crystallized form, plaintiffs, defendants, and lawyers stumbled over terms, correcting themselves and using different terms interchangeably.[130] The fact that there were so many misunderstandings and disagreements reflects the rise in labor tension in the early 1960s, a result of both the increasing pressure on workers and (as I will discuss in the next chapter) the political climate.

Striking for Lower Wages

This was the context for the 1963 strike at the Usina Roçadinho, which appears to have been a story of competing payment systems and workers' attempts to exert a measure of control over their own labor rhythm. The *Diário de Pernambuco* noted that workers on some *engenhos* of Sampaio's Roçadinho mill were striking, but the reporter commented incredulously that the workers actually sought *lower* pay. Work on Roçadinho was remunerated by a task of 625 square *braças*, the reporter explained, paid at a rate of two thousand cruzeiros. Since a worker could complete a task in two or three days, he should earn six hundred to seven hundred cruzeiros a day. But the strikers preferred payments based on the "count," understood as ten *braças* square more or less, depending on the terrain (hilly, rough areas might be nine by nine, while clear, level fields where work went easier could require a twelve by twelve measure). Generally, workers completing "counts" earned five hundred cruzeiros a day. "One sees, by the numbers," the author concluded, "that the strikers 'demand' the reduction of their salaries from seven hundred and even eight hundred cruzeiros to five hundred a day."[131]

Sampaio held a similar view of rural workers. In 1965, he wrote a long letter to Brazil's president, General Castelo Branco, describing in great detail the challenges Pernambuco's sugar industry had faced in the previous years. Referring to political gains and worker unionization in 1963, he wrote, "Lamentably, when progress was made, putting an end to an extremely exploitative process and rehabilitating the rural man, productivity was greatly lowered by class conflict, disorder, disharmony, the poisoning of spirits, and the exaltation of small production and minimum effort."[132] This statement directly reflected Sampaio's analysis of the people who had kept his mill going; he believed that rural workers lacked initiative as a matter of culture. "Since [the cane worker] earns by the task," he said, looking back in a later interview, "if he put forward a little more energy, instead of cutting one ton, he would cut three, he would earn three minimum wages. But he is resigned to poverty."[133]

More likely than either the reporter's interpretation or Sampaio's was another, revolving around the workers' desire for control. Just as other workers around the region had incorporated efficiencies into their own work habits, the Roçadinho workers probably wanted control over their own work rhythm. This was something more easily achieved with a job designed to be completed in a single day. Tasks were too vague, too easily manipulated by foremen, who would inflate their measurements or refuse to approve the workers' completed tasks. Though in theory the Roçadinho workers should have been able reliably to make the wages calculated by the reporter, in practice they likely did not. Tasks also only provided payment every few days, keeping workers tied to the job for that period. Rather than striking "for the simple fun of it" (another of the *Diário* reporter's theories), the Roçadinho workers no doubt sought to maximize their flexibility and their options for splitting wage work with subsistence farming.[134]

As it happened, they were swimming against the tide. As we have seen, the area planted in cane in 1962 was nearly two-thirds larger than it had been in 1948.[135] While well over half of workers still had access to land in 1961, seven years later that number dropped to 46 percent. And of those who still had access, only half were able to use it.[136] In 1963 those workers wanted a payment regime that would facilitate a daily schedule under their control. They battled the simple physical advance of cane (facilitated in many instances by the hardy and adaptable 3X), which displaced garden plots and increased labor demands. Planters moved toward tasks to get more labor from workers, and they changed tenantry agreements to withhold land access.[137]

Rural workers did not have a culture of sloth, nor were they collectively resigned to misery. They did, however, see the world they lived in as unjust, and amid changing circumstances they developed strategies of work that would best approximate freedom from the captivity of the boss. Workers in some places devised means of increasing their efficiency; in the case of the Roçadinho workers, they protested for an older payment method to which they were accustomed and which could be balanced with subsistence farming. Across the region, workers also changed employers with greater frequency. Planters, too, made choices on the basis of their landscape discourse, as their experience with 3x demonstrated. They followed scientific advice selectively, serving their own desires for commanding their laboring landscapes. Planters also wanted their labor supply to be reliable and stable, a project that had begun in an organized way with Barbosa Lima Sobrinho's ELC.[138] This chapter has shown the connections between persistent landscape discourses, on the one hand, and agricultural change and *reactions* to agricultural change, on the other. The clear oppositions between the workers and the planters produced contradictions and conflict during the 1950s and into the 1960s. These came to a head in an explosive political context I describe in the next chapter.

FIVE

The *Zona da Mata* Aflame
Political Upheaval, Strikes, and Fire

In January 1962 a public security investigator hurried to an *engenho* in São Lourenço, northwest of Recife, to look into reports that an airplane had dropped a red object into a field and set canes on fire.[1] Investigator Oliveira stayed in São Lourenço for days, visiting twelve *engenhos*, interviewing "the greatest possible number of rural workers, for greater clarity," and soliciting the opinions of *senhores de engenho* before writing an exhaustive report. Though Oliveira eventually concluded that the story had no basis in fact, local and national newspapers wrote that an apparent firebombing had taken place.[2] This nonevent provides a window into the climate of the early 1960s, when authorities and observers could be taken in by the idea of "outside agitators" engaging in aerial warfare against the sugar industry.

Josué de Castro captured the temper of the time in the title of his book *An Explosive Zone of Latin America: The Brazilian Northeast*, written in 1963 and 1964.[3] In describing his home region as explosive, Castro joined many other residents and observers who resorted to incendiary metaphors to characterize the cane zone. In a recent article, a history professor from Pernambuco examines his own memories of the time, stirred by seeing a cane fire at night and experiencing a resurgence of old feelings from the 1960s. He relates the anxieties and fears associated with fire and, most important, its politicization in the context of labor unrest.[4] Many others, both those involved with the sugar industry and those with no connection, saw events in the cane zone as he did: tinted flame-red. The political drama that unfolded during these years continued the process of opening up the *zona da mata* that agronomists and social scientists had begun. Indeed, the region became a focus of worldwide attention because of popular political mobilization.

Many outside observers reacted ignorantly to the dynamics of this mobilization. A *New York Times* reporter reduced the entire Northeast to a dry wasteland, for instance, but even better-informed commentators missed local specificities. An agro-environmental history must attend to variation even within an agricultural region, and in the case of Pernambuco's cane zone, environmental variations shaped different forms of mobilization. Worker experience dictates its own sort of geography, and in this chapter I examine variability across the cane zone to help understand this pattern. I also explore a key theme of many discourses from the period: fire. After describing its material realities and the process of its politicization, I describe how fire emerged from the period with a new use in the cane zone.

Realities on the Ground: Regional Specificity in the Cane Zone

The *zona da mata*, despite its powerful regionwide association with sugarcane, is not homogeneous. Climate, topography, trends in land use, and patterns of landholding varied over time, with a discernible split between the northern and southern parts of the region. Development and the expansion of cane cultivation in the second half of the twentieth century brought a "homogenization in the landscape, and the small differences [between regions] became less perceptible."[5] But despite the uniformity of monoculture farming and the increasingly rationalized and regularized production methods introduced by the big mills, certain regional specificities persisted, and these must be grasped in order to understand worker experiences in the north and south and the subregions' different political trajectories. Though greater burdens were common throughout the industry, workers' experiences of the pressure differed around the cane zone.

The most stable regional differences were rooted in the environment. In the northern area, for instance, rainfall drops sharply from the coast toward the interior (from 1,981 millimeters per year on the littoral to 920 farther inland). This earned the north the name "dry *mata*," while in the "wet *mata*" of the south rainfall ranges from fifteen hundred millimeters to well over

Municipalities of Pernambuco's *zona da mata*

Source: Manoel Correia de Oliveira Andrade and Sandra Maria Correia de Andrade, *A cana-de-açúcar na região da mata pernambucana: Reestruturação produtiva na área canavieira de Pernmabuco nas décadas de 80 e 90: Impacto ambiental, sócio-econômico e político* (Recife: Editora Universitária, 2001).

two thousand. When cotton mounted its brief bid for economic primacy in the province early in the nineteenth century, it was much more widespread in the drier north.[6] Cane planters had always sought out the well-watered valley floors for cane, and they associated good land with the highly prized *massapê* soils found in these low-lying, clay-rich areas, more abundant in the south. Topographically, the north tends toward broad, sandy ridges, while the south has more consistent and steeper hills.

Because it had a less single-minded tie to sugar work, the north supported more agricultural diversity as well as a denser population. In 1914, for instance, the northern region had 391 *engenhos* while the south had 973. The south also converted to large mill (*usina*) production earlier and more emphatically than the north. Of the thirty-six central mills built before the beginning of the twentieth century, thirty-four were located in the southern part of the sugar region. This trend continued in the next decades, as mills continued to be built more frequently in the south, while small-scale *engenhos* persisted in the north. In 1935 the south had fifty-four *usinas* and 180 *engenhos*, while the north had only thirteen *usinas* and 316 *engenhos*.[7] With the lack of diversity and the dominance of *usinas* in the south came much higher rates of land concentration. Palmares, Catende, Gameleira, and Ribeirão, all in the south, were among the top municipalities in the entire sugar region in terms of concentration of land in the hands of the mills and neighboring Escada, Joaquim Nabuco, and Água Preta were not far behind. Catende, the southern municipality that was home to the empire built by Antônio da Costa "Tenente" Azevedo, had 80 percent of its lands controlled by mills. A northern municipality such as Vicência, by contrast, had only 9 percent of its land controlled by mills.[8]

Along with the *usinas*, research support was concentrated in the south. Less likely to employ new technologies, the north did not have the same experience that the vast expansion of cane brought to the southern *mata* in the 1950s. The sandy ridges, or tablelands, were "empty expanses, like dead frontiers between the domains of cane extended through the valleys," wrote an agronomist in 1958.[9] Only in the 1970s did the north make serious strides to push cane into these formerly unviable areas. To take two municipalities as exemplary of their respective regions, Catende, in the south, had twenty thousand hectares devoted to cane in 1960, or 80 percent of its total area. Vicência, in the north, had seven thousand hectares of cane, occupying about 55 percent of its land. From this land investment, Catende produced more than twice as many tons of sugar as Vicência, or a difference in yield of forty tons per hectare to twenty-eight tons.[10] These discrepancies flowed from the differences outlined above: greater rainfall in the south along with more emphasis and attention given to the sugar industry. As another index of sugarcane's dominance in the south, one could look to the production of a staple food. While Vicência had 1,170 hectares of manioc under cultivation in the 1957, Catende in the south had a meager twenty-six. This last statistic also points to the fact that northern families produced far more food for their own consumption.[11]

The divergences visible in these statistics from the 1950s were probably accentuated by 3x's rapid spread. The variety that facilitated the spread of cane because of its hardiness was "widely accepted" in the northeast and northwest sections of the *zona da mata*, but it was "dominant" in center, southeast, and southwest.[12] The variety compounded existing differences in work conditions. For example, more children and women performed wage work in the south than in the north.[13] A worker interviewed about this period said, "in this region of the northern *mata* the woman didn't work; she worked at home, took care of the children."[14] This conforms to the general trend of older, more patronage-mediated tenantry relations persisting for longer in the north. The "yoke," which required unpaid labor in exchange for access to land, lasted into the 1960s there.[15] A northern worker who brought his complaint to a labor court in 1963 testified to this, saying that he worked two days of the week without pay and earned a daily wage for another three days.[16] As part of the erosion of patronage in the south, workers changed jobs more frequently. Mobility had been an aspect of workers' lives in the past, but before midcentury most workers remained on an *engenho* from birth to death.[17] By 1961, fully half the respondents to a large survey listed a place of origin elsewhere in the cane zone; the south had significantly higher rates of transience.[18]

Despite these differences, there were obviously broad areas of shared experience for workers in the two subregions. In 1961, for instance, workers in north and south still overwhelmingly lived in houses provided along with their jobs. In fact, only about 10 percent of families in that year reported living in a house that was *not* ceded to them by a planter as part of their labor agreement. And the conditions of those homes were generally consistent: dirt floors, wood fires for cooking, a potable water source some distance away. In addition, only about 4 percent of children regionwide finished elementary school.[19] Both the generalized experience of exploitation and the differences between north and south were important to the patterns of politicization of the region, even if they were not widely recognized by observers from outside. This disjuncture between the geographic understanding of the region from within and without was discernible in the heated discourses of the 1960s.

Mismapped: Popular Politics in the Cane Zone

Josué de Castro began his 1965 book on the Northeast with an incisive commentary on the persistent misperception of the region's history. He argued that the region was discovered twice, both times by mistake. The original Portuguese colonizers arrived in Brazil through an error of navigation, as Pedro Cabral swung too far west on his way around the African continent. Then, in the early 1960s, anxious North Americans made another discovery through error, this time "an error of interpretation."[20] Castro cogently identified the mismapping to which the cane zone was subject during the period of its extreme politicization, when observers in the United States took the region for another Cuba rather than a place with specific historical patterns, challenges, and conflicts. The North Americans were not alone in their failure to understand the social, political, and physical geography of the region; some Brazilians also failed to grasp the nature of the events there because of their misinterpretation of realities on the ground.

Politicization Begins

Political activists, especially Communists, turned their attention to the cane zone in the 1940s. Pernambuco had participated when the Aliança Nacional Libertadora (ANL, or National Liberation Alliance) attempted to foment revolution in 1935. The ANL drew a mass following of antifascists, liberals disaffected with Vargas's rule, and Brazilians sympathetic to the charismatic former army officer and Communist leader Luís Carlos Prestes. In Pernambuco, the ultimately fruitless uprising of November 1935 was led by Gregório Bezerra, who from his rural worker upbringing had become an Army sergeant and a Communist Party member.[21] Despite this failure, the Communists maintained an active organization; they took notice when the work of social scientists like Josué de Castro drew more attention to the working class and led more people to explore the connection between urban and rural labor. At the early stages of the region's politicization, the north-south differentiation in the *zona da mata* mattered less than the urban-rural differentiation. Although they set out to organize rural workers and forge an alliance with the urban proletariat, Communist activists were concentrated in the city, so their "rural" organizing took place largely on *engenhos* close to Recife.[22]

The Party's politics reacted as much (or more) to national-level dynamics as to local concerns. In 1944, for instance, Getúlio Vargas's regime opened an avenue of organization with the passage of the Rural Unionization Law. In theory, this policy aimed to bring rural labor in line with the oversight norms governing urban labor. Vargas had created the labor judiciary in 1939 as part of the federal justice department and then, in 1943, promulgated the Consolidação das Leis do Trabalho (CLT, Consolidation of Labor Laws) that outlined all regulation of urban labor. Labor courts took shape under the

CLT, with local juntas presided over by four *tribunais regionais de trabalho* (TRTs, regional labor tribunals) and the Tribunal Superior de Trabalho (TST, Superior Labor Tribunal) at the top. The CLT governed labor standards and workers rights and established a minimum wage, opening to urban workers a crucial new means for contesting their exploitation. But the CLT's relationship to rural workers remained an ongoing question, and it was considered null by most employers and courts.[23]

With the redemocratization associated with the end of Vargas's Estado Novo in 1945, the Communist Party was legalized, and in 1946, two years after the rural union law passed, the first rural union appeared in the cane zone. An agitated *engenho* owner from close to Recife wrote to police in 1946 that Communists were spreading word that he did not own his property and that it would soon be theirs "because the leader Luiz Carlos Pretes [sic] was working for this."[24] The Party also attempted to create peasant leagues to protect the interests of smallholders and nonsalaried rural workers, circulating boilerplate copies of charter documents for founding leagues.[25] However, the unionization effort and the attempt to create peasant leagues went nowhere, especially after the Party's brief period of legitimacy came to an end in 1947.[26] In the fifteen years following passage of the rural unionization law, only a single union gained legal standing in Pernambuco's sugar zone (Barreiros).[27] And legal recognition did not save that union from manipulation by the Ministry of Labor, which intervened in its leadership during 1957 and 1958.[28]

Communist leaders reestablished momentum in the 1950s and made another push into the rural areas. They released a document in 1950 that came to be known as the "August Manifesto," which again emphasized molding rural workers into a viable ally for the organized urban workers. In 1954 they convened a "peasant congress," trumpeting the meeting as "a new and decisive step toward the future in the struggle of peasants for their own organization."[29] The conference took place under the aegis of the União dos Lavradores e Trabalhadores Agrícolas do Brasil (ULTAB, Union of Brazilian Peasants and Agricultural Workers), which inspired a multistate unionization effort and convened a national conference in São Paulo.[30] The ULTAB conference briefly reinvigorated the movement for rural organization, but the state Party structure suffered another defeat in 1956, when a lack of security among activists in the rural areas led to its complete collapse. Five hundred and fifty suspected militants were arrested, the national Party cut ties to the state, and the central and regional committees were disbanded.[31]

By the time of the Communists' setback, other actors had begun to get involved. A group of tenants on a defunct *engenho* called Galiléia, on the western edge of the cane zone, started a mutual benefit organization to pay for burials and protect their small plots of land from the whims of the landowner. The tenants called their organization the Sociedade Agro-Pecuária de Pernambuco (SAPP, Farmer's and Cattle Breeder's Association of Pernambuco), and they sought legal help from a socialist politician and Recife lawyer named Francisco Julião, who came from a sugar planting family.[32] With Julião taking on a larger role, Galiléia's owner grew concerned and tried to expel his tenants. In the publicity surrounding the ensuing conflict, a sensation-hungry journalist foisted the label "peasant league" on the group to link it to the Communist organizations from a decade earlier. The name stuck and "peasant leagues" spread around the cane zone under Julião's guidance, growing most on the region's peripheries among smallholders and subsistence farmers.[33]

The same year that the SAPP formed, a socialist alliance called the "Recife Front" that had been building over a long period in the capital managed to elect as mayor a progressive candidate named Pelopidas Silveira. Silveira had run for governor in 1947, losing to the popular and powerful Barbosa Lima Sobrinho just as the latter finished his influential stint as director of the IAA. Though Barbosa Lima Sobrinho won that contest, his failure to win Recife was a signal of the growing leftist strength in the city. Three years after Silveira's victory in 1955, a similar coalition backed the *usina* owner Cid Sampaio in the 1958 election for governor. With the odd spectacle of an *usineiro* using Communist support, Sampaio won on a platform of support for the working class. Recife gave him a large majority of votes, and he enjoyed strong support in the cane zone, in a sign of the growing electoral importance of the region.[34]

In his first address to the state assembly after his election to the governorship, Sampaio declared that he had reached office "after a movement of opinion that had the characteristics of a true revolution. The aspirations of the people defeated resistances and social, class, and partisan prejudices, to appear as an obstinate will." The victory "signified that which is most constructive in the spirit of a people: the reaction against poverty, underdevelopment, misery, and migration."[35] Soon, faced with the "obstinate will" of the SAPP, Sampaio expropriated the Engenho Galiléia and distributed land to the members of the peasant league.[36] This impromptu, state-level land reform measure drew attention and whetted the appetite of Julião's followers.

At the same time, the founding of the Superintendência de Desenvolvimento do Nordeste (SUDENE, Superintendency for the Development of the Northeast) in 1959 raised hopes for further reform based on the plans of its teams of technicians. The agency received an ambitious mandate and the freedom to operate in a broad realm.[37] Led by the influential economist and dependency theorist Celso Furtado, SUDENE carried the responsibility of fomenting development throughout the Northeast from its headquarters in Recife. Aiming to create viable enterprises funded with federal, state, and private capital, SUDENE's model revolved explicitly around planning—assessing, critiquing, and ultimately shaping the economic direction of the entire region. The agency employed economists, geographers, agronomists, and other specialized technicians, as well as sociologists to chart interlocking plans for regional development that would decrease the socioeconomic divide between the Northeast and the rapidly developing center-south of the country.[38]

Drawing Broader Attention

With Sampaio taking office and SUDENE setting up shop, national attention turned toward Pernambuco. This interest was fueled by breathless spectatorship and commentary from activists, politicians, and journalists converging on the state. The number of pages devoted to such a short period in the life of a region is extraordinary.[39] Antônio Callado, a writer for the Rio de Janeiro newspaper *Correio da Manhã*, led the charge, spreading the word around the country about the heady events in Pernambuco. His book based on a stay in Pernambuco between September and December 1959 was titled *The Industrialists of the Drought and the "Galileus" of Pernambuco* and appeared in 1960 with the ink barely dry on the *Correio* reportage it reprinted.[40]

Callado's vigorous, insistent, and hopeful accounts spoke to a generation of young people caught up in the push for reform. His writing communicated the contagious energy of the workers he covered; he had clearly been affected himself. The resonance of his reports demonstrated the emergence of a new Left out of the crisis of populism. Military pressure on Getúlio Vargas in the early 1950s, followed by Vargas's suicide in 1954, had had a politically polarizing effect. Juscelino Kubitschek's energetic presidency had overseen economic growth, but the same dissatisfactions with inequality remained that had propelled Vargas's populism. Jânio Quadros's bungled resignation of the presidency in 1961 after only seven months in office was not followed, as he had hoped, by a return to power through the acclamation of the people. Instead, João Goulart, the vice president and heir to Vargas's legacy, had assumed the office with restricted powers. After consolidating his position, Goulart began pushing an agenda for "basic reforms," which though vague included agrarian reform.[41]

The hope associated with "Jango's" presidency crept into Callado's writing as he repeatedly claimed that the Pernambuco sugar workers had decided to take their future into their own hands; they did not share the rest of the country's "shame to make history." These rural masses had begun to organize, agrarian reform was on its way as Sampaio's solution for Galiléia showed, and political allies occupied the highest offices. This was emphasized with Miguel Arraes's election as Recife's mayor in 1959, feeding the hopes of reformers and radicals even further. While working for the IAA, Arraes had advised cane producers in 1945 to improve conditions for their workers. He went on to succeed Sampaio in the governor's palace, winning the election in October 1962 with tremendous support from the *zona da mata*. Callado called the events in Pernambuco a "pilot revolution," and he felt that the country as a whole could do worse than be dragged along toward true democracy by "the most democratic state in the Federation."[42] The rest of the country had "an excess of common sense and not enough passion," Callado wrote, whereas northeasterners—and really the workers of the cane zone—had inverted the formula, putting passion first and letting their enthusiasm carry them along as they began forging a new society.[43]

Callado and other journalists who picked up on the political effervescence in Pernambuco starting at the end of the 1950s saw a parallel group develop in the social sciences—it was composed of heirs to the efforts of Josué de Castro. Under the auspices of the Joaquim Nabuco Foundation, Telmo Maciel carried out a survey of more than four thousand workers in the cane zone in 1961. He noted in his report interpreting the data that little research had been carried out on the standard of living in the cane zone, and social scientists like him found in the sugar fields propitious terrain for posing questions with national implications. In his 1964 follow-up of Maciel's study, Fernando Antônio Gonçalves remarked that the region was "one of the most dynamic laboratories of social changes in the Northeast—and maybe in Brazil." Maciel's conclusions, self-consciously pitched to policymakers, tied substandard living conditions directly to the mode of production in the cane zone, and he pointedly observed that the mortality rate in the *zona da mata* was far higher than that in agriculturally diversified areas.[44]

Watching progressive electoral victories and signs of mobilization among rural populations in Pernambuco, Brazilians on the left found reason to hope that they could address these problems and make radical change on a large scale. Many traveled to Pernambuco to put their hopes to the test. Among these were Catholic radicals, associated with one or another of the many activist strands within the Church. Spreading out into the cane zone alongside Communists, peasant league organizers, and parish priests were members of Juventude Universitária Católica (JUC, Catholic University Youth), Juventude Obrera Católica (JOC, Catholic Worker Youth), Ação Popular (AP, Popular Action), and, especially important in Pernambuco, the Juventude Agrária Católica (JAC, Catholic Agrarian Youth). The Church had released a statement of concern for rural workers only a month after the Communists' "August manifesto" in 1950.[45] The Catholic radicals also shared with their Communist colleagues a commitment to the unity of theory and practice and the idea that "learning [conscientização] could not occur unless it was accompanied by a testing out of the newly acquired knowledge."[46] Consciousness-raising was central to the Basic Education Movement, a group connected to AP and active in Pernambuco.[47]

Many observers from within and outside of the state saw in the struggles of the cane zone's workers an illustration of the plight of a country full of landless and disenfranchised laborers. And just as Gilberto Freyre had expanded the sugar region's experiences to a national scale in his sociological and historical work, so this new generation portrayed the small region as a microcosm of Brazil—in this case of Brazil's problems. Manuel Correia de Andrade claimed that his influential 1963 study *The Land and the People in the Northeast* "is not just a regional study, it is also a national study, since it projects the region in the country as a whole and conducts the discussion as a function of national systems and realities."[48] But while Correia de Andrade wrote from within the region looking outward, others like Callado looked at it from the outside and inevitably saw the national perspective first.

As Castro made clear in his indictment of the U.S. error of interpretation, observers in the United States also had their eyes on Pernambuco. The *New York Times* correspondent Tad Szulc put Pernambuco on the Cold War map when he wrote in October 1960 that "the makings of a revolutionary situation are increasingly apparent across the vastness of the poverty-stricken and drought-plagued Brazilian Northeast." In this article and another the next day, Szulc's language put readers at the time instantly in mind of Cuba; if they missed the connection, he made it explicit by dubbing Julião the "maximum leader" of the peasant league movement and calling the leagues "the closest thing to an organized 'Fidelist' movement in Latin America outside Cuba."[49]

Though he correctly identified a vital movement of the rural poor in

Pernambuco, Szulc's reporting clearly went over the top. "The bulk of the Northeast residents are not consumers or producers in the economic sense," he claimed. "Physical survival is their only concern, and it becomes desperate when the periodic drought hits."[50] The articles left the impression that all of the Northeast, and especially the miserable peasants being organized by the Marxist Julião, lived in dusty, sun-baked squalor. Not only did Szulc fail utterly in giving a sense of the region's diverse environment and variation in agricultural scale, he never mentioned the sugar industry or the pressures facing sugar workers specifically. The simplistic view from abroad probably derived from a distracted and homogenized view of Latin American hotspots. Peasant leagues and unions had made news elsewhere since the 1940s, and U.S. observers no doubt lumped Pernambuco's versions with other examples.[51]

The international press coverage along with other forms of intelligence caught Washington's attention. Echoing Callado's language, a U.S. diplomat wrote in a 1961 report, "This part of Brazil is presently making history." In July of that year, President Kennedy sent his brother Edward to visit the Galiléia peasants. After receiving a briefing on the trip, the President donated a generator to the SAPP, a gesture that Sampaio tried to leverage into further aid.[52] In 1962 Goulart visited Washington, and Kennedy promised $131 million of aid specifically for the Northeast. When the Alliance for Progress was formally introduced in 1963, Kennedy made special mention of the region as an example of where the program needed to concentrate. Still, the U.S. perspective was of a homogenized Brazilian Northeast that simply represented a new hot zone in the battle against communism in this hemisphere.[53]

Conservative Brazilians feared the significance of events in Pernambuco for many of the same reasons that U.S. leaders looked on with concern. Even the leftist activists observing events in the state from afar read into them their own political dreams. All of these reactions derived from perceptions and representations of the *zona da mata* different from the landscapes held in the minds of workers and planters in the cane zone itself. The actual borders and the specificities of terrain mattered little in these representations, while national and international politics mattered a great deal.

Boiling Over: Unionization in the *Zona da Mata*

External political influences played their roles, but as Callado had observed it was the workers who drove events in the cane zone in the early 1960s. The pattern of organization from this period shows us the importance of regional differences and worker reactions. The new friends workers found in the form of social reformers and radical activists, as well as the election of a progressive governor and the presence of a reform-minded president, provided a critical sociopolitical opening. It was the expansion of cane, however, and pressures on land access combined with the increasing labor rationalization and the shift in payment regimes that pushed workers toward action. These were the catalysts that produced the "strike for lower wages" at the Usina Roçadinho, and they had broad effects across the region. Workers' opportunities were greatly improved when the sociopolitical opening produced the legalization and recognition of rural unions.

In the first years of the 1960s, a movement to unionize rural workers began to take over the momentum that the leagues had enjoyed over the previous six or seven years. In 1962 Goulart's federal Ministry of Labor signaled a willingness to recognize rural unions within the structure of Vargas's rural labor decree of 1944. Unions began to proliferate in particular areas of radicalism (notably in São Paulo's sugar areas) and in June the ministry released a decree outlining the recognition process.[54] The movement in Pernambuco drew strength from Arraes's election in October 1962, which followed a campaign in which he promised support to the sugar workers. One of Arraes's most famous and widely remembered acts as governor was his explicit order to the state police not to persecute rural workers. In his first address to the state legislative assembly after taking office, Arraes said, "What is certain is that the Secretariat of Public Security will not, in this Government, be an instrument of pressure, much less of oppression."[55] A decade later, a worker succinctly observed, "They stopped beating us after Dr. Miguel Arraes took over."[56]

Workers responded eagerly, and Julião himself exhorted workers to join the unions and took credit for creating an atmosphere in which they could flourish.[57] Paulo, who had worked the cane fields for thirty years, said he signed up the first day he heard of the union. Severino noticed how worried some planters had become. He heard one *senhor* say, "The time is coming when the people are going to get what's theirs. . . . The people who will be in charge [*mandar*] will be you workers, and there will be wages, and people will pay wages, and holidays, the thirteenth wage."[58] It is not coincidental that the planter framed the change in terms of who would be in charge—who would command.

In the midst of the popular mobilization, though, most planters clung stubbornly to the assumed inheritance of command over their domain in the *zona da mata*. Not long after Arraes's election, an *usineiro* named José

Lopes da Siqueira Santos killed five workers who had come to request their back pay. Then, gunmen for the Caxangá *usineiro* Júlio Maranhão shot and killed a union delegate and critically injured his companion. Maranhão hurriedly buried the delegate directly in front of the *usina*.[59] Siqueira Santos and Maranhão (both in the southern *mata*) were lashing out against the workers' challenge to their authority. Union leaders rallied rural workers around their slain comrades, casting the killings as an additional impetus for organization.

Their cause received strong support with passage of the Estatuto do Trabalhador Rural (ETR, Rural Worker Statute) in 1963. Besides clarifying the legalization process for rural unions, the ETR extended labor legislation to rural workers, giving them an analogous law to urban labor's CLT. The ETR stipulated a series of benefits from paid holidays and indemnification in the case of dismissal to the year-end bonus known as the thirteenth wage (the list of measures to which Severino's boss referred). Before, workers had rights to none of these things and instead received unpredictable favors (or nothing) from bosses. Paulo, for instance, after being expelled from a long-term tenantry arrangement in the 1950s, received a goat as recompense for his years of service. And Severino earned "whatever the boss wanted to pay," though the boss occasionally supplemented his wages during harvest time with molasses from the mill.[60] The ETR also gave rural workers access to a system of labor courts for lodging complaints, from the local *juntas de conciliação e julgamento* (JCJs, boards of judgment and conciliation) to the TRT and TST.

The push for unions accelerated with promulgation of the ETR and took on a bifurcated character as the Communists and the Church competed to bring workers into the fold. The *New York Times* announced in April 1963 that "a race to organize rural workers into unions is under way here between Roman Catholic and Communist-oriented labor groups." At that time, eight unions had been officially recognized, of forty-nine formed. As they organized local unions, Communists and Catholics vied for a larger prize—control of the state federation of rural unions, once such an organization was recognized by the government.[61] The battle took on a geographically specific character, as the Communists built a large, twenty-one-municipality umbrella union in the densest part of the cane zone around Palmares in the south, and the Church established municipality-based unions in the north. Gregório Bezerra led the drive for the Communists, bringing an impressive thirty thousand members into the Palmares union and establishing sophisticated organizational techniques.[62] The north fell under Padre Paulo Crespo's leadership and followed the directives of the organization he guided, the Serviço de Orientação Rural de Pernambuco (SORPE, Pernambuco Rural Orientation Service). His effort had begun in earnest with a July 1961 meeting of twenty-five priests in Jaboatão, Crespo's parish just west of Recife. The group resolved to help organize unions at the meeting and several weeks later participants founded the SORPE.[63]

The unions adopted from the leagues the key goal of land reform, and they gave land redistribution a durable place in their discourse. This was an easy case to make, given the extreme concentration of land in the sugar zone. It also matched the tenor of the time, with President Goulart repeatedly advocating passage of a nationwide land reform. The unions also adopted tactics basic to their structure, such as strikes. Small strikes became commonplace starting in the middle of 1963, as workers on an *engenho* or from several *engenhos* owned by one mill would stop work until a particular demand was met. In May, for instance, twenty-five workers from Engenho Oriente in Rio Formoso stopped work and demanded that the foreman be expelled for measuring tasks unfairly.[64] The pattern of this combativeness provides one index of the differences in mobilization between workers in the north and south. Of thirty newspaper articles the *Diário de Pernambuco* published on rural worker strikes during the tumultuous months between May and November 1963, only three covered action in the north.[65]

With the north under Catholic control and the south on the Communists' side, the central *mata* around Recife became a battleground. Unions from five municipalities in this central area declared in mid-May 1963 their desire to stay with the group organized by Padre Crespo. The Communists threatened to organize a strike to force the labor ministry to revoke these unions' certification, but the Crespo-led unions retained their registration.[66] On September 2, thousands of rural workers reportedly made their way toward Recife to demonstrate their opposition to Communist organizers and their support for union leaders who were not "government stooges." (The fact that Communist leaders could be cast as government stooges, even as a gross exaggeration, shows that the political conditions had strayed from the norm of oligarchic control. Even Cid Sampaio, who had some Communist support, was a mill owner. Arraes was not.) They decried the state labor delegate's supposed bias toward the Communists. The *Diário de Pernambuco* noted that the protesters came from "various municipalities" and listed nineteen. Again demonstrating the regional division in the union movement, every single one of those listed lay in the *mata norte* or the central region around Recife.[67] With the decision of the central region's unions, the competition turned in the

Church's favor and by November 1963 most recognized unions were under Church control, and they controlled the state rural union federation.[68]

What is interesting in observing the course of the struggle between the Communists and the Church for control of the rural unions is not who ended up with control but the marked geographic divergence to the political battle. The regions' differential development and the earlier consolidation of the *usina* system in the south related to the more complete adoption of 3x in that region and therefore the greater expansion of cane. The more widespread use of the task system and the disappearance of traditional tenantry relations also contributed to conditions that made workers in the south more amenable to the Communists' more radical rhetoric and more likely to strike for their demands. The Church established more unions, while the Communists built one large union, and the Church managed to control the federation. The way the process of unionization took place, with regional differences highlighted by the presence of two different political groups, owed its shape to environmental and agricultural factors.

As the 1963–64 harvest season approached, federation leaders demanded an 85 percent increase in the minimum wage (most of which would go toward simply catching the wage level up with inflation). Outraged planters denied the request out of hand and the unions announced a major strike.[69] (In keeping with Szulc's skewed coverage earlier, the *New York Times*' reporting on the strike was limited to a three-sentence squib under the headline "Brazil Strike Brings Violence," though the four days of the stoppage were actually relatively peaceful.)[70] The strike officially began on Sunday, November 17, with unions maintaining close coordination and organization of pickets. Officials at the rural union federation claimed proudly that 90 percent of the region's workers—throughout every municipality—had joined the mobilization, which included an estimated two hundred thousand people. The 1963 strike eclipsed any other demonstration to that time in the history of Brazil's rural labor movement. With its organizational core in the Palmares union, the strike achieved its aims. The workers were granted their raise and planters promised to abide by the stipulations of the new labor laws.[71]

The economist and theorist Caio Prado Jr. noted several months before the military coup of 1964 the dramatic scale of the improvements workers were experiencing, "a true transformation of their conditions of life." "It is sufficient to cite some statistics," he wrote. "Less than a year ago, they earned 80 to 120 cruzeiros a day. Today they earn 900!"[72] Workers associated their major victory and the mobilization that produced it with Miguel Arraes (or "Pai Arraia") and remember him as the deliverer of "rights."[73] A worker named Benedito, who was young at the time, said "The bosses don't like to pay wages, but [Arraes] argued against the rich and in favor of the poor."[74] Workers invariably credit Arraes with increasing their standard of living enough to afford a bed (rather than a cheap hammock).[75] Considering the stakes, it is little wonder that Arraes inspired such strong emotions in everyone with an interest in the cane zone.

Arraes faced the challenge of managing the social and economic changes unleashed by the opening up of the *zona da mata* over the previous several decades. In a sense, the dilemma produced by these processes boiled down to a question: How do you rationalize a landscape characterized by captivity? Arraes arrived at this crucial moment equipped with experience in the implications of this question, since he had worked as an IAA technician as a young man. In 1945 he had given a speech to Pernambucan sugar producers in which he argued that "the rationalization of production" should be pursued in part for "the advantages that it will bring, without doubt, to those who work in cane cultivation." "Save those who already enjoy the benefits given them by more progressive producers," he continued, in a bow to mills such as Catende, "our rural man constitutes a species apart from human beings, given the conditions of hygiene and health and education with which he lives."[76] But rationalization could not progress without freeing workers from the captivity that they traditionally endured. The convulsions of the cane zone in the 1950s and 1960s came as a product of challenges to planter authority and their discourse of the laboring landscape brought by new actors' "opening" of the region, as well as unionization among the workers. Workers faced worsened conditions while simultaneously finding increased opportunities for mobilization. The rationalization of the landscape of captivity was experienced as a great clash between classes.

Arraes became a political lightning rod because he pushed the transformation forward. Arraes's secretary of state, Hélio Mariano, interviewed twenty years later, contended that Arraes was partly to blame for what some perceived as a climate of anarchy during his tenure, since he wasn't a man of "great firmness." "He didn't, let's say, *assume the command,*" Mariano said. Instead, he "was placed at the whim of circumstances, forced into the position of fireman, to put out fires here and there."[77] No doubt Mariano expected Arraes to act as a traditional man of power, commanding over and controlling those below. Arraes's agriculture minister, Dorany Sampaio, responding to the indictment of Arraes as a "lunatic or crazy," said that Arraes "was a conscientious firefighter who used water and not the gun."[78] Again, a traditional leader in Arraes's position would have "used the gun," perhaps quite literally—directing the police to open fire on protesting workers. (After all, planters such as Siqueira Santos did not shy away from such tactics.) As the repeated references to fire in comments about Arraes indicate, one effect of the impressive politicization of sugar workers at this time was the specific politicization of fire.

Fire in the Fields: Lived Realities and Rhetoric

A burning field of sugarcane gives the disturbing impression of being alive—it crackles and pops loudly enough to be heard hundreds of yards away, and it heaves and waves as the fire's heat disturbs the air, bending and twirling the canes. Once started, fires move quickly and can be hard to stop. Because of their destructive potential, cane fires have occupied a powerful place in the anxious imaginations of planters and the rebellious dreams of workers—slave and free—across many different regions devoted to cane agriculture.[79] From the end of the eighteenth century, the danger of cane fires evoked for planters images of vast burning fields during the Haitian Revolution. Fields of cane represent so much capital, and fire presents itself as such a ready tool, that it makes sense that people from multiple classes set fire to cane when retaliating or going on the offensive against planters.

Fire was as common as it was feared in sugar fields. For one thing, it had long been used as a matter of course to clear new land and burn the "trash" off cut fields to prepare for new planting. The Jesuit André João Antonil in his 1711 text recommended that planters burn the cut leaves after harvest to "make the land more fertile," and that practice was widespread. Also, sugar growers have always confronted accidental fires. The engines traveling mill-owned railroad tracks threw sparks from their stacks, occasionally setting cane alight. Workers in the fields throwing away a cigarette butt or lighting sticks for their pipes risked starting a fire.[80] Bird hunters firing shotguns, children playing with matches, and an unlucky lightning strike all could catch a cane field on fire.

During the dry part of the year, which coincides with cane's reaching maturity, planters sometimes placed sentinels in the fields to provide early warnings of fire.[81] The outbreak of a fire forced planters to initiate a sequence of urgent tasks, mustering workers first to put the fire out, then to cut the burned canes. Fire triggers a biochemical process in the cane that leads eventually to the degradation of the canes' sucrose—turning it into glucose and levulose. The threat of lowered sugar concentrations and the eventual spoiling of the juice make the harvest of the burned canes urgent.[82] Mills

Burning cane to prepare the field for harvest. (Photograph by the author, 2003)

only accepted burned canes at a discount, and they rejected canes if they had sat for long in the field after a fire.

Though often accidental, fire could obviously be deliberate, frequently a sign of conflict, whether between rivals or between planters and their dissatisfied workers. In January 1947, for example, representatives of the Olho d'Agua mill, in the northern cane zone, contacted the secretary of public security in Recife to complain that two fires had been set in their cane.[83] The secretary dispatched an investigator, who was told by the local police delegate that the fires were probably linked to political rivalries. The elections for governor had just taken place and the municipality had voted for Barbosa Lima Sobrinho. The local judge had campaigned for Lima and the delegate surmised that some people who had voted "correctly" were taking advantage of the judge's gratitude, and therefore protection, by setting fires in rivals' cane fields.[84]

While electoral politics stoked people's emotions, more personal disputes could also escalate to arson. A man named Antonio Tomaz was arrested for a fire in late 1959 after burning several fields outside the town of Carpina belonging to the São José mill. The case involved property near an *engenho* owned by a certain José Aymar. Tomaz had allegedly set the fire in a field in retaliation for a fine levied against him by an administrator from the mill. But he was eventually freed from jail "through the interference of sr. José Aymar." As well he might, the investigator wondered, "What interest does Mr. Aymar have in the liberty of an arsonist, keeping him on his property?"[85] Tomaz surely owed his freedom to some intricacy of the patron-client relations so common to the region. Aymar, as a cane grower who sold his cane for processing, likely supplied the mill (the closest to his property). He may well have nursed a resentment involving this relationship and could have arranged for the fires as a means of settling his score.[86]

Cane suppliers also set fire to their own canes to shift a mill's harvest priorities. Since mills depended on suppliers to fulfill specific cane quotas, they did not want to see canes go to waste. Therefore, if a fire broke out in a supplier's field, those canes would be bumped to the top of the mill's list for cutting. By surreptitiously lighting his own field on fire, a planter could be assured of a prompt harvest and, therefore, a payout for the cane supplied.[87]

With most fires, planters and authorities were forced to guess whether the blaze was deliberate or accidental, an ambiguity that fed the anxiety about fire. Shortly after Communists' unsuccessful attempt to establish unions and peasant leagues in the 1940s, planters became concerned about radical elements, "agitators," and above all the peasant leagues. But the new wave of better-organized leagues established in the 1950s made a much larger impact than the 1940s version. Planters made their fears known in correspondence with police officials and through the press, which increased its coverage of fires from the mid-1950s. A February 1956 article in the Pernambuco cane growers' magazine *Boletim canavieiro* denounced frequent fires as "a new form of banditry" that threatened the "destruction [of] a patrimony of the Pernambucan economy."[88] The Usina Tiúma owner wrote a similar letter to the police at the same time, claiming that "the incidence and volume of fires in the cane fields is creating a disastrous situation."[89] The next year, the mill again appealed to the police, requesting their "closest attention to the scourge that still at the moment affects us: the burning of canes."[90]

In the face of this sort of self-reinforcing rhetoric and the anxiety it produced, the routine aspects of fire in the cane zone—the regularity of accidental fire and the fact of occasional arson—were forgotten at the end of the 1950s and beginning of the 1960s. In his 1956 letter to police, the owner of the Tiúma mill also noted that 1935, 1946, 1949, and 1952 had been particularly bad years for fires, indicating that his was not a new problem.[91] As the political situation in the region heated up, it became more and more difficult to separate the rhetoric from the reality. Livid planters spoke as if the entire *zona da mata* were going up in smoke, but their outrage seemed directed as much toward the political changes afoot as toward the actual burning of cane. Incendiary language—the use of flame-filled metaphors to describe both political mobilization and supposed "agitation" in the fields—drew on the charged history of fire in the cane-growing region and lent power and resonance to descriptions of the events unfolding in the region.

Planters' entreaties to police for assistance shifted what had been matters of private power into the public realm. As we have seen in preceding chapters, planters did not shrink from wielding power within their domains. And the treatment of suspected arsonists in 1944 fits the pattern of how they addressed threats to their authority. In a report from that year, secret police investigators acknowledged to their superior that they had delivered men suspected of setting fires in cane fields directly to the offices of the Usina Santa Therezinha. No doubt a form of justice was meted out after their departure.[92] Police involvement until the 1950s had occasionally involved arrests of alleged arsonists, but they were accustomed to deferring to the true forces of order.[93] However, steady complaints from planters in the late 1950s and early 1960s, along with heavy press coverage, impelled police to direct more energy toward investigating reports of fires. Early in 1960, Governor Sampaio responded to planter demands by creating a special mobile police unit dedicated to preventing and investigating cane fires, a measure cane suppliers publicized with satisfaction in the *Boletim canavieiro*.[94] The unit quickly began producing extensive documentation of the distribution of fires and the progress of investigations into the causes. The squad's priorities showed clearly in the organization of its reports: they tracked fires not by municipality or jurisdiction but by mill ownership boundaries.[95]

The politicization of fire increased as newspapers denounced every reported fire as arson and political terror. An editorial from around the time Sampaio created the patrol captured the tone of the coverage: "It is generally known that a wave of fires has recently swept through Pernambuco's cane fields, set by pernicious elements and led by agitators, with the aim of fomenting indiscipline and agitation in the formerly peaceful cane zone of the State. Unfortunately, though these acts of sabotage against cane planters have intensified ever more, our authorities have not given, it seems, the attention that the disorder has demanded." The editorial called for punishing the "malefactors" with "all of the rigors of the law."[96] An article published in 1963 under the headline "Agriculture Could Halt in [the Municipality of] Vicência: Arsonists Act" warned that "the climate of agitation continues to worsen in the face of the demands of workers making unwarranted claims."[97]

The anxieties of planters and police were stirred up by, and in turn continued to feed, sensationalist reporting on fire. The reporting was also influenced heavily by sugar bosses, who besides informal patronage ties had some reporters and news directors on their payroll.[98]

In the 1980s, when the Joaquim Nabuco Foundation interviewed people about the events of twenty years earlier, a remarkable number of subjects discussed the issue of fire. Many of those interviewed, including the port worker and union leader Amaro Valentim do Nascimento, the *usineiro* scion Artur de Lima Cavalcanti, and the activist priests Paulo Crespo and Wanderley Simões, suggested that many fires were started in an attempt to destabilize or delegitimize the Arraes government. Cavalcanti stated "already in that period there were soldiers disguised as peasants, radicalizing in order to destabilize. This is an absolutely concrete fact, known in every political area." The late geography professor Joaquim Correia de Andrade felt that cane planters themselves set most fires, for a variety of reasons, including the desire to implicate the radical Left.[99]

The basis for the authorities' anxieties and concerns must be questioned (especially considering the manipulation of the newspapers by planter interests.) Investigator Oliveira's 1962 report on the alleged aerial firebombing provides a case in point. Oliveira found that the story originated with a child who accompanied his father to work in the fields. The boy dutifully retold his tale of a low-flying plane and a falling object to Oliveira, though his father and other workers dismissed it as "worthless child's talk."[100] Most of those interviewed said they first heard the story on the radio, and all were skeptical of it. Oliveira elicited the most loquacious responses from local planters, one of whom said a similar rumor had circulated in 1947 (perhaps not coincidentally the year the Communist Party was banned). Another said he had not had any fires in his fields that year but added that it was "just a brief breath of luck." And two planters admitted that at the end of the harvest season they burned all of their cane to expedite cutting so that as much as possible could be rushed to mills for grinding.[101]

Oliveira knew from experience that fires were often lit for prosaic reasons. In a 1959 report he explained to his superior the basic calculus of workers who set fires knowing that they could make more money despite being paid less for cutting burned cane: "In the same space of time that a worker finishes a bundle of raw cane, he can cut three or four bundles of burned canes." Oliveira also noted that burned-over fields were free from "inconveniences" like reptiles, insects, and ants, as well as the leaves and "hair," or abrasive spines, of the cane being cut.[102] Other cases revealed that workers sometimes resorted to fire because of household demands. A man arrested in 1960 for setting two fires in cane fields near Maraial (at a time when fires were becoming increasingly politicized) confessed that he did not make enough money cutting raw cane to support his eight children, two brothers, and mother-in-law. Even though he earned five cruzeiros less per bundle of burned cane, it was worthwhile because he saved the considerable amount of time he had to spend trimming the leaves from "raw" cane.[103] At the end of his aerial bombing report, Oliveira dismissed the airplane story out of hand and summed up his findings. "The great number of fires in the cane zone," he wrote, resulted from a range of factors, including planters' burning their cane, accidents or negligence, and the presence of psychopaths and criminals.[104]

Albert Rabida, the U.S. consul in Recife, offered a similar analysis in a dispatch to Washington about the supposed aerial firebombing. After mentioning the fears of a conspiracy among politicians, journalists, and cane planters, Rabida wrote that "those who found confirmation of their suspicions in this report noticed or cared not that the sole witness was an eleven year old boy." He said cane fires were an annual occurrence, likely the work of careless smokers, "cane planters themselves seeking to force rivals to sell their properties at low prices, and some by grudge-bearing field workers wishing to avenge themselves of real or imagined injuries suffered at the hands of the cane planters."[105]

The police did not share these opinions, nor did they seem to take Oliveira's sensible conclusion to heart despite the fact that his interviews in São Lourenço offered leavening to the fears of numerous activists hiding among the workers. In an earlier fire investigation, the investigators (including Oliveira) had demonstrated the depth of their suspicions. Though they acknowledged that the "infiltration of teachings by the rural unions and peasant leagues" was not a factor, they still warned that this possibility "should not be discounted in future observations."[106] Eudes de Souza Leão Pinto, who was the state secretary of agriculture at the time, continued to claim in the 1980s that the peasant leagues taught rural workers methods of burning cane during their meetings.[107] And Armando Samico, chief of Pernambuco's public security agency for parts of this period (1958–59, 1964–65, 1969–70), argued, unsurprisingly, that Communists were responsible for the fires in the cane. He says: "initially [the Communists] didn't intend to disrupt the work, there were very few strikes. But after seeing that strikes were opportune they began to burn cane fields, to reduce production."[108]

True, some "outsiders" set fires occasionally. Investigators looking into a rash of fires around the central *mata* town of Jaboatão in 1955, for instance, found that migratory laborers were responsible: forced by drought to come from the interior looking for work, they had been dissatisfied with their treatment by planters and had set fire to cane as they left *engenhos* in the middle of the night.[109] But these were probably not motivated by Communist politics or even the idea of unionization, and few other police reports record the presence of "outsiders" in the sense intended by planters.

Workers did set fires, as many admitted to police, but they did so because of the daily needs that Inspector Oliveira had mentioned in his reports and not necessarily for the unionist political reasons assumed their employers. The increased work demands and time pressures I discussed in the last chapter were motive enough. Agriculture Secretary Eudes de Souza Leão Pinto acknowledged that many cane fires were set by "chronically hungry" workers.[110] A worker arrested in 1960 admitted to setting a fire, saying he was at the end of his rope. Deeply in debt to his planter boss and trying to provide a respectable baptism for his new baby, Severino da Silva visited the *engenho* store to ask for an infant's dress. The proprietor bluntly denied the request, saying, "A poor person's child is rolled in a rag when it's born and taken off to be baptized." Infuriated, perhaps as much by a sense of patronage rules violated as by the store minder's harsh rebuff, Silva set fire to a cane field near his home. He readily admitted his crime to the police and seemed eager to tell his story as a justification for his action.[111]

During the November 1963 strike, workers were aware of the danger associated with fire in the prevailing political climate. They were also savvy to the ways employers could manipulate public opinion and took care to avoid potential stumbling blocks. During the strike, the mill administrator Severino Moura visited the *engenhos* under his direction and in each one found organized pickets of ten to twenty men. The union ordered that workers follow him, to ensure that he not set fires and implicate the strikers. Though offended that the union would treat him as a suspect, Moura sympathized with the workers' aims.[112] Gregório Bezerra also mentioned the scrupulous organization of the strike, describing a *senhor de engenho* who went to the rural union asking for help fighting a large fire in the cane. The union officials first made him sign a document saying that the fire was accidental and not purposeful because they did not want him complaining to the authorities the next day that the fire had been set by agitating workers.[113]

Whether or not the pace of burning actually increased over this period is a difficult question to answer. The preoccupation with fire in press and police reports from this period probably derived as much from the anxieties of planters and security personnel as from actual experience. About the Caribbean, Bonham Richardson claims that "cane fire frequencies indicated—indeed marked and underscored—levels of social satisfaction and economic well-being among an island's working peoples."[114] But it is not clear that such a direct causal relationship existed in Pernambuco, and there is no convincing evidence for the period. It is true that state government data on incidences of fires for the years 1954 to 1958 reveals a steady increase of fires in "forests, agricultural establishments, etc.," from eleven to seventy.[115] But these numbers are actually suspect because they are so low—even seventy fires during a given year is a small number for a region of fifteen thousand square kilometers and thousands of agricultural establishments, large and small. Since the official statistics probably amount to an underreporting of fires, it is even more difficult to assess whether planters' and security officials' preoccupations were justified by an increased incidence of arson.

A doctor from the region, discussing the politics of the period in an interview twenty-five years later, dismissed the notion that fires became more common during the period. "I was born on a sugar *usina* and I lived my whole childhood and life in the sugar zone," she said. "I always saw fires in the cane fields."[116] Padre Crespo, the activist priest deeply engaged in organizing workers in the northern sugar zone, explained in a 1985 interview that mills "cut lots of burned cane, [and] most of the fires were probably scheduled." Accusations of setting fires in the cane fields, he said, gave the police an excuse to terrorize the region's workers.[117] We are thus faced with a question that routinely bedevils environmental and other histories: What relationship does a widespread discourse bear to a material reality? Whatever the actual incidence of fires, they became an apt metaphor for understanding or expressing the shock of events during the 1960s. Yet ironies of fire in the fields run deeper. As the chaotic social conflict faded, fire found a place in agricultural practice as a deliberate strategy.

During the same period that planters posted armed guards in their fields and political discourse seemed combustible, the sugar industry began to see fire as a potential harvesting tool. The process clearly had to overcome all of the anxieties about fire that have been outlined, but research elsewhere helped the case. Not only did people associate fires with dangerous radicalism and sabotage, they also saw it as a destructive agricultural practice. Writing about the alleged increase in cane fires in 1960, a journalist for Recife's *Jornal do comércio* claimed that "burning cane to facilitate its harvest would be an absurd and fruitless practice."[118] The reporter's unequivocal statement would, in not so many years, be belied by widespread use of fire in the fields.

In 1945, an agronomist writing in *Brazil açucareiro* captured the prevailing scientific view of fire in cane fields. He referred to fire as a "terrible enemy of the land" and warned that it "destroys all of the organic material, eliminates the fertilizing nitrogen and kills the insects that protect against *broca*, the worst insect pest of sugarcane!"[119] During the 1950s, *usinas* grudgingly accepted delivery of fire-damaged canes at a reduced purchase price. Investigators looking into fires in Nazaré da Mata in 1959 noted that *usinas* accepted delivery of "damaged" (i.e., burned) canes for eight days after the fire.[120] (Considering the pace at which sucrose breaks down, cane is viable for about forty-eight to seventy-two hours after a fire, so this was actually a generous policy.)

But the same year as the article condemning the "terrible enemy" appeared, another scientist—from the respected agricultural research station in Piracicaba, São Paulo—took it on himself to systematically study the effects of burning on sugarcane. Jaime Rocha de Almeida published his results in 1946, beginning with an acknowledgment that, "from the agronomic point of view, the practice of burning cane before cutting it, for any reason, is contraindicated."[121] However, he continued, this practice was actually increasing among São Paulo mill owners. He surely knew, too, that other sugar-producing areas of the world were experimenting with preharvest burns. At the end of the 1940s, when Sidney Mintz carried out fieldwork in Puerto Rico's sugar fields, his informant don Taso mentioned that cane cutters were making more money because they were cutting burned cane.[122]

In light of the fact that some people were using preharvest burns despite received notions of fire's harmful effects, Almeida decided to test the effects of precut burns on the cane juice itself. His results must have been somewhat encouraging to *usineiros* eager to speed the process that brought cane from field to mill. He found that juice remained sucrose-rich into the third day after burning, if the canes were left standing. Burned and cut cane degraded quickly, and rain was also bad for burned cane. Assuming no transportation bottlenecks, the scientist concluded that burning cane was an acceptable practice.[123] Almeida's research made little apparent impact among Pernambucan *usineiros*, as there is no evidence of planters' intentionally burning cane before cutting in the years immediately following his research.

Just before the beginning of the 1958 harvest, however, the journal of the Pernambuco cane planters' association excerpted parts of a statute proposed by the São Paulo cane planters' association that outlined regulations for burned cane. The two provisions of what must have been a larger agreement regulating relations between cane suppliers and mill owners stated that suppliers would be required to provide the exact hour when the cane burned, with corroborating witnesses. *Usinas* would have no obligation to receive cane that had been burned more than two days before. Cane burned through "the fault or negligence" of the supplier but presented to the *usina* between twenty-four and forty-eight hours after burning would be remunerated at 10 percent less than the normal level. Finally, suppliers to *usinas* that burned their own cane for milling would not receive lower payment levels for their burned cane.[124] This piece in the sugar bulletin tells us several things about the incorporation of fire into the harvest regime. First, the practice seems to have begun in São Paulo, though the Pernambucans' reprinting of the statute indicates their interest in emerging techniques. Second, cane was still seen as being compromised by fire, so suppliers were penalized for delivering burned cane to *usinas*. Qualifying this point, however, was a third issue: some Paulista *usinas* seem to have already been burning cane intentionally. Suppliers feeding into those mills, of course, received full payment for burned cane.

At the beginning of the 1960 harvest, the IAA released a resolution about relationships between cane suppliers and mill owners for Brazil as a whole, establishing a new set of guidelines for burned cane. "In cases where *usinas*, for their own convenience, adopt the practice of burning cane fields to accelerate and facilitate the harvest," the resolution stated, generalizing the São Paulo rule, "suppliers will be assured of the freedom to adopt the same process." If suppliers burned in the same proportions as *usinas*, and delivered their cane within twenty-four hours of burning, they would receive full payments. Deliveries between twenty-four and forty-eight hours old would receive 90 percent payment and those over two days old could be turned away by the *usinas*.[125] The IAA promised to review the rules once they had been in effect for a year. The sequence of regulatory modifications as they appeared in Pernambuco, from the publication of a model from São Paulo to the IAA's decree, clearly indicates the gradual incorporation of fire into accepted agricultural practice precisely during a period when workers appeared to be lighting more and more fires on their own.

In a 1965 article, Bento Dantas noted that still, among Pernambucan planters, "burning is roundly condemned, under the pretext that it makes cutting difficult and accelerates the inversion of sucrose." However, he said, cane fields were burned before being cut in most other cane-growing regions in the world, and studies completed in São Paulo demonstrated that losses in sugar yield occur only if the cane is harvested a full three days after burning (probably a reference to the Almeida study). Workers would also have pointed out that far from making cutting difficult, burning eased their task. Because other research questioned the deleterious impact of fires on soil quality, Dantas concluded, and since fire reduced labor demands, "the use of fire in the cane fields before cutting will be inevitable."[126]

Dantas was right; the practice was already spreading quickly. The same year as his article, *Brasil açucareiro* reprinted a chapter from a British scientist's study on cane burning that claimed the practice "rarely results in damages" and "no problems present themselves at the mill."[127] Like so many other changes, the incorporation of fire was incremental and regionally specific. A northern *mata* worker interviewed in 2003, for example, remembered pre-harvest cane burning arriving in the mid-1980s.[128] However, the general trajectory remains the same: the burning of mature cane had been transformed from a weapon of pressure, resistance, or revenge to a standard procedure. By 1985, 99 percent of workers queried in a survey said they only cut burned cane.[129] And it appears that workers may have helped drive the transformation, since they used fire as a way to ease their work demands and help them keep up with quotas. With their incorporation of fire into the harvest routine, however, planters gained control over the practice. In this respect, then, they reasserted a degree of control over something they saw as a weapon used against them. This progression also points to the realities of labor costs' (and savings') trumping the violent discourse associated with fire.

Already in the 1970s, however, the detrimental effects of large-scale cane burning was attracting attention. The state environmental agency highlighted the extent of air pollution caused by burning in the cane industry. The agency could not separate the ash and gas production of cane field burns from that produced in mills by bagasse (cane pulp) burning, but the two practices together accounted for a huge amount of overall pollution. In the southern cane zone the industry sent 117,000 kilograms of ash into the air per day, in addition to 109,300 kilograms of carbon monoxide. The numbers for the northern cane zone were not far behind, and in both cases the cane industry accounted for the large majority of these types of airborne pollutants as a whole.[130] Cane burning also contributes to pollution of the region's waterways. Burned cane requires thorough washing before being ground at the *usina*, and the ash-, fertilizer-, and herbicide-soaked wash water generally gets pumped directly into regional waterways.[131]

By the 1990s, a growing number of people saw burning as destructive and inefficient and planters' views of the practice of cane burning have nearly come full circle; while still a common practice, it is increasingly called into question. Even considering the labor savings, fire has long-term consequences, since burning off leaves translates into less organic matter covering the ground after harvest, exposing the soil to the full force of the sun. At the immediate level, recent research indicates that fire can significantly reduce sucrose content; a long-term study on preharvest burning published in 2005 demonstrated that abandoning the practice increased cane yields by 25 percent.[132] In addition, 40 percent of *zona* residents recently said that they regularly lived with ash from fires, and research has demonstrated the harmful effects of pollutants from cane field fires on respiratory health.[133]

Conclusion

Most scholarship on the cane zone during the late 1950s and early 1960s emphasizes organizations and organizers, the clearly *political* components of the region's transformation. The leagues and unions, Francisco Julião, Gregório Bezerra, and others deserve scrutiny, but their stories unfolded in a context of great change that is not fully captured by political analysis. A labor and agro-environmental history opens up important perspectives on the broader context of the period. The specificities of workers' experiences in different parts of the cane zone reveal subtexts to the pronounced subregional divergences in political mobilization. The electoral and organizational activity that stunned Brazil and pushed the Northeast onto the international stage were linked to changes in agriculture and labor relations revealed in the pattern of unionization.

Similarly, fire's "career" during the third quarter of the century cannot be understood without grounding it in agricultural processes and labor relations. A careful assessment of the range of motivations compelling some people to light fires in cane puts in perspective the hysterical, politicized treatment of fire. Some workers had set fires in an attempt to speed their labor and increase earnings. Fire's development into a prosaic tool of preharvest preparation reflected planters' search for savings on labor costs (perhaps in part a reaction to the increased costs brought by the ETR). These distinct aspects of cane fires intertwined and informed one another, a fact that points to a larger argument: for regions such as the cane zone, historians cannot separate agro-environmental processes from labor relations, and both have political dimensions as well.

Fire's trajectory during this period tells us that we need a historicized sense of agricultural science that situates decisions and techniques as products of their times. There is no single, rational way to harvest sugar that lies outside history, nor for that matter is there a way to match labor efficiencies with the best environmental practices. Elements of agricultural professionalization as early as the 1930s contained the seeds of an ecological consciousness that would reflect back on agriculture many decades later. An agronomist in 1946 wrote in a Pernambuco agriculture department pamphlet that "the tendency of modern agronomy is toward scientific work, done in accord with the principles of nature. The farmer should work with nature, and not wrench it or attack it."[134] By the 1990s an emergent environmental movement articulated a version of this approach when it denounced the practice of preharvest burning that emerged in the 1960s.

PART III

The Dictatorship Commands the *Zona da Mata*, 1964–1979

SIX

The Only Game in Town
Workers, Planters, and the Dictatorship

In late March 1964 rumors flew that the military planned to overthrow President João Goulart and Pernambuco Governor Miguel Arraes. The Communist union leader Gregório Bezerra went into hiding after learning that he was the quarry of an armed band assembled by the murderous *usineiro* José Lopes de Siqueira Santos. The military did seize power on April 1, and several days later an army unit found Bezerra before Siqueira Santos could. Transported to a military barracks in Recife, Bezerra was stripped to his underwear and paraded around city streets, led by a rope tied around his neck. In full public view, officers beat him nearly to death with an iron bar.[1] The story of Bezerra's flight and capture represents the transition in authority that stands at the center of this chapter, a shift from planters' unchecked dominance to a concentration of power by a federal regime that promised an end to radical threats. The dictatorship thought that the task of rationalization was best accomplished by minimizing the uncertainties of democratic politics.

The planter-style justice meted out to Bezerra signaled the dictatorship's emphatic taking of command, even in this area of entrenched traditions. The regime dedicated itself to rationalizing and modernizing Brazilian society through state intervention. James Scott characterizes the regime's approach as guided by high modernist ideology, a creed adopted by many twentieth-century authoritarian states. Scott argues that states seek to improve the "legibility" of societies by devising what he calls "simplifying instruments," such as censuses and cadastral surveys, for clarifying technocrats' vision of the natural and social order. When states seize far-reaching power and confront a cowed civil society, they can place these mechanisms in the service of large-scale transformative projects.[2] The Brazilian military policymakers had full confidence in the capacity of state planning to address and conquer crucial problems. In the cane industry, the state set out to revolutionize export agriculture and arrogate to itself control over labor relationships.

For decades, the state had been represented in the *zona da mata* as a police presence and by the more distant presences of the Instituto do Açúcar e do Álcool (IAA, Institute of Sugar and Alcohol) and the Estatuto de Lavoura Canavieira (ELC, Sugarcane Farming Statute). The goal of a strengthened central state, which arguably had roots in the *tenentes*' movements of the 1920s, became a reality with President Getúlio Vargas after 1930 and especially under the Estado Novo. But in the 1940s police still delivered suspected arsonists to *usina* offices, a sign that in the sugar zone the state continued to defer to planters. In the 1960s workers became politicized through electoral politics and empowered by unionization and the Estatuto do Trabalhador Rural (ETR, Rural Worker Statute), while planters sought assistance from the government in extricating their industry from the crisis of 3x and the increased labor costs associated with the workers' victories. After 1964 the dictatorship took responsibility for corralling conflict and directing change. The sharply increased presence of the military state—guiding industrial expansion, attempting to quantify and regulate the agricultural environment, and setting the rules for the labor bargain—directed the floodlight of state intervention on a region that was only haltingly moving out of the shadows of patronage and individualized domination.[3]

Terms of Command: The Dictatorship in the *Zona da Mata*

Workers had made clear progress toward gaining a collective voice before 1964 through unionization. Their material gains were more modest. Fernando Antônio Gonçalves's 1964 labor force survey found that only 2 percent more workers lived in a house not ceded by their employers than had in 1961, before the establishment of unions. The improvement in workers' diet was more encouraging: between 1961 and 1964 their estimated average daily calorie intake rose by 20 percent, from 2,320 to 2,760. During that same period, however, food prices almost doubled. The *engenho* stores (*barracões*), which traditionally held workers in a form of debt peonage, began to lose their place as the primary node for worker consumption. But in a cruel irony of the transition, the vicious pricing and the controlling hand of debt were replaced by long, difficult trips to the small towns of the region, where prices were only marginally better.[4] The minor gains that workers achieved likely related to the bargaining capacity of the rural unions and the efforts of the Arraes administration; this is certainly how the workers themselves perceived the situation.

In workers' collective memory, the arrival of their "rights"—the protections and privileges accorded them through the application of the ETR and the ability to challenge planters—represented a major rupture in the region's history. Before 1963, the worker and union leader Paulo said, he and his colleagues had "no one to complain to."[5] They accord much more importance to the moment of gaining an avenue of recourse—a power from above to rival the bosses'—than they do to the transfer of state power that took place with the coup.[6] The climate obviously shifted with the military takeover, but in terms of the trajectory of workers' struggles there was continuity across the divide of April 1, 1964—the ability to press complaints and seek improved conditions remained, if in an attenuated form. The numerous strikes of 1963 ended due to repression and the dictatorship's strict Strike Law, but the number of cases brought before the labor courts grew. (The Cabo union did mount a series of strikes between 1965 and 1968, with workers demanding adherence to the Task Table and payment of back wages.)[7] But the lack of demonstrative confrontation can also be seen as a marker of the government's success in channeling worker dissatisfaction through new structures, particularly the rural labor courts.

The best evidence that the regime's rise was not an unequivocal victory for planters was its decision not to abolish the ETR. Another illustration from which lessons can be drawn about the dictatorship's approach was its retention of a tool that planters detested: the Task Table. In June 1963 representatives from the newly influential rural unions had met with representatives of planter organizations under Arraes's mediation to discuss payment norms in the fields. The ETR's shake-up of labor relationships and extension of labor legislation to rural areas impelled the two class groups to meet under the state's sanction to begin establishing a new modus vivendi. The *Diário de Pernambuco*, covering the talks, noted that the parties started with the most basic questions, getting to the heart of tensions that had simmered in the fields for decades. "In what manner should tasks be measured? How should the terms 'weeding' [or even] 'task' be translated into a current system?" The *Diário* reporter understood the significance of the debate, noting the numerous daily disagreements provoked by these issues, since "rules peculiar to each zone [of the region] guide tasks." The abundant discrepancies were "an inheritance of the origins of the agro-industry, when each *senhor de engenho* directed work according to his personal point of view and his convenience." Sensing the scale of the changes afoot, the article concluded that "the agro-industry has evolved in all senses, and the measures currently being taken will overcome" the problem of multiple sets of standards.[8]

Through extended debate, the unions and planters hammered out the Task Table, a unified set of standards. The table stipulated norms and expectations for each job associated with cane agriculture. Since the payment of wages depended on the completion of tasks, the table played a critical role in setting the boundaries for tasks. The labor system's drift from time-oriented remuneration toward spatial tasks during the 1950s had opened room for manipulation, as field managers and foremen could simply set larger and larger task units for the same wage.[9] The table established how much work was to be accomplished in a given job during an eight-hour period. "If you don't have a Task Table," Paulo summarized, "you don't have anything."[10] Bezerra called the table one of the most important conquests of the entire series of struggles from 1963. Though they did not yet know to demand it when they "struck for lower wages" that July, this was the tool the Roçadinho workers needed (see chapter 4). The negotiations that produced it contributed to Miguel Arraes's legend and became known as the "Accord of the Fields." The table was reaffirmed in the collective agreement following the November 1963 strike, another moment when workers demonstrated their newfound strength and forced planters to make a series of concessions in a state-sanctioned negotiation.[11]

With the military's ascent after the coup, planters saw an opportunity to undo the accord and insisted that the table be reformulated. They argued that the table was unfair to them and that they had been "forced to respect it by the Communists." In response, the dictatorship made its first major incursion into the nitty-gritty of labor in the *zona da mata*. Its method said a great deal about how it would approach the region and its dominant industry. Representatives from the new military-appointed Ministry of Labor, the state government, the army, and other "qualified technicians" were assigned to research work norms in the region in order to formulate a new table. This

Stipulations for Weeding Cane Fields, 1964 Task Table

Terrain	Task (in square *braças*)
Plowed	100
Unplowed, thick brush, hard ground	50
Unplowed, thick brush, soft ground	60
Unplowed, thin brush, hard ground	70
Unplowed, thin brush, soft ground	80
Unplowed, thin brush, sandy ground	100
Quick weeding, abundant brush	80
Quick weeding, sparse brush	100
Cultivating (running the hoe)	200

Source: "Contrato coletivo de trabalho, na lavoura canavieira de Pernambuco," in SSP 28688: Federação dos Trabalhadores Rurais de Pernambuco, Acervo DOPS, APE.

thorough and heavily bureaucratized approach set the tone for how the military state would address labor issues and agricultural planning alike.[12]

The dictatorship's representatives were probably attracted to the table's explicit acknowledgment and accommodation of environmental specificity. The table's very structure—the fact that it sought to mold a set of work guidelines to the infinitely variable and changeable environment through a set of simple categories—matched the state's ambitions to bring rural work in line with an industrial model of regularity and organization. The table recognized six different types of terrain and prescribed different task measurements for each under the category of weeding, for example (see table). In the stipulations for cane cutting, the Task Table included the category "good cane." Although such a category gives the orderly grid an inherent subjectivity, the table did offer a short explanation: "Good cane to cut is understood as thin cane on clear land."[13] A worker interviewed in 2003 clarified that good cane meant "thick and growing close together, and good and strong, so a person can cut more. . . . It's good in a flat area or a recently planted area."[14] In addition to this confusion, it should be pointed out that good cane to cut is not necessarily the same thing as good cane for making sugar.

The fact that the table, with its task sizes carefully calibrated to different conditions, still failed to reflect the full diversity of the *zona da mata*'s environment did not bother its designers or those charged with its revision. Continued conflict was to be expected over such basic questions as what constituted good cane or thick brush. The table nonetheless had definite effects. Along with other symbols of the state's intervention in labor relations, the table and its implied ordering of the environment represented a new landscape discourse. It also challenged the planters' laboring landscape discourse, according to which quibbles about thick and thin brush were beside the point since workers were commanded to labor. The table also gave workers a framework for contesting their assignments, and its status as an instrument of state power and a tool for negotiation shifted the terms of the workers' engagement with the landscape. In this way, the dynamics of the table show that "simplifying instruments" can sometimes offer opportunities to those with less power. The simple presence of a competing authority in the planters' realm was the most important change, a chance to modify the terms of captivity. As we will see the state's authority was embodied in other institutions as well. The closed, subjugated world of the *engenho* was pried open by the Task Table and other instruments of the state. Workers in the fields still toiled the land of the planter, but with a layer of protection—tools for negotiation and complaint.[15]

Union participation sagged following the coup, but by 1967–68 the organizations were back to the membership levels of 1963. To some, this illustrates the regime's corporatist philosophy, an idea supported by the fact that the same approach was taken with rural employer groups.[16] It is also a lesson about the impact of the coup and the fact that it did not crush the labor movement despite the repression.[17] In 1966 Euclides Almeida do Nascimento, the president of the Federação dos Trabalhadores na Agricultura do Estado de Pernambuco (FETAPE, Federation of Workers in Agriculture in the State of Pernambuco), sent an appeal to the national minister of labor. Nascimento described the difficulty of life for workers in the cane zone—high food costs, the constant need to press complaints in the labor courts, and "disrespect" of the Task Table. He made a simple request of the minister: "Recognize the right of the worker to the wages of fifty-four thousand *cruzeiros* for eight hours of work, according to that which is prescribed by law."[18] Rural workers continued to struggle, but Nascimento's letter to the labor minister captured their changed approach—appealing directly to the state for aid and support.

Workers' new tools for negotiation and complaint clearly did not ensure their rights, and state enforcement of wage levels had limits. A U.S. consular official wrote bluntly in 1966 that "the worker now finds himself worse off than before while his Revolutionary leaders prattle about high-minded inflationary controls, planning agencies for economic development and labor laws that 'guarantee' the workers' rights."[19] Despite the ETR's stipulation that workers choose their own payment method, almost 80 percent of workers in the *usina*-dominated southern *mata* were paid according to tasks or production levels in 1968, probably not by choice (they likely preferred a daily wage). As a military investigator observed, most workers needed help from family members to finish a task in one day, meaning they were failing to make the minimum wage.[20] The northern *mata* had a somewhat higher percentage of workers paid by the day, but another consular report mentioned that only one *usina* in the region consistently paid the minimum wage.[21]

Faced with these challenges, but cognizant that they still had some room for action, workers increasingly looked to the state. They were reinforced in this strategy by the state's own strategies. The Brazilian anthropologist Mauro Koury argues that the state "recuperated certain battle standards of the precoup unions, including agrarian reform and the fight for labor rights and social security, and transformed them into its language in order to return them to the workers in a paternalistic form."[22] Key to this process was the ETR, which in addition to the benefits I have already mentioned established a welfare program funded by a tax on agricultural products. The program was streamlined and better funded in 1967, the same year that another law established an administrative structure for rural medical and dental services (the Fundo de Assistência ao Trabalhador Rural [Fund for the Assistance of the Rural Worker], which became known as FUNRURAL).[23] Once workers received benefits and privileges as concessions from the state, they began to play the roles envisioned for them by the state. That is, to the extent that they accepted and abided by the rules established by government, the workers fit in the state's social framework—they were "legible," in Scott's terms.[24]

While the state translated union goals into its own language, it also appropriated producer initiatives and took a larger role in the agro-industry. The mid-1960s brought the Pernambucan sugar industry a three-pronged crisis, and it needed support badly. The rise of southern growers had continued as rapidly expanding producers in the state of São Paulo left northeastern producers a shrinking percentage of the national market share.[25] The (mis)use of 3x created a sharp decline in industrial yield (sacks of sugar per ton of cane). And the workers' "social conquests" in 1963 resulted in steeply increased labor costs. The multitiered reactions to these problems represented the beginning of fundamental changes intended to address the "crude technological backwardness" of Pernambuco's fields.[26]

The first united response from state-level Pernambucan producer groups to their crisis was the formation of the Grupo de Estudo do Açúcar (GEA, Sugar Study Group) in 1963. An ad hoc emergency response team, the group released a report in 1964 enumerating the challenges facing sugar producers and emphasizing the need for change. "We know it is not easy to execute agricultural tasks that stray from routine, we know everything is quite difficult and expensive, but we cannot continue as we are, tied to these problems without putting forth the effort required to conquer them." The members of the GEA pledged to convince their own colleagues of the imperative to adopt scientifically sound agricultural practices, and after April 1964 they looked to the new military regime for help.[27]

In 1965 Pernambucan sugar producers formed the Grupo de Trabalho para a Indústria Açucareira (GTIA, Working Group for the Sugar Industry), with the aim of boosting production as well as discussing land reform. Only seven months after the coup, the dictatorship released the Land Statute, an agrarian reform law. The law stipulated a progressive land tax intended to punish owners of unutilized land and spur a broad redistribution, but the government failed to provide the funding for the evaluations and compensation the program required. The military's apparent strategy for the sugar-producing Northeast involved a reform of property holding designed to cultivate a rural middle class that could produce food for the urban centers. The GTIA was essentially just a reconfiguration of the GEA, but it reflected planters' growing understanding of the dictatorship's approach—that it saw some value in land reform but had clearly prioritized increased sugar production. They tailored the GTIA's initiatives to these parameters, announcing plans to increase industrial efficiency, cut down the bloated labor force, and release unused and underused land for new small-scale producers.[28] Among its early recommendations, the group proposed an increase in the task size guidelines of the Task Table, a measure aimed at increasing the productivity of the labor force.[29] In 1966 the apparent synergy between producers and the dictatorship produced another organization: the Grupo Especial para a Racionalização Açucareira do Nordeste (GERAN, Special Group for the Rationalization of the Northeastern Sugar Industry). Releasing reports that combined guidelines for increased productivity with a stated concern for the future of rural workers, GERAN secured widespread support among planters.[30]

In a 1983 interview, Eudes de Souza Leão Pinto, Pernambuco's secretary of agriculture in the late 1950s and early 1960s and later an official in the federal Ministry of Agriculture, made a curious observation. Declaring his commitment to the importance of professional, state-sponsored support for farmers, he argued that "60 percent of the progress registered in the United States of North America is owed to the fact that farmers never feel alone in the fields, never feel helpless." Instead, "they are always oriented and guided, not by adventurers and opportunists, but by technicians or people with proven professional training, who give the farmers every chance to apply their physical capacity, financial and land resources, and intelligence."[31] This rosy model of state-proprietor relations—with government technicians keeping farmers company in the fields—captures the aspirations of the tightening relations between producers and state during the 1960s.

Courts and Contracts: Planters and Workers Maneuver for Advantage

In the days following the military takeover, a rural union official named Amaro Pedro escaped persecution by seeking refuge on an *engenho* belonging to José Bezerra (no relation to Gregório), a man known for maintaining good relations with his workers, which is to say, he was known for being a skillful patron. The next year, Pedro filed a labor court complaint against Bezerra seeking payment of back benefits. He eventually won a monetary judgment, after which an indignant Bezerra ordered Pedro off the plantation. Later, the repentant worker went to his employer, returned the money, and tearfully begged forgiveness for violating the trust of their relationship.[32] The story shows two men caught between systems of social relations—the old, patronage-based employment relationship of *morada* and the emergent system governed by labor law and brokered by the courts. Many workers and employers in the sugar zone confronted similar situations in the 1960s and 1970s.

When the ETR replaced Vargas's defunct twenty-year-old rural unionization law, employers had to issue work cards (*fichas*) to workers, a mode of documenting the labor relationship that helped make sugar workers visible to the state. Workers refer to this status as being *fichado*—having a card. The terminology associated with undocumented workers underscored the importance of the state's vision. Workers with no work card acquired the label *clandestinos* (clandestines), since they held no legal employment relationship with a planter and were therefore "hidden" from the labor law. With a legal, visible employment tie, workers became entitled to benefits, union membership, and social services.[33] They also were given access to the labor courts—the Juntas de Conciliação e Julgamento (JCJs, Boards of Judgment and Conciliation) and the state and federal labor tribunals. These courts were crucial to the new system of labor relations, since they became the privileged forums for conflict between workers and employers, with their own norms and a language that was used increasingly self-consciously by both parties.

The courts were far from perfect instruments for the mediation of disputes, not least because of employers' efforts to undermine their effectiveness. Employers dissembled, feigned ignorance, and leveled outrageous accusations in order to slow cases in their already ponderous progress through the courts. The dockets quickly became unwieldy following the courts' establishment in 1963 and 1964. In 1967, for instance, the Palmares JCJ received 6,181 complaints, but only 252 hearings were held, and judges handed down only 93 judgments (though the next year, when the government intervened in the Palmares rural union, the JCJ received only a third as many complaints). Another 1,718 of the complaints reached conciliation. Demands for fair payment accounted for the largest percentage of complaints brought before the JCJs, since workers earned an average of 20 percent less than the minimum wage.[34]

Many workers complained that their own union lawyers tried to force them to accept settlements with planters. Some workers also sought a settlement because they saw how many cases were ahead of theirs on the docket and how slow the courts were to rule. This drove average settlement sizes to a level far below the initial demands. For their part, union leaders noted that the large number of cases required them to spend nearly every afternoon in court.[35] These patterns became embedded in the culture of the courts, and workers' dispositions toward the JCJs remain subject to manipulation by employers. A worker interviewed in 2003 mentioned an episode from the early 1990s when the Catende mill fired every worker who had brought a complaint before the JCJ. "It was a holy cure," Cícero said. "No one wanted to make trouble anymore [*ninguém quiz mais questão*]."[36]

Already in the first years of the labor courts, the judges heard cases from workers without cards petitioning for recognition of their "rights," while planters coolly denied having any relationship with them. In one 1964 case from the southern *mata*, a planter disclaimed any connection or knowledge of a worker before allowing that the man "might have worked" on the *engenho* but was never *employed*; that is to say, he was never *fichado*.[37] Another planter's lawyer noted that his client was an "illegitimate or inappropriate party to respond to the obligations of the [work gang] contractor, who, by law, ought to be called to justice to respond to said obligations."[38] Another lawyer for a *senhor* found no "contractual work connection" between his client and the petitioning worker, since a particular contractor "is the one who contracts the labor force of sugar industry workers in this municipality," and the workers "only receive orders from the contractor himself."[39] Understanding these battles requires grasping the crucial difference between holding a work card with an *engenho* or *usina*—with the benefits that implies—and working for a contractor (*empreiteiro*), a middleman between the workers and the planters who musters gangs of informal workers with no legal tie to any boss.

These arguments over the status of employment took place in the labor courts as a result of planters' disinclination to comply with the stipulations of the ETR, whether these be paid holidays, the thirteenth wage, or anything else. From the moment of the legislation's passage, and especially after the coup, employers severed official employment relationships with increasing numbers of workers. The case files for the Palmares JCJ (in the heart of mill territory) include thirty-six hundred cases for the year 1965, every single one a contract rescission delivered by an employer. This JCJ served parts of other municipalities in addition to Palmares, but to give a sense of proportion, Palmares's total population was only thirty-eight thousand. The firings represented a considerable percentage of the economically active population.[40] Most files contain a worker's thumbprint signaling his or her acknowledgement of the termination, though many are blank, meaning the worker did not appear at the JCJ to accept or contest the rescission. The scorched earth firings continued into the following years, as *usinas* and *engenhos* dumped work cards at a vertiginous rate.[41]

Obviously, most of these fired workers continued to perform their jobs. Instead of holding work cards, they forged short-term agreements (just for a harvest, for example), while the majority began work with contractors.[42] By hiring gangs of workers through contractors, planters could dispense with their responsibilities toward resident workers and many of their overseers, since they no longer held work cards. Contractors took a cut of an established payment from the planter or mill and distributed the rest among their workers, short-circuiting the state-mandated benefits system. Contractors had appeared as early as the 1930s, but after passage of the ETR their numbers grew rapidly. In the 1970s, contractor-organized labor became a standard aspect of work in the cane zone.[43]

Since contractors arranged jobs on a daily or weekly basis, workers could play the field, moving from contractor to contractor looking for the best pay or the best working conditions. But having two pools of workers, one with work cards and the other an eminently flexible reservoir of labor mustered by contractors, was very much to the planters' advantage. The two groups could be opposed to one another; the extra expenses of holidays, paid leave, and bonuses were not due to workers without cards, and in hard times no obligations bound this group to planters. Growers made sure to keep their specialized workers on the *engenhos* and otherwise found a sort of equilibrium (variable from employer to employer) between *fichado* and informal workers—a balance that oscillated with sugar prices and other considerations.[44]

The shifting norms of work relationships further eroded the already weakened traditional tenantry system. The ETR regulation of work cards started the process, as a state-sanctioned system began a new phase of informal agreements forged between the parties. Planters then reacted against their obligations as employers and, seeing no need to continue maintaining houses for workers on *engenhos*, began to physically expel them or compel them to leave. Planters' techniques included refusing to renew work agreements, bodily removing workers from their homes, and treating *clandestino* workers better than *moradores*. In 1972, for instance, an Usina Santa Tereza official ordered several houses destroyed and recalcitrant workers beaten; several workers were killed. All of those targeted had worked for the *usina* for long periods; this was not a coincidence, since employee-paid retirement benefits rise with years of service.[45] In another example, a card-holding worker from Moreno told police that thirty *clandestinos* on his *engenho* had been given new hoes to clear brush, but he had gotten nothing, despite being assigned the same task. When he asked for a new tool, the foreman shot at him and beat his wife with a shovel.[46] The aftermath of this episode is undocumented, but it seems likely that the altercation resulted in yet another of what a union report referred to as the "incalculable expulsions" from *engenhos*.[47] By the end of the 1970s, when the French journalist Robert Linhart interviewed a series of rural union leaders, all of them told him that most cane workers operated without work cards and lived in the region's small towns, having been expelled from *engenhos*.[48]

The drop in the proportion of workers with work cards and their accelerating departures from the *engenhos* changed the human geography of the region by increasing general mobility and shifting the concentrations of habitation to small towns. In 1961 almost all workers still lived in homes ceded as a condition of their tenancy on an *engenho*. But there were also high rates of job turnover and worker movement, an indication of growing instability in work relations. Half of workers at the time had worked less than a year with their current employer and nearly half were from a different area of the *zona da mata* than where they worked.[49] In 1963 and 1964, over half of the labor court cases from two sample municipalities (Palmares and Nazaré da Mata) involved workers who had only started working on their *engenhos* within the previous five years.[50] A 1968 union federation survey also documented high mobility rates; only 37 percent of workers surveyed by FETAPE had spent more than a decade at their current job.[51]

As workers increasingly moved away from *engenhos*, they occupied shacks "on the street" (*na rua*, the phrase invariably used to describe movement off the *engenhos*) in the *mata*'s small towns. By 1985 this population accounted for about 56 percent of the workers, with the remainder still on *engenhos*.[52] The basic trajectory of "deruralization" began, as we saw, in the 1920s and 1930s and then grew by the decade. Between 1950 and 1960 the percentage of rural residents in the northern and southern areas of the region dropped by 6 percent and 8 percent, respectively (from 78 percent to 72 percent and from 75 percent to 67 percent). Over the next ten years, to 1970, they dropped another 8 percent in the north and 9 percent in the south.[53] Accelerating through the 1970s, this movement of people finally tipped the regional balance between urban and rural; 1980 was the first year that the *zona da mata*'s urban population exceeded the rural.[54] This was not unlike the trajectory for the country as a whole. Indeed, not all of the workers leaving the *zona da mata* were going to Recife; many were joining the heavy flow to São Paulo and Rio de Janeiro.

The importance of work cards, the dynamics of labor courts and their bearing on work relationships, and the role of the Task Table were all relevant to a 1970s case from Nazaré da Mata. In 1971 José Panciano was admitted to tenantry on the Engenho Boa Vista. When he moved in, his wife Severina Rosa da Conceição and their four daughters accompanied him. Six years later, José complained to the JCJ that he had not received some wages due him, and the court ruled in his favor. Shortly thereafter, the *senhor* refused to allow Severina and the daughters to work, which José alleged was in retaliation for his disloyal behavior. The five women brought a complaint before the JCJ, claiming they had been fired and petitioning for back benefits as well as indemnification for the firing.

The planter's lawyer called the case "an adventure without precedent," explaining that Severina and her four daughters helped José work and augmented his production, but that this did not constitute a labor relationship. No wages were ever paid to the women directly.[55] Though he probably ought to have rested his case there, the lawyer continued, "just for the sake of argument." If they held cards, he pointed out, the women must not have been fired by the plantation, because they continued to live there. Here the lawyer mixed the logic of patronage and tenantry with the logic of labor law. The women could not have been fired because they continued to live on the plantation; yet they did not hold work cards, so they must be prohibited from working in the fields. The rules of the inherited system of labor tangled with the emergent one to catch Severina and her daughters in a trap. According to the old tenantry arrangements, habitation signaled employment. But the new laws required that employment be accompanied by a work card. The women lived on the plantation and therefore had not been dismissed, but they also lacked cards and therefore could not work. As for their acknowledged presence in the fields with José, the lawyer described workers' habitual use of family labor to augment their production as a "peculiarity of work in the cane zone."[56]

The lawyer accurately characterized the culture of labor in the cane zone in terms of women's work. A union report from 1968 found that only 8 percent of women in the Pancianos' subregion (the north) earned their own wages. In the south, as much as 27 percent of women held waged work. A large proportion of remaining women worked alongside husbands or fathers, but without their own direct compensation.[57] This was precisely the sort of "peculiarity" that the state sought to stamp out with the new labor regulations and social service programs, since work performed by women in support of male family members, and remunerated through payment to the men, was invisible to the state. By the time Severina brought her case, it was no longer uncommon for women to work for their own wages, though they had to wait until the 1980s for full incorporation into the rural unions.[58]

The JCJ judges decided that the question of the women's labor relationship to Boa Vista revolved around how much work José regularly accomplished. If more than one task were produced every day but only one worker received wages, the judges argued, that would imply fraud by the employer. Essentially, the judges' approach was to literally apply the Task Table. According to the table, any given task is equivalent to an eight-hour day's work. Therefore, the completion of more than one task in eight hours would imply multiple laborers and therefore multiple employees. Through careful parsing of contradictory witness testimony, the judges decided that José regularly completed more than one task per day, an accomplishment clearly made possible by the help of family members. Then, using witnesses' statements about the frequency of the women's attendance, they decided that three of José's daughters had provided labor to the plantation on a consistent enough basis to warrant their classification as employees. Therefore, they ruled, the planter's refusal to offer them any more work must be understood as dismissal without warning. They ordered that he pay them restitution.[59]

The courts clearly did not always agree with planters' protestations that they had no legal relationship to clandestine workers. Though judges privileged the importance of a work card, they also recognized continuity and consistency of service and a personalized relationship between worker and employer, even if a contractor stood between the two. And if they felt a pattern of consistent labor could be demonstrated, they required planters to pay benefits to *clandestinos*. Notwithstanding its flaws and imperfect application, the Task Table was a tool the judges had at their disposal for cutting through the informalities of labor in the sugar zone (and perhaps battling the routine exploitation of assigned tasks too large to be completed in a day). The Nazaré da Mata judges' use of the table emphasizes the crucial role it played for workers as a benchmark for negotiation in the fields with the foremen and as a legal device in the courts. Even as planters sought a fully "flexibilized" workforce without leverage of its own, workers resisted this process with the resources at their disposal.[60]

Severina and her daughters' experience in the courts also demonstrates the changing role of women's work. The "peculiarity" of family labor came under the judges' scrutiny in their case and others, leading to a growing recognition of women's work as something individually rendered rather than an occasional addition to a man's labor. In 1975, for example, a woman named Emília Maria da Silva took the Usina Catende to court for unpaid benefits. She worked occasionally with her husband and also on her own through contractors. Witnesses testified to seeing her regularly in both roles, though when she worked with her husband he "received the fruit of her work." In their ruling after hearing testimony, the judges emphasized "the individuality and continuity" of work and the fact that it must not be "sporadic, occasional." From the testimony heard, they declared, it appeared that this woman had worked consistently and personally for the mill (that is, not solely with her husband) over an extended period. Even though she worked for a contractor and had not been *fichado* for years, her service indicated she should be paid benefits.[61] This was not just an issue of informal work versus work with a card, it also signaled the judges' respect for the labor women performed. As with the Task Table and its aim to regulate task work, the state sought to bring "irregular" aspects of the region's work regime such as women's supplemental work under its oversight.

Land, Proletarianization, and the Metaphor of Captivity

The reconfiguration of patron-client ties and the decline in the number of workers holding work cards spelled an important transition for all parties. For planters, the increased flexibility of the workforce implied the loss of control over *engenhos* populated by "their" workers, but it offered new opportunities for a form of control without strings attached. Workers faced divergent opportunities and difficult choices. The personalized relations of patronage had offered the hope of support in times of need. Workers remaining on the *engenhos* could appeal to the planters for aid (not that it would necessarily be forthcoming), or they could try to supplement their diets and possibly incomes with personal crops, if they had access to land. *Clandestinos*, in contrast, frequently enjoyed better working conditions and the freedom to seek the closest or highest paying job on a given day, but when labor demands dropped they were the first to feel the pinch. These complexities were tied to the multiple ambiguities of the shifting labor relations—formal, card-sanctioned work versus work with a contractor; law-governed relations versus patronage.

Driven by these changing circumstances, workers' definitions of captivity and freedom shifted. This change in perspective would obviously impact the way workers conceived of improvements in their lives and made decisions regarding the political struggles they pursued. Through the 1950s at least, workers generally identified freedom with a broad range of work options (knowing how to do every job but not having to do them all) as well as access to land and few restrictions on its use. They achieved these by building a good relationship with a *senhor*, and the system as a whole led to an intricate geography of power on *engenhos*. With passage of the ETR, workers were supposed to enjoy a minimum wage with labor benefits and access to land. In addition, their unions argued vociferously for a restructuring of economic relations through sweeping land reform. This section treats the two aspects of changing worker views separately: first work and living relations, then the question of land.

Paulo associated the transformation of tenantry, and especially expulsions from *engenhos*, with the ETR. He said that bosses began expelling workers out of "hate" for the workers (a common claim) and from the frustration of having to pay labor benefits. Of the workers' plight, he made this broad and distinctive statement: "They don't want us to have liberty."[62] The deterioration and transformation of tenantry relations and the regulation of labor relationships made the optimal situations of regular work and stable access to land—the circumstances associated with greater "liberty"—rarer. In this context, captivity increasingly came to be defined in terms of a simple duality between living on an *engenho* and living "on the street," though workers did not necessarily agree about which was better. Subjection to employer control had always been associated with captivity, but a good work relationship could earn a worker greater freedom. Now that life on the *engenhos* offered narrower margins for improvement, many preferred "the street."

Workers' preferences for the *engenho* or the street tended to vary according to how they left the *engenhos*. The rapid termination of work cards after the ETR and especially after the coup did not necessarily imply expulsion from an *engenho*, though the pressure planters exerted on many *moradores* to leave was often extreme. A worker named Bento described his family's expulsion straightforwardly: "The boss needed our plot to plant cane. They didn't let us work there anymore because they wanted the plot to fill with cane, because it made more profit for them. So we left and came to live on the streets."[63] Despite the frequency of coercion or the apparent lack of choice, workers generally refused to say they had been expelled. They acknowledged that planters kicked *others* off of *engenhos* ("botado para fora"), but individuals all claimed to have themselves left of their own "free and spontaneous will" ("livre e expontanea vontade"), a heavily used phrase lifted from the language of a work-card rescission form in the JCJs. Union officials found that they had to read statistics about worker firings carefully, since workers would not even admit to their union officers that they had been fired.[64]

Apparently, workers described getting kicked off as something that happened to people in general, while leaving freely was what individuals did. This view is still generally held. A Catende worker I interviewed in 2003 said that the owners expelled twenty-three hundred people in the early 1990s: "They fired everyone; completely fired. They kicked everyone out." But when asked if he, or anyone from his *engenho*, had been expelled, he said only that some people had left "of their own spontaneous will."[65] The pattern of workers' interpretations of expulsions suggests their reaction to the operations of power involved. They resisted acknowledging the planters' ability to coerce them and emphasized instead their own self-determination. The stakes were high on leaving: no *morador* could leave an *engenho* and hope to be admitted to another as a *morador*.[66]

Once on the street, workers identified the *engenho* as a space of captivity. A union official described the "suffocating" conditions of life on *engenhos*, and workers said they went to the street because they "didn't want to be prisoners anymore." Many focused on the issue of control, of subjection to planter demands. "The *engenho* was worse than the street. The boss made us work in the plantation all week, from Sunday to Sunday."[67] Workers told of having to load cane carts in the middle of the night, in the rain, or while sick. Their complaints were confirmed by an unlikely source, a 1977 report from an inspector in the regional labor office about conditions on the São José plantation and mill, just north of Recife. After determining that workers were being bused several hours to and from their work sites and were being denied time to work on their subsistence gardens, the inspector gave the plantation management a harsh review. "The rural man does not live," he wrote, "because he is always subject to the caprice of the *senhor de engenho*, who does not see him as a human creature, but probably sees him as a thing (a slave)."[68] Workers justifiably described such conditions as captivity, compelling their constant availability to the boss and their inability to supplement their diets or incomes with gardens.[69]

Yet life on the street also offered only a thin space to maneuver, with few options for the application of individual enterprise, no recourse to someone "above," and no guarantee of regular work. Paulo said of the mobility of expelled workers, "At least a person has liberty, and can work wherever he wants." "That is to say," he corrected himself, "he knows how to *look* for work wherever."[70] And there were those who found life on the street inferior to life on the *engenhos*. They said life was better "in the *mata*" (on an *engenho*) where "we could plant a little for ourselves."[71]

Gaining access to land for family production was at stake in policy developments from the 1960s that changed workers' views toward land and political struggles that revealed these changes. Since their inception, rural unions had placed calls for land at the core of their rhetoric. They pressed their demands for reform in petitions to the state and routinized their views through popular pamphlets and songs sung at union meetings. Their supporters in government or the academy or international work, such as Josué de Castro, also took up their call.[72] The socialist politician Francisco Julião had pushed vocally and persistently for agrarian reform, winning more followers after Governor Sampaio's expropriation of the Engenho Galiléia in 1959.[73] A major conference, "The Agrarian Problem in the Cane Zone of Pernambuco," held at the Joaquim Nabuco Foundation in May 1963, featured detailed debates about land reform among participants ranging from Julião to General Humberto Castelo Branco (then head of the Fourth Army, based in Recife).[74]

The next year Castelo Branco became president and promulgated the Land Statute, but the promise of the debate over agrarian reform did not translate into programs on the ground in the *zona da mata*. In 1965 rural unions attempted to force the issue, trying to use the apparent enthusiasm of government and producer organs for the potential synergy between agro-industrial modernization and land redistribution. Padre Crespo and his Pernambuco Rural Orientation Service led a campaign for federal recognition of rural workers' right to land access based on a forgotten piece of the 1941 Sugarcane Farming Statute that had guaranteed rural workers access to land for subsistence crops. Crespo and FETAPE persuaded the federal regime to establish a new law confirming this provision. The new law became known as the "law of the plots" or the "two-hectare law." Essentially recapitulating Barbosa Lima Sobrinho's guideline from the ELC, it obliged employers to cede two-hectare plots to all employees with more than a year's standing.[75]

Crespo's success in getting such a law passed provides another sign of the marginal but not insignificant power unions retained following the coup. And considering the context—the appearance of an emerging consensus on the importance of land reform and a newly protected workforce in need of ways to supplement its unpredictable income—one might think that the measure had a bright future. The law could have begun the process of fostering the small-scale producer class envisioned by the military's planners. Not surprisingly, planters demonstrated little enthusiasm for granting the land required. But the workers also participated in a lackluster way in the campaign for the law, and the ensuing push for its enforcement and their reticence must be factored into an analysis of their desires. In the Brazilian phrase, the measure "never got off the paper." Padre Crespo pushed doggedly on, advertising the law and demanding that it be applied, but the labor system in the *zona da mata* was moving in a different direction—with ever more labor contractors, fewer workers with work cards, and less and less stability on *engenhos*.[76]

Other signs of ambivalence toward land and agrarian reform—an issue that occupied such a central place in rural union discourse—offer some insight into Crespo's failure to popularize the two-hectare law. In March 1967 a regional meeting of rural cane workers convened in Pernambuco. Among other things, the delegates grappled with the discussion question "Proprietor or proletarian?" Many delegates "preferred the acquisition of land, provided it was of good quality for economic exploitation and if adequate assistance were secured." However, others maintained that correctly paid wages and duly apportioned benefits would obviate workers' need for land.[77] The northeastern federations of rural unions met the next year and conformed to their historical position: the most urgently worded of their demands to the federal government called for an immediate, efficient national land reform that prioritized the Northeast.[78]

Several months later, however, FETAPE issued a frank self-assessment that drew attention to the gap between this call from the leadership and the desires of the rank and file. Questioning "the emphasis union directors [gave] to the organizing principle: 'fight for the two hectares,'" the authors expressed reservations about whether it should be a central union goal, since it did "not reflect the anxieties and aspirations of the masses." Indeed, a survey of union workers found that very few expressed interest in land. Instead, workers showed far more concern for prompt wage payments, health care, and other benefits than with access to a subsistence plot.[79] Cícero, interviewed in 2003, said he left his family's small farm in the 1970s to work on an *engenho*, where he is now foreman. He preferred regular, waged work to owning land of his own; likewise, none of his siblings had shown an interest in continuing to cultivate the family farm.[80]

FETAPE's 1968 report effectively acknowledged that unions had evolved to mirror the state's vision of worker organization when it noted that the only effective gains workers had made in the late 1960s involved social assistance. This was the state's focus, not work conditions or access to land.[81] Mulling the unions' activities in the two decades following the coup, Severino said, "I think the union movement talked about one thing, but advanced in another. . . . [The movement] retreated a bit on agrarian reform and advanced on the question of wages."[82] The anthropologist Anthony Leeds, studying the region in the 1970s, confirmed the shift in worker feelings, saying he had "seen no evidence that [workers] actually *want* the land to be divided among them into small peasant holdings, other than in the slogans of the 'leaders.'" What did seem clear, he reported, was that workers wanted higher wages and improved living conditions.[83]

Workers' changing views of land coincided with the declining use of subsistence plots. As I discussed above, workers lacked the time to make use of land even if they enjoyed access, and the ETR did not improve this problem. The statute granted a series of benefits to workers only if they provided six days of work a week, but consistently working six days left little time to care for crops. Even greater pressure was put on some workers, as documented by a union's 1968 complaint to police; hours in the fields were extended and unfairly remunerated even when workers finished their tasks. Besides not having enough time, workers lacked resources such as seeds and tools.[84]

The story of the two-hectare law could be interpreted as a brilliant balancing act by the state that helped it consolidate control over all sectors in the *zona da mata*. The promulgation of the law put planters on notice that the dictatorship would not shy away from measures that threatened the interests of the entrenched landowning elite. For workers, the law's total lack of enforcement and application, combined with the state's focus on employment-associated benefits, signaled that they should stay within the bounds of struggle established by the state. Work should be rendered consistently, and complaints should be funneled through the labor courts. (As a salutary side effect for the federal government, the law showed the U.S. government that the new military regime appreciated the importance of reform to pacify what was seen as a dangerous leftist hotbed.)[85] In a sense, the set of interactions and discourses from the earlier period of struggle made land into a symbolic but not a real goal, a transformation accomplished through the unions' rhetoric (ineffective, if not empty) and the state's law.

Looking back at the previous decade in 1971, the agronomist Bento Dantas wrote in a GERAN report that agricultural modernization "resulted in the widespread discharge of the labor force and [considering] that the production of foodstuffs was historically ignored by the traditional planter class, it was made clear that a land reform was needed that would fix the liberated laborers to the land." The regime did seek to "fix" workers, in the sense of rendering them available on the flexible terms preferable to producers.[86] The dictatorship's approach did not lie in a systematic land reform; in fact, the regime abolished GERAN in the same year that Dantas published his report. This producer-state partnership that had actually supported land reform had only existed for five years. It was replaced by the Programa de Redistribuição de Terras e Estímulo à Agroindústria do Norte e do Nordeste (PROTERRA, Program for Redistribution of Lands and Stimulus for Farming in the North and Northeast), a land reform agency that accomplished almost nothing, while producers received support on a massive scale.[87] Unsurprisingly, this transition coincided with the beginning of a new phase of industrialized and incentivized production.

Conclusion

The central theme of the postcoup period was the systematic application of state power and the conscious recognition of that power by workers and producers. The military regime retained the legal framework for managing rural labor established under Goulart's government—the unions and courts sanctioned by the ETR—and even endorsed the Task Table.[88] The dictatorship handled the much more explosive question of land reform through the ultimately ineffectual Land Statute. Producers found relief from an economic crisis, facilitated by partnerships with the state to centralize and rationalize agro-industrial planning. The dictatorship's careful stewardship would in the 1970s produce a new boom.

The metaphor of captivity remained ubiquitous during this period. Escape less frequently depended on the power-soaked relations with a *senhor* than on a new, evolving relationship with the state, one still marked by asymmetries of power. The state recognized workers as, and expected them to be, card-carrying individuals with legal ties to particular employers and drawing specific benefits in exchange for consistent work. Interpellated by the state in this way, workers saw themselves, their hopes and their aims, differently. Land was no longer the sine qua non of freedom. Mobility was now important, along with reliably paid wages, and the terms of captivity now were understood in the dichotomy between the *engenho* and the street. Together, these changes disrupted the meanings of landscapes in the region. When this happens, according to Thomas Greider and Lorraine Garkovitch, "it is our conceptions of ourselves that change through a process of negotiating new symbols and meanings."[89] As freedom came to be associated with engaging the bureaucracy of the state, workers less frequently saw themselves in "peasant" terms, as even aspiring to ties with the land. They were wage earners first and foremost; one might say they were proletarianized.

SEVEN

An Agricultural Boom and Its Unexpected Consequences

Although they had begun earlier, the state's attempts to intervene in the *zona da mata* accelerated once the military took command. The state interacted with producers and workers through representatives both familiar (technicians from the Instituto do Açúcar e do Álcool [IAA, Sugar and Alcohol Institute], the police) and new (labor court judges, economic planners). Though they had separate briefs, these actors contributed to a common state thrust toward bureaucratic solutions to problems at multiple scales, both natural and social. The modernist ideology guiding state action privileged the power of science, planning, and technocracy to rationalize social relations and industry.[1] This value system produced the Brazilian Miracle, a period of economic acceleration between 1968 and 1974 that averaged nearly 11 percent annual growth. Warren Dean describes the fetishization of development during this period as a millenarian system of belief: "Backwardness would cease, traditionalism would give way to modernization, and the nation would attain development, an edenic plateau."[2]

Export agriculture was a key engine of the growth, even though the "miracle" focused more on urban areas and industrial production. Sugarcane received special attention, especially after planners recognized its potential as a fuel.[3] The rapid increase in cane alcohol production, facilitated by credit and agricultural support and a further expansion of cane fields, was a remarkable example of agro-industrial mobilization. It appeared to be the realization of the modernization dreams nurtured by IAA pioneers, state-level planners, and progressive planters like Antônio da Costa "Tenente" Azevedo of Usina Catende. Even as Pernambuco's cane sector had grown across the twentieth century, it had lurched from one crisis to another. The dictatorship and the emphasis on alcohol ushered it into a boom.

Visions of Agricultural "Revolution"

A 1966 *Brasil açucareiro* article described the backwardness of Brazilian agriculture and the need to grant more recognition and power to agronomists. J. Motta Maia, an IAA divisional director, predicted that agronomists, suitably empowered, would carry out a revolution. Putting the effort in grand terms he wrote, "This revolution will strengthen [Brazil] through the real and not superficial identification of man with the land, through telluric love. . . . The urgent task, which is a technological revolution, will be accomplished by the hands of the agronomists, elevated to the condition of practical scientists."[4] As seemingly extravagant as Maia's rhetoric was, the transformation achieved through government initiative in the 1970s nearly matched his tone. And he captured the ethic of the revolution, that the embrace of modernization would produce a sense of dominion and control. His article also contributed to an emerging landscape discourse associated with the state that would displace the planters' laboring landscape discourse and reframe workers' ideas of captivity.

A complex combination of legislative initiatives, with interlocking bills promulgated over the course of the 1970s, engineered the boom in the cane agro-industry. Together these initiatives formed a broad program of support for the industry through incentives, direct loans, and research aid. The ruling party, the Aliança da Renovação Nacional (ARENA, Alliance for National Renewal), passed the Sugar Agro-Industry Rationalization Program and the Programa Nacional de Melhoramento da Cana-de-Açucár (PLANALSUCAR, National Program for Sugarcane Improvement) through federal laws in 1971. PLANALSUCAR's primary goals for Pernambuco included distributing industrially productive cane varieties and exorcising the 3x cane variety that had been so disastrous in the 1960s (though retaining some of it exclusively for use at the end of the harvest).[5] In 1973 a site in the northern *mata* town of Carpina became PLANALSUCAR's headquarters for northern Brazil. The rationalization program and PLANALSUCAR were followed by the Support Program for the Sugar Agroindustry, in 1973, and the Programa Nacional do Álcool (Proálcool, National Alcohol Program), in 1975.[6] With these programs, every important facet of the industry received attention: cane varieties, agricultural research, funding and credit support, and incentives for alcohol. Their impact was striking, described by agronomist Gerson Bastos as a "light turning on in the countryside."[7]

The state pursued its aggressive support for the agro-industry under the influence of several factors. First, it wanted to help this large export sector take advantage of a rise in world sugar prices. Cuban and Indonesian sugar production had dropped earlier in the 1960s due to revolution and independence struggles in those countries, and the United States eventually reallocated Cuba's sugar quota to other Latin American producers. The price of sugar in 1970 reached a staggering $1,400 per ton (though it dropped to $164 by 1978). Brazilian exports rose from 800,000 tons in the 1960s to a high of 2.8 million tons in 1973 and remained around 2 million tons for the rest of the decade.[8] With the oil shock of 1973 sugar prices rose further, crucially important for cane growers since this spurred government support for alternative fuels. The central thrust of Proálcool was the large-scale production of fuel alcohol, with the goal of reducing the country's dependence on imported oil.[9]

The impact of the new national legislation can hardly be overstated. It helped trigger a sugarcane boom in the midst of a broader process of accelerated modernization in Brazilian agriculture in the late 1960s and 1970s.[10] This agricultural transformation was part of the worldwide "Green Revolution." Brazil experienced greater change than other Latin American countries, a fact clearly visible in the explosion of fertilizer use. In 1961 Brazil used only 8.4 kilograms of fertilizer per hectare, but by 1974 it used 40; the rest of Latin America, by comparison, increased fertilizer use only from 6.5 kilograms per hectare to 12.[11] Forty years later, few careful studies exist of the shape and consequences of the Green Revolution in Brazil. Such studies are far outnumbered by assessments of the country's current agricultural boom, led by soy and sugar.[12]

Even in the context of this impressive nationwide growth in agriculture, nothing approached the incentives given to alcohol production. One of the more concrete measures of the program was the generous opening of credit markets for sugar and alcohol producers; credit made available in rural areas increased between 1971 and 1977 by 1,900 percent.[13] Over the course of the decade, outstanding agricultural loans actually exceeded the total value of the annual agricultural product.¹⁴ All of the easy money was necessary for producers to buy the proliferating "agricultural inputs": fertilizers and protective chemicals. Sugarcane consumed 85 percent of all fertilizers used in the Northeast, and in Pernambuco the crop accounted for a fifth of all herbicides used, a proportion that continued to grow.¹⁵

As an earlier generation of frustrated agronomists could confirm, Pernambuco cane growers had not made heavy use of fertilizers before midcentury. Regular use of nitrogen—one of the most basic and widespread fertilizers—had only begun in 1955, when "Nitrocal" became available. Phosphorus, little used for cane agriculture in other areas of the world, also saw increased use in Pernambuco for encouraging root system development in the cane region's steep hillsides. Potassium, the element directly linked to the synthesis of sucrose, drew interest even more recently than phosphorus and nitrogen. In 1961, the Northeast used 20,000 tons of the standard nitrogen-phosphorus-potassium fertilizer. The next ten years brought a substantial increase, to 95,000 tons. Then in just the next five years—to 1976—it reached 239,000 tons. In percentage terms, fertilizer consumption in Brazil as a whole only increased 200 percent between 1950 and 1964, but it skyrocketed by almost 1,500 percent over the next dozen years, from around 200,000 tons to 3 million tons. By the mid-1970s, northeastern growers used more fertilizer than the entire country had a decade earlier. Similar rates of growth were seen with herbicides and pesticides. By 1974, northeastern cane planters were using 101,057 tons of these, an increase of 600 percent over the previous decade.¹⁶

The substantial shifts in agricultural practice represented a sea change for the *zona da mata*. Expanded education, experimentation, and extension efforts translated into steep industrial growth.¹⁷ The research and industrial support structure that had been so tenuously funded from the 1920s into the 1960s received full funding and came to enjoy a great deal more stability. Staffing the flush cane research centers were the agronomists who had vented their frustration at planter intransigence in the 1950s and early 1960s. With considerable state backing, they acquired greater influence over practices in the fields, approximating the lofty vision of Motta Maia.¹⁸ Planters began to routinely fertilize their fields, a practice whose absence had long been criticized by agronomists, and the increased attention PLANALSUCAR brought to varietal specificity made it unlikely that an episode like the overuse of 3x would be repeated. Planters' new willingness to submit to the wisdom of agricultural science was signaled in a contract between Usina Catende and a planter renting one of the mill's *engenhos*. The agreement explicitly dictated that the renter obey the *usina*'s agronomists in any decision relating to crop cultivation.[19]

The agro-industry even grappled with secular and apparently intractable problems such as topography. Almost three-quarters of Pernambuco's cane-growing lands lie in hilly terrain, and more than half of the state's cane grew on slopes of 20 percent grade or steeper. Grades steeper than 15 percent made it difficult to use tractors, so northeastern producers had to employ more manual or animal power.[20] Because of increased fertilizer use and the continued rapid expansion of cultivation, however, cane could spread to areas in the *zona da mata* traditionally seen as unsuitable. Planters showed enthusiasm for the dry, sandy tablelands of the northern *mata* because the flat ridge tops made it possible to mechanize production. As late as 1958 Mário Lacerda de Melo found planters only tentatively experimenting with the tablelands, though they had trouble devising a fertilizer regime for the areas, telling him that "the [only] fertilizer there is rain."[21] In a 1964 report the Grupo de Estudo do Açúcar (GEA, Sugar Study Group) expressed some ambivalence about this direction of growth. It projected that the expansion would require extensive irrigation, "an over-ambitious objective in the cane zone where, in general, the simplest agricultural techniques are still not observed or implemented." Despite this, it recommended expanding onto the tablelands to the greatest extent possible.[22]

The increased availability of fertilizers and access to the capital to buy them brought a decisive extension of cane onto the tablelands in the 1970s. Traditionally occupied by smaller-scale operations than the south, the northern *mata* saw the establishment of multiple large mills.[23] The tablelands were heavily populated by the already relatively dense standards of the *zona da mata* as a whole, and, traditionally excluded from cane cultivation, these areas had provided food crops for the region. Expansion onto broad ridges, then, further aggravated the region's ongoing struggle to feed its population.[24]

The expansion into dry and formerly inappropriate areas captures several aspects of the impact of this decade's cane agriculture. Cane had been creeping outward for decades, but the availability of credit, the rise in sugar and alcohol prices, and the technology to extend cultivation meant that fields expanded much more in the 1970s. The industry continued to grow, putting 332,412 hectares under cultivation by 1974, almost double the 1952 hectarage, and more than doubling output to 13,244,324 tons of sugar. In the next harvest, planters added 40,000 additional hectares, and by the mid-1980s, they had 542,000 hectares planted in cane.[25] Sugar production increased from

944,000 tons in 1970 to 1.2 million tons in 1980 and 1.4 million in 1986. Alcohol production, which had dropped during the 1960s, rose from 81 million liters in 1970 to 236 million liters in 1980 and 666 million liters in 1986. During the same period production became further concentrated; the number of *usinas* dropped from fifty-three to thirty-eight. In a 1965 article written when the industry still suffered from declining yields and a pinched credit market, Bento Dantas had optimistically predicted that the state's cane industry would one day reach higher levels of production with reduced acreage (through intensive cultivation and more efficient production technology). His hopes were unrealistic; by 1978, well over half of the total land area in the southern zone belonged to *usinas*, while the mills controlled nearly a third of the northern zone.[26]

Planters' awareness of the difficulties presented by Pernambuco's topography and its soil deficiencies grew with the expanded research infrastructure, along with a concomitant faith in and reliance on agricultural science to solve such problems.[27] This marked an attitudinal shift toward energetic state planning for agriculture. The aggressive support for the cane agro-industry stemmed from the state's efforts to expand and fine-tune Brazil's gigantic agro-industrial machine, which highlighted the cane-producing Northeast.[28] The serial construction of new agencies and bureaucracies exemplifies the state's high modernist convictions.[29]

Unintended Consequences for the Agro-Environment

One problem with high modernist projects of economic growth was that they failed to factor in the environment. The boom increased cane acreage and production levels through impressive escalations in fertilizer use, varietal engineering, and funding, but it also increased the destruction of the native forest and the pollution and siltation of rivers and streams, to name just the most obvious ways the environment was affected. *Usinas* cleared between five and ten thousand hectares of forest each year, about 0.3 percent of the entire cane zone. Such a rate clearly compounds rapidly: a decade would suffice to eliminate 3 percent and fifty years 33 percent of the region's entire forested area. A federal law requiring landowners to preserve a minimum of 5 percent of their land in forests went almost entirely unenforced.[30]

The destruction of forest was progressive, constant, and repetitive. One worker interviewed in 2003 had moved to an *engenho* in the 1970s and remarked that an entire valley planted in cane had been forested when he arrived. But the term he used—*capoeirão*—indicated secondary forest rather than first-growth, meaning that at some point decades before, the valley's original forest had been cleared.[31] The GEA, in the same report that recommended expanding onto the northern tablelands, acknowledged that some of these previously uncultivated lands—and especially hilltops—held most of what remained of the region's original forest. The report's authors suggested that planters leave some scraps of the forest but without compromising the cultivation of choice lands and easy access for planting and harvesting.[32] The Pernambuco Development Council, a state agency, began a reforestation program in 1978. But fifteen years later the region's total forest represented less than 5 percent of the original cover. This had been a centuries-long process, accelerated by the rise of *usinas*, but according to environmental researchers it was Proálcool and the 1970s expansion that carried deforestation almost to completion.[33]

Cane expansion and incentives for alcohol also increased mill waste and wastewater. The pulpy cane waste, or bagasse, churned out by the mills' rollers was frequently dumped into rivers. Mills burned as much pulp as possible to generate energy, but expansion and especially the use of the fibrous 3x, meant huge excesses of bagasse.[34] Sugar and alcohol production also demands a great deal of water—water boiled off during the crystallization process and wash water for dirty and charred canes. Over the course of the 1970s, as agricultural inputs rose, the canes also arrived at the mill covered in herbicides and insecticides, which when applied threatened food crops and when washed during the industrial process entered the streams. For every liter of alcohol an *usina* produced, it generated about fifteen liters of wastewater (or *vinhoto*). During the 1964–65 harvest, almost 1 billion liters of mill waste water was pumped into the region's rivers, roughly equivalent to the annual wastewater production of 38 million people.[35]

Some scientists expressed increased concern for water degradation in the mid-1960s. Discharging wastewater had been banned in the 1940s, but (as with forest protection) the measure had never been enforced.[36] Gilberto Freyre directed researchers at the Joaquim Nabuco Foundation to look into the issue, and several studies warned of the dangers of *usina* byproducts, fish kills, and the spread of disease. Schistosomiasis was a particular threat, as the snails carrying the parasite lost many of their natural predators when fish died from high levels of organic material and low oxygen levels in streams and rivers. One study from 1965 claimed that the parasite killed more workers than tuberculosis, then a chronic epidemiological problem in the cane zone. The author accused the mills of spreading "germs that kill the people who gave the mill owners their fortunes" and demanded increased regulation of the industry.[37]

Despite the clear threats to public health associated with degradation, water quality in regional rivers deteriorated further over the next twenty years, reaching critical levels with increased alcohol production. By the 1970s the state's numbers showed a staggering increase in wastewater production, up to 35 billion gallons every year. Encouraging mill owners to find solutions for the pollution problems had little effect, and even a new agency founded in 1976—the Companhia Pernambucana de Controle da Poluição Ambiental e de Administração de Recursos Hídricos (CPRH, Pernambuco Company for the Control of Environmental Pollution and Administration of Water Resources)—struggled to make an impact (the institution is now called the Agência Estadual de Meio Ambiente [State Environmental Agency]). In conjunction with Proálcool, the government took the small step of requiring new distilleries to hold wastewater in reservoirs during the harvest season. These, however, often poured into the rivers if the water level rose with rain. An activist organization called the Associação Pernambucana de Defesa de Natureza (ASPAN, Pernambuco Association for the Defense of Nature) emerged in 1979 and helped put pressure on producers.[38] CPRH and ASPAN persistently advocated for practices such as spraying a wastewater-water mixture directly on cane fields (called "ferti-irrigation"). Tested as early as the 1940s but without taking hold, the process finally began to spread in the 1980s.[39]

In addition to harming water quality, mill waste transformed stream flow. Loads of bagasse dumped in streams, together with the silt eroding from deforested hills, made waterways shallower and less regular. The disrupted regularity of streams aggravated problems associated with increased water consumption, especially during the region's periodic droughts. In 1962 the *Diário de Pernambuco* lamented the region's radical contrasts: "When it is not burdened by drought it suffers the horror of the inundations, the floods that have no mercy in their uncontrollable avalanche and make of the region a permanent theater of trials and tribulations."[40] The newspaper failed to point out the human role in creating, or at the very least aggravating, these problems.

Spikes in water volume from heavy rains were more dangerous in shallow, silt-filled rivers, increasing the frequency of flooding. The region suffered from especially damaging floods in 1940, 1960, 1962, 1965, 1966, 1969, 1970, and 1975.[41] Severino Moura recounted a flood that rolled down the

Sirinhaém River in 1961. At the time he was in charge of agricultural activities for a large part of Usina Pedrosa, and the flood impacted all of the land under his responsibility. "This is how we know the severity of Nature," Moura writes, detailing the destruction of more than five hundred houses, numerous bridges, and several hydroelectric dams belonging to *usinas*. The next day dawned "clean and dazzling, with the greatest indifference, as if nothing had happened with the children of God." The mill director Fernando Cavalcanti was apparently just as indifferent, visiting the site where the flood had swept an *usina* locomotive from its track. He asked Moura the name of the *engenho* he was visiting and in which the direction the *usina* lay from where he stood. Moura wrote incredulously, "How wonderful to be so rich you don't know what you own!"[42] Ignoring the broader drama around him, Cavalcanti concentrated on getting his mill running again.

In July 1975 a large flood arrived in Recife with punishing force, killing one hundred people and leaving sixty thousand homeless. Losses to the cane industry amounted to 60 million cruzeiros. Concerned by the increasing incidence of floods, the state government convened a symposium on the issue with independent technicians. Despite the fact that the cane industry was by far the largest landowner in the region and had exerted the greatest long-term environmental impact, the group's official report never mentioned the industry specifically, limiting itself to criticizing, in one brief passage, the deforestation and erosion associated with "tropical agriculture." Instead, the report sought a large-scale technological solution—a complicated flood control system that required the damming of three major rivers—the same sort of approach that had fomented the massive expansion of the cane industry in the first place.[43]

On the one hand, a flood is simply a flood: any people living near fluvial features will face some threat of floods. On the other hand, changed conditions—the transformation of the environment—can increase their frequency. Floods represented a distinct, "natural" consequence of the cane agro-industry's manipulation of the environment through rampant deforestation, bagasse dumping, and river siltation, not to mention the pollution of waterways associated with sugar and alcohol manufacturing. The state's response to the problem spoke to its vision of the environment and contributed to the development of its landscape discourse. Rather than questioning the agricultural model or enforcing and expanding conservation measures that would mitigate the agro-industry's impact on the environment and connection to flooding, the state chose to dam multiple rivers. The idea that control was possible through technology and systems of oversight and management also extended to the state's management of labor relationships. Water was a resource (and potential energy) to be managed; workers were energy to be managed.

A troubling example of the ethic of control and management that linked the sugarcane agro-industry's environmental and human impacts comes from a 1976 issue of the Pernambuco state development journal. A planner named André Cavalcanti predicted that if *usinas* decreased their discharge of wastewater into streams fish populations might recover, a development that would expand workers' subsistence options and therefore draw them away from the cane fields. The industry would then suffer from increased labor costs, he argued, until rural residents fished enough to thin the fish populations again.[44] Cavalcanti's cynical calculation—weighing environmental destruction and workers' diets against an industry's reliable access to labor—captures the technocratic perspective that envisioned human, "natural," and industrial components of big agriculture as so many columns in a report on productivity.

Unintended Consequences for Labor

The labor component of the agricultural machine actually registered another set of "downstream" consequences unleashed by the 1970s cane boom, some of them linked to the unintended consequence of environmental degradation. Changes in workers' lives were visible both in the context of the *longue durée* of their experiences in the *zona da mata* and in their immediate labor relationships and conditions. Water was crucial to workers for consumption, cooking, and washing. According to 1960s numbers, individual families in the region used about four hundred liters a week. Most of these people retrieved water from an uncovered well, but more than 20 percent used water directly from a river or spring. Since three quarters of families did not have toilet facilities, human waste generally found its way, untreated, into the region's rivers; the same was true of common household garbage. Virtually no families had water running into their homes, and one in five had to walk more than half a kilometer for water.[45] The agro-industry's degradation of the region's river network threatened worker health and basic subsistence.

Also, as was clear from Cavalcanti's 1976 article, workers traditionally had relied on rivers for fish. Gregório Bezerra wrote in his memoir that early in the century his family survived the "dead" months of the year in between harvests partly by fishing in the creek of the Engenho Brejinho.[46] But as Cavalcanti implied, fishing was no longer viable because of pollution. Around the same time he wrote his report, a group of mill workers grumbled to the anthropologist José Sérgio Leite Lopes about declining fish and game. As children and young men, they said, "we went down to the river and the river had a lot of fish; we went to hunt in the forest and there was a lot of game." But things had changed: "We go to fish, all day long; it's better to be home sleeping—you won't catch anything."[47] Workers I interviewed in 2003 complained repeatedly about their impoverished environment and referred to the past as a "time of abundance." One worker said of his childhood, "everything was plentiful. . . . And now, my friend, every effort we make is like a waste." He blamed the overuse of herbicides and pesticides and a long-term drying that he associated with deforestation.[48]

The mill workers' testimony, in conjunction with Cavalcanti's Malthusian analysis and the retrospective testimony of old workers, points to the impact on workers of pollution from the cane industry. Leite Lopes offers a different, cultural interpretation of their words, describing them as a "transposition, for symbolic ends, of social relations onto the field of relations between man and nature." He suggests that this connects to "a specific way of legitimating the organization of labor of the *usina*—a legitimization that is realized in [the industry's relations to nature]—given the subordination of this type of large industry to 'nature,' represented by the seasonal agricultural cycles." Workers make of their social location a naturalized fact. The agricultural process that their work propels is subject to the rhythms of "nature," so they can subtly blame that broader subjection (a form of helplessness) for their own powerlessness vis-à-vis their superiors.[49]

The mill workers' view, as presented by Leite Lopes, is similar to the argument that Gilberto Freyre advances in his framing of systemic inequality in *Northeast*. When he argues that the entire region has been marked by hierarchies that all flow from the dominance of cane over all other plants, he similarly subsumes social inequality into a "natural" relationship.[50] His characterization of cane monoculture as inflicting the "deepest wounds" on the region rests on the same premise. This perspective toward the natural order of things depends on a conflation of environmental processes (and their capricious nature) with the social order.

A perspective that naturalizes hierarchies can similarly naturalize the place of workers at the bottom of those hierarchies. This partly explains scattered comments from workers about the importance and inevitability of being "in one's place." Some of the disgruntled mill workers envied their rural counterparts for supposedly leading freer lives less subject to the captivity of the mill but felt the respective roles were part of the order of things.

One said, "For whomever is accustomed to the field, it's better in the field, no? Because they garden, right?" And more emphatically, "Those from the field don't want to come to the mill, and those from the mill don't want to go to the field."[51] Although cutting cane is difficult, said a field worker interviewed in the early 1980s, "I'm satisfied with my job, this is what was meant for me. God put me here, I can only be satisfied."[52] These comments recall the assertion of a worker quoted in chapter 3 who said of his family's tradition of working the fields, "The people who live here, these people are part of this land. . . . This is why people say, 'Each person in his place.'"[53] In each of these quotations, we can discern the worker's recognition of his place in established hierarchies. Yet the remarks do not link the presence and enforcement of hierarchies to the power of employers or the state; they are natural facts, not engineered ones.

Motta Maia was familiar with this type of worker discourse, "the mentality of those who live on and work the land," he called it, characterizing it as "their passive and almost fatalistic attitude toward the land." In his 1967 paean to agronomists, he argued that part of the realization of their revolution would be the transformation of their perspective in order to "give them a consciousness of dominion over nature."[54] He felt that agronomists carried the power to seize dominion over the natural world, and that they could even teach workers to assume that power. The first part of that formula, at least—the faith in scientific knowledge to control nature—typified the state approach. As for the second part, agronomists might not have taught workers to feel a dominion over nature, but the new structure of their labor relationships did provide them an opening to achieve a greater dominion over their own roles. The labor courts became forums for the expression of these new worker identities, in the same way that they arbitrated women's work and the Task Table.

On March 7, 1980, Usina Catende auctioned two black donkeys, male and female, both nicknamed "Blackie" (Pretinho and Pretinha). The mill eventually remitted proceeds from the sale to the local *junta de conciliação e julgamento* (JCJ, board of judgment and conciliation) to pay the judgment awarded to Emília Maria da Silva in a case that had begun three years before—the same woman whose case from 1975 had paralleled the legal battles of Severina Rosa da Conceição and her daughters. Since 1977, when Emília had returned to the JCJ with a new set of complaints, the case had wound its way through the labor court system, eventually finding its way to the top at the Tribunal Superior de Trabalho (TST, Superior Labor Tribunal). The TST judges supported her claims and ordered Catende to pay restitution. The

15,544 *cruzeiros* deposited for Emília represented back payment for labor benefits she had never received. The court also sanctioned a remarkable request Emília had made to be recategorized as an industrial worker. Even though she continued to perform the same field duties as before, the court compelled Catende to amend Emília's work papers and pay scale.[55]

When Emília had signed a new work card with the mill in April 1976 (which identified her as a "rural worker" for a "cane cultivation establishment") it marked the first time in a decade she had worked *fichado*.[56] Less than a month after she brought her claim in 1977, a flood of other workers arrived at the Catende JCJ with the same request for reclassification. Several of them worked on the same *engenho* that she did (Cana Brava) and most were represented by the same lawyer—Floriano Gonçalves de Lima. Like Emília, many of the plaintiffs had filed a labor complaint against Catende in the past.[57]

After five hearings through the spring of 1977, the Catende JCJ ruled in Emília's favor on July 28. The first part of the judgment—winning back benefits—was the same as the victory she won against the *usina* two years earlier, which had given her back benefits from 1974. But the second part of the ruling approved her new work classification. Gonçalves de Lima based his petition for her reclassification on a 1963 Supreme Federal Tribunal summary ruling stating, "Even if performing a rural activity, the employee of an industrial or commercial company is classified in accordance with the category of the employer."[58] Sugar mills were industrial operations, Gonçalves de Lima reasoned, and therefore *all* of their workers, including those in the fields, were industrial workers. He also drew on a ruling about sugar workers from 1974, when the TST issued a summary ruling dictating that "the agricultural workers for sugar mills form part of the professional category of industrial workers, benefiting [therefore] from the regular wage increases obtained by said category."[59]

Catende refused to accept the decision regarding Emília's classification. The mill's lawyer argued that workers involved with planting, weeding, and cutting cane were manifestly rural workers and had nothing to do with industry.[60] The lawyer's stiffly formal communications with the court barely conceal the mill's incredulity. For planters, rural workers were historically a malleable and utterly disempowered group. It was a matter of keeping things, and people, in their proper places. Catende tried a variety of arguments to convince the court not to reclassify the waves of workers filing complaints. Besides the argument that field tasks did not relate to industry, the lawyer sought refuge in the state's own expanding social service system. He pointed out that social services had legally prescribed links to particular categories of workers; a structure existed for rural workers (the Programa de Apoio à Pequena Produção Familiar Rural Organizada [PRORURAL, Program of Assistance to Organized Small-Scale Rural Family Production]) so they had no need of the agency that took care of industrial workers (the Instituto Nacional de Previdência Social [INPS, National Institute for Social Welfare]).[61]

After losing to Emília in the JCJ, Catende appealed to the regional tribunal in Recife, which in late November affirmed the ruling in her favor. The judges rejected the mill's argument that the 1974 TST ruling was just an opinion "and, being [just] a summary judgment, can be struck down."[62] The mill's appeal was denied, so Catende filed a motion forcing the TST in Rio to consider the case. The TST declared its final refusal to reopen the case in October 1978 with a statement conveying irritation with Catende's reiteration of the same arguments at each stage of the case's progress through appeals.[63] Emília had officially triumphed. The installments of her payment were released at a miserly pace through 1980 (helped by the sale of Blackie and Blackie).In 1981 her new paperwork was finally signed identifying her as an industrial worker. Many of the other petitioners won their cases, too, and the mill was forced to sell off a series of donkeys and even a jeep in order to pay the difference in the workers' salaries.[64]

The distinction in work categories mattered materially, since it raised Emília and her colleagues to a higher pay scale.[65] And on a symbolic level, it helped pull rural workers out of the area of traditional planter dominance and into the modernity implied by industrialization. These cases raise important questions about the form of modernization the state envisioned for the sugarcane agro-industry and for the agricultural sector in general. In petitioning to be recognized as industrial laborers, the workers called attention to recent rulings in the labor judiciary that classified agro-industrial businesses as industries, decisions that finessed the discrepancy between the military regime's focus on industrialization and the reality of a largely agricultural economy.[66]

Emília's case also demonstrates the savvy of the rural unions' lawyers. Floriano Gonçalves de Lima cut his teeth on this work; listed as a legal intern in the court papers for Emília's initial case, by the end of the decade he had his full certification and was successfully ushering dozens of Catende workers through hearings at the JCJ. He had his arguments so clearly outlined that he eventually began submitting mimeographed copies of the demand for industrial reclassification with his clients' complaint forms. The individual unions, supported by their federation (the Federação de Trabalhadores

Agrícolas de Pernambuco, [FETAPE, Federation of Agricultural Workers of Pernambuco]), pursued every potential legal avenue to advocate for the workers.[67] The overall sector had been deemed an industry, allowing some workers to claim the professional category appropriate to this classification, a decisive shift away from the laboring-landscape discourse of the planters. The Estatuto do Trabalhador Rural (ETR, Rural Worker Statute) began a shift away from the norms governed by that discourse, and the 1970s boom completed its displacement by the state's discourse of an ordered, industrialized landscape.

A New Strike

If planters had gained allies in the form of agronomists and agricultural agencies, the unions newly established in the 1960s gained allies in the form of lawyers. FETAPE's legal sophistication showed in the late 1970s, when the rural unions embarked on another bold challenge to the state-built sugarcane juggernaut. In some ways, it appeared that the state securely commanded the cane zone with its regulations and orderly management. Sugar and alcohol production boomed under the oversight of state support agencies. Workers channeled complaints through closely monitored unions, and disputes were controlled by the labor courts. At the end of the decade, however, workers in their orderly unions and lawyers who had honed their skills in the labor courts made surprising use of the space allowed to them within the state's controlling grid.

The state had constructed a dense network of restrictive laws for the rural unions to navigate. The repressive Strike Law was intended to keep a tight lid on worker mobilization. The military regime thwarted an attempted strike of one rural union in 1968, as the dictatorship's regional labor delegate declared the strike illegal and took actions against the union.[68] Undaunted, the rural workers in 1979 mounted a careful movement to prepare a strike at the beginning of the harvest. The organization of the strike took place on several tiers, with the northern *mata* unions of São Lourenço da Mata and Paudalho leading the way and voting for the strike before other unions in the region met.[69] In August the *Diário de Pernambuco* published an admonitory article under the headline "Illegal Strikes Will Be Rigorously Repressed." The president, the article read, had instructed the minister of labor to strictly apply strike legislation in order to prevent stoppages in the so-called essential sectors of the economy.[70]

Having learned from their earlier experiences, the unions took a meticulously legalistic approach and planned a careful route through the "arid terrain" of the Strike Law.[71] They also mounted the strike in the context of widespread labor mobilization during this period of major political transition in the country as a whole as the military regime slowly moved toward democracy. In 1979 there were at least sixty-eight strikes by workers in various sectors, with as many as 2 million total participants. Some of the rural unions began planning for their strike months earlier, during the preparatory meetings for the Third National Congress of Rural Workers, held in Brasília in May 1979. Under FETAPE's leadership, the rural unions organized *engenho* by *engenho*, training and preparing delegates elected by the workers. The unions explicitly included both contracted workers and *clandestinos* in their preparations for the strike, a key strategy if they were to mobilize the workforce as a whole. And during August and September the unions held the series of votes required under the Strike Law.[72]

The sugar workers used the restrictive strike code to their advantage, demonstrating publicly and ceremoniously their adherence to the letter of the law. Their "political utilization of the labor law"—putting public pressure on both producers and the state—helped them reach a collective agreement during the strike.[73] Workers remember the strength of the unions and the exhilaration of the struggle. Cícero, now a foreman, remembered the period as a time of constant union vigilance: "Whatever little thing, the union would act," he said. About the period leading up to the strike, he said, "Everyone stopped working in that damned pressure I remember; many were working and others came in and stopped them. Everyone went to the mill gate to fight for wages."[74] The head of the rural workers' union in Ipojuca told the *Diário* that workers were bringing back "the true sense of unionism" with their efforts. Adding to the political drama of the growing movement, Miguel Arraes returned from fifteen years of exile under a blanket amnesty offered by the federal government. In a triumphant appearance in front of his mother's house, the sugar workers' old friend declared "we will be the thread that will stitch together the unity of the exploited masses."[75]

Though Cícero and his colleagues marched to Catende's gates, they and many other unions did not technically go on strike. In mid-September, about seventy thousand workers from around the cane zone voted to strike, but as with the initial order of the union assemblies, the São Lourenço and Paudalho unions struck first. Around twenty thousand workers from the two municipalities refused to work and carefully organized pickets and *engenho* "commands" to maintain unity. Union and FETAPE leaders began negotiations with producers and after four days of these, an accord favorable to the workers was signed. The strike won a 52 percent wage increase, guarantees from producers that they would adhere to provisions of labor law, and the adoption of an improved Task Table (the workers had hoped to simply revert to the table of 1964, but the bosses refused). The anthropologist Lygia Sigaud helped organize during the strike. She writes that after the announcement of the agreement, worker families "could hardly contain their happiness. Some cried, others danced, others shouted, others simply laughed."[76]

As the union organizing for coordinated action gathered steam in August and September 1979, the government anticipated problems with the state-level labor department and replaced the hated regional labor representative with a less polarizing figure. As a whole, though, the labor ministry, the secret police, and the military police apparently did not expect the organized movement and strike with which they were eventually confronted. In subsequent years, though, they monitored union activities extremely carefully. As it turned out, they faced a predictable routine. These were years of redemocratization, when Brazil was transitioning away from military government and back to direct elections, transparency, and popular engagement. The sugar workers surely benefited from this "opening," as the period was called.

The 1979 strike established a pattern for conflict and negotiation that proved quite durable. Each year, the unions planned a strike for the beginning of the harvest season, timed to occur just before the end of the last year's contract in the first week or so of October. Supported by the Confederação Nacional dos Trabalhadores na Agricultura (CONTAG, National Confederation of Workers in Agriculture), FETAPE, and volunteers, the movements continued to be characterized by strong solidarity among the workers and excellent coordination at the Federation level.[77] Police files show that the authorities tried to match this level of organization. During the 1980 strike the Departamento de Ordem Político e Social (DOPS, the political and social police) sent out investigators to compile reports, supplementing their intelligence with numerous newspaper clippings. They tracked the unions obsessively and carefully divided responsibilities for monitoring the strikers. Photocopied pages from a text on labor law were circulated to officers to ensure they were familiar with the regulations governing strikes and could pounce on any breach by the unions.[78]

Police files covering the 1984 strike include a large map of the *zona da mata* used for organizing surveillance of union activities and enforcement of the Strike Law. Five color-coded regions have been highlighted on the map, with each color corresponding to a particular agency's assigned territory.

AN AGRICULTURAL BOOM   195

The DOPS, the federal police, the Fourth Army, the military police, and the civil police each covered an assigned list of mills. The map also represents each of the rural unions with a dot, colored either black or green in accordance with the given union's stance toward the government—green for "favorable" and black for "contrary" (with one yellow "undecided"). The police map showed a dense cluster of black dots in the *mata norte* and a more heterogeneous black and green pattern throughout the south. For the region as a whole, black dots outnumber green by a ratio of two to one.[79]

The political geography represented by the map is not surprising. That the strength of the unions lay in the northern *zona da mata* makes sense when one considers that the 1979 strike had originated in the north, with the São Lourenço and Paudalho unions, and the next years followed the same pattern. But this distribution diverges from the 1963 strike—the last great mobilization of the cane workers. The pattern then had been the opposite; while the entire region was paralyzed in November 1963, adherence to the movement was strongest in the large umbrella union based in the southern municipality of Palmares. Most scholars have interpreted this as a political fact, only acknowledging in passing that the southern *zona* was the more "industrialized" cane region. As I have pointed out, this shorthand description stands for a great deal: in the aggregate, workers in the south faced pressure of a different level than those in the north. In some ways, 1979 was the same story, but inverted.

The northern *mata* had been the region organized by the Catholic Church and had suffered less intervention after the military coup, partly due to Padre Antônio Melo's intervention and mostly because of the military's greater distrust of the Communists. But in addition to this political reality, the north experienced greater transformation during the 1970s agricultural revolution than the south—the extension of cane onto the tablelands shifted long-term patterns of mixing wage work and subsistence farming. The greater change from gendered work customs also separated the north from the south. Finally, the northern region, receiving less rainfall, has less reliable water flow and lower stream volume than the south. The north experienced more severe flooding during the 1960s and 1970s (the three rivers dammed following the 1975 flood were in the north). Interestingly, São Lourenço da Mata was the municipality most affected by destructive floods in 1977.[80] These factors together—agricultural, cultural, and environmental more than political—explain the regional picture of union participation and political alignment.

The "cycle of strikes" begun in 1979 played out on two stages—the *engenhos* throughout the cane zone and the bargaining table in Recife. On the *engenhos*, workers bore the responsibility for maintaining pickets and preventing colleagues from returning to work. Strikers were subject to threats and violence by planters' gunmen, but the strength of the negotiators' position at the bargaining table in Recife depended on the unity of the strike commands in the fields.[81] The strikes ended in one of two ways: either the negotiators reach an agreement within the stipulated five days of negotiations, or the question was sent to the regional labor tribunal and a contract was imposed on both parties. The 1979 and 1985 strikes, for instance, resulted in accords, but in each of the five intervening years the negotiations failed.[82]

While the 1963 strike represented the culmination of a nearly decade-long period of progressive victories at the polls and political ferment in the *zona da mata*, the 1979 strike reflected careful planning and a scrupulous attention to the regulations stipulated by the labor ministry. The workers saw the successful negotiation with (or against) the producers as "one more blow against the private power of the bosses."[83] Indeed, by the 1980s disputes ranging from the common complaint about back pay to the annual contract negotiations were hashed out in the public arenas of state power: the local labor courts and the labor delegate's office in Recife. This does not mean that planters abandoned the use of violence. During the early 1980s, an average of more than four rural workers were killed every year in the sugar zone, almost all union leaders.[84] Compared to the impunity of previous eras—especially the time before Arraes and the ETR—the hearing of disputes in public forums must be considered a victory. Certainly the ability to muster a strike and force a negotiated agreement was an achievement, considering the repressive power the state brought to bear in the aftermath of the coup. Workers voted overwhelmingly with their feet in supporting these achievements. In 1985 nearly 83 percent of surveyed workers belonged to their rural unions.[85]

Sigaud invests the annual negotiations between the unions and industrial associations with transformative potential. The unions, she argues, demonstrate power by forcing planters to the table. Once there, each time the proprietors agree to a point, "such acceptance signifies not just the registering of a claim, it implies symbolically the recognition of the workers' power to take the initiative and propose/impose norms relative to the structure of relations between employees and bosses." These moments of recognition can consecrate or normalize de facto rules or bring new measures into existence. Sigaud does hint at the limits of the negotiations when she says that the planters misread the dynamics of the 1979 strike, agreeing to an accord only because they were "betting on the inefficacy of the instrument they were approving."[86] The planters may have been surprised by the lengths to which the unions would go to ensure enforcement, but they largely followed through on their threat to ignore the measures established during the negotiations.

Marcelo Rosa analyzes the strikes from a similar perspective, situating them in their historical context. Though the unions' activities during the military period were basically "juridical battles against the large property owners who disrespected their legal obligations to their employers," Rosa suggests that they were radically defiant given the limitations of the political opening. Leftist leaders continued to endure persecution, he writes, and "their actions for the maintenance of labor rights unfolded along the tenuous limits of the legal standards established and imposed by the government."[87] That the workers and their union leaders managed to carry off successful annual strikes, abiding by the constraints imposed on them but not becoming discouraged by those limits, should be considered a great victory. Indeed, Sigaud published her 1980 book *Greve nos engenhos* in part because she felt Brazilians did not realize the significance of the rural workers' accomplishment. Many Brazilians were focused on the large urban strikes taking place, including the pioneering effort by the metalworkers in São Paulo (famously led by Luiz Inácio Lula da Silva, who was elected president in 2002).

The planters had their own strategy to adapt to the strikes as the new mode of addressing labor relations. Since the dictatorship commanded the situation, the planters sought to find their own advantage. The U.S. political scientist Anthony Pereira argues that the strikes served planters interests, since mill owners used the yearly mobilizations to draw attention from the state and apply pressure for higher sugar prices.[88] Federal Deputy Marcus Cunha, of the Partido do Movimento Democrático Brasileiro (PMDB, Brazilian Democratic Movement Party), looked into precisely this issue in 1984, opening a parliamentary commission of inquiry to investigate allegations that the sugar industry was in crisis. Cunha alleged that planters used trumped-up claims of economic difficulty to extort higher prices for their commodity from the government. He further accused planters of using the "false crisis" as an excuse for the violent and inhuman conditions suffered by their workers.[89]

Crisis may have been the wrong word, and the planters' blaming their difficulties on high labor costs and appealing to the state may have been misguided, but the agro-industry's boom was coming to an end. Despite the impressive numbers surrounding Pernambuco's sugar and alcohol production in the 1970s, it continued to lose ground to the state of São Paulo. São Paulo's efficiency perpetuated discrepancies in labor costs between the two states, and the flatter fields, better management of productive cane varieties, and easier access to capital in the south further separated the two. In 1972 São Paulo planted nearly twice as many hectares of cane as Pernambuco; by 1986 that ratio rose to nearly four.[90]

By the state's own later evaluations, the 1970s programs that fueled the boom had dismal results. The industry had plenty of warnings. In the same period from the early 1950s to the early 1970s, during which sugar production nearly doubled and per hectare cane yields rose 15 percent, for instance, the industrial yield (sacks of sugar per ton of cane) dropped by 20 percent.[91] The 1961 Expansion Plan of the National Sugar Industry had aimed to achieve sixty tons of cane per hectare, a goal it still had not met twenty-four years later (when Pernambuco's rate stood at fifty-three tons per hectare). In the decade between the 1972–73 harvest and the 1984–85 harvest, the Northeast's sugarcane agro-industry did increase its per-hectare cane yields by 10 percent, but the center-south increased by more than 25 percent. Worse, the industrial yield dropped between 1950 and 1985, from 104 to 95 kilograms of sugar per ton of cane. Less than 10 percent of funding from the grand sugar aid program begun in 1977 went to the agricultural side of the industry.[92]

Besides the comparative lack of attention to the field sector as opposed to the mills, the disappointing progress likely had to do with the marginal areas into which cane expanded. In spite of the poor results of many aspects of the 1970s programs, the industry had continued its growth. Pernambuco's acreage planted in cane grew more than 70 percent between 1970 and 1980. By the end of the 1980s it had grown another 18 percent.[93] By that point, though, a new economic model was on the ascent in Brazil as the democratically elected President Fernando Collor ushered in an age of neoliberalism. Business regulation and price supports did not fit the new development model and the Collor administration dissolved the IAA and PLANALSUCAR in 1990.[94]

Conclusion

In 1919 the wife of new *usina* owner Pessoa Queiroz recommended that he change the name of the stream running through their property to Cruanjí, which in the indigenous Tupí-Guaraní language means "Stream of Constant Waters." In a superstitious effort to defeat the drought threatening his *usina*, he followed her advice and also took more practical measures. He dammed the stream where it rolled out of wooded hills on his lands, collecting the water in a reservoir. Then he rerouted the stream down a valley adjacent to its historic course and directly toward his *usina* and a power generation station. Pessoa Queiroz's grandson José Guilherme Azevedo Queiroz now runs the much expanded Usina Cruanjí. He has hydraulic operations of his own underway, including a large new reservoir to increase the supply of water available to his growing mill. Talking about his plans in 2003, José Guilherme repeated several times the lesson he gleaned from his grandfather's efforts: "He who controls water and energy controls the world."[95]

José Guilherme's refrain captures the essence of the state's aspirations during the 1970s as it confronted environmental obstacles such as water scarcity and flooding and sought to control a particular type of energy: labor power. But the security of control can be illusory. A cautionary example comes from José Lins do Rego's novel *Usina*, set around the same time as Pessoa Queiroz's efforts, with a plot that bears an interesting similarity to his story. The narrator Carlos's uncle Juca dreams of building an *usina* on the site of the family's old *engenho*. He engages in an epic struggle to redirect the flow of a creek for power and irrigation, sending it to his mill and to cane fields instead of through his workers' garden plots. Juca "would spend a fortune" on the stream, Lins do Rego writes, "to place it in servitude, at his orders, like a submissive and useful prisoner."[96] At the end of the book, however, the water strikes back; Juca watches a flood destroy his proud new mill. In standard Lins do Rego style, the story serves as an argument in favor of maintaining the traditional social structure. In the first of his sugar novels, *Plantation Boy*, a flood rolled through Carlos's grandfather's *engenho*. But this wise patriarch, with his single *engenho*, his family, and his loyal workers, reacted with equanimity, recognizing "the wealth the river had brought him in the form of rich slime to fertilize his fields."[97] Juca's profiteering hubris led him away from his father's smaller-scale paternalism. The self-consciously modernizing 1970s state had sweeping ambitions to control industry, development, even the environment. But like Juca, it found that its capacity for control had limits.

In this chapter I have described the ambitious state-led growth of the 1970s, which pulled the sugarcane agro-industry out of its economic funk with easy credit and grand production incentives. The state's confidence in the efficacy of technocratic approaches to agriculture extended also to its approach to labor, and I have analyzed here the management of labor disputes. The various plans, decrees, and pronouncements emanating from agencies and organs of the state amounted to a new landscape discourse that bore some similarities to the planters' discourse even as it displaced the latter. The state's apparatuses for making agriculture and the labor system legible produced some blind spots. Their approach degraded the environment and sparked worker protest. Measured purely by the occupation of hectares by cane or the production of sacks of sugar and liters of alcohol, the agricultural revolution that Motta Maia had hoped for did take place. But his intrepid agronomists did not achieve their accomplishments through the enhanced "telluric love" that he predicted. Instead, cane expansion came at the cost of fetid streams, ash-filled skies, slashed forests, and increased floods. And the workers' mentality was transformed, not necessarily in relation to nature but in terms of their collective sense of power. Working within the state's own rules and structures, they achieved greater control over their own roles (though not the power to guarantee the security of their jobs). Their clear-eyed sense of the limits constraining them and the possibilities available to them facilitated a broad mobilization and a successful battle against producers and the state.

CONCLUSION

Power, Labor, and the Agro-Environment of Pernambuco's Sugarcane Fields

The *zona da mata* is approaching five hundred years of sugarcane cultivation. Remarkably, the most dramatic environmental change in the region during the twentieth century was the expansion of cane. Remarkable considering that this was likely the case for the three preceding centuries as well. Cane occupies over a third of the region's land, and clearing for the crop has destroyed all but the smallest fragments of the native Atlantic Forest and eaten into former food crop land and pasturage. The expansion of railroads and the rise of the *usinas* starting in the 1880s brought a new level of environmental impact to the *zona da mata*, opening more and more land to cane. Writing as a witness to this period of transformation, Gilberto Freyre lamented the wounds that cane culture had inflicted on his region. In the 1980s, the impact reached a new peak. Faulting the national alcohol program (Proálcool) for alarming levels of alcohol wastewater in the region's rivers, the chemical engineer Antônio Geraldo Brandão Alves declared to the *Diário de Pernambuco* that "we are living a biological cataclysm, total devastation."[1] His comments seemed prescient several months later when the largest distillery wastewater spill in the state's history took place, a nationally reported ecological disaster that affected the Capibaribe and Pirapama Rivers.[2]

The discourse of landscape produced by Freyre and his generation privileged planter power in commanding the agro-environment and labor. The *usinas* spread under the hands of powerful families confident of their authority, and the agro-industry grew impressively into the 1940s and 1950s. In the early 1960s, however, the planters faced radical challenges from workers and an agricultural crisis partly of their own making. The state took command in 1964 and directed a new cane boom in the late 1960s and 1970s, guided by its own discourse of landscape. The map used to coordinate surveillance and repression of the 1984 cane workers' strike provides a concrete as well as metaphorical example of the state's vision. The large map assigned territories to various police forces and showed municipal boundaries, the locations of mills, and dots colored to indicate unions' degree of political friendliness to the dictatorship. As instruments of representation that reflect a particular way of seeing, maps carry no power in and of themselves, but they can show how political actors who "deploy the perspective of [a] particular map" conceive of and wield power.[3] In the landscape of the late-twentieth-century Brazilian state, technocratic control mattered, from the small scale of Task Tables that guided workers' daily assignments to the large scale of Proálcool. The state operated with the conviction that workers could be contained by bureaucracy and the agro-environment could be perfectly managed through improvement programs, statistical analyses, and maps.

As it turned out, the state's approach failed just as the planters' traditional approach had. The carefully annotated map itself marked this failure, since it was produced as a tool for containing a strike that challenged the legitimacy of the regime's system of labor control. Alves criticized Brazil for being "bound by the belief that progress means unlimited growth," and he cautioned that "the unrestrained search for maximum growth makes us forget the importance of achieving good growth."[4] Far from achieving "good growth," the state's rationalization efforts polluted the environment and sparked the backlash of a massive labor mobilization. Like the planters before it, the dictatorship thought it commanded the landscape. In both cases this control was an illusion. These failures indict the very notion of command as an approach to environmental or labor management.

In this book I have told the story of workers in Pernambuco's sugar zone and their changing agro-environment. By exploring these two themes together, I have sought to enrich our understanding of politics, state dynamics, and labor mobilization. Beginning with a long view of the region's history, I have shown how sugar agriculture (and to a lesser degree cotton) progressively transformed the environment. The *zona da mata* historically was taken as a model of quasi-feudal labor relations, in which planter power and control had no rival.[5] The state's attempt to create central mills in the late 1800s failed, and even the transition away from slavery was managed by the planters rather than the government. Then railroads and *usinas* spread and extended cane's reach, transforming the environment and the social relations of agricultural production.

Divergent discourses of landscape crystallized during the transition away from slavery and the emergence of the new production model. Planters described a laboring landscape meant to bend to their will, while workers spoke of a landscape of captivity, in which they struggled to achieve a degree of freedom for themselves. The planters' discourse, an aspect of their class *habitus*, encompassed a terrain in which they exercised unquestioned control over both the natural and human worlds. Workers found themselves captured by a landscape redolent of planter power, a predicament confirmed in their daily experiences of subjection to planter demands and reflected in regional folklore, which portrayed cane fields as sites of fearsome power. They associated freedom with a relationship of trust and loyalty with the boss, which facilitated access to land and the privilege of using it as they pleased. The two discourses obviously helped structure each other, locked as they were in a tense mutual opposition.

Outside groups such as agricultural professionals, state officials, social scientists, and reformers came to the region to advise, oversee, study, or intervene. They gradually opened up the region to new influences and an uneven process of rationalization in the 1940s and 1950s. The state's first meaningful intervention came in 1933 with the creation of the Instituto do Açúcar e do Álcool (IAA, Sugar and Alcohol Institute), which set the tone for decades of halting agricultural modernization. Evidence of the disjuncture between the ideas of agronomists and planners, on the one hand, and planters' traditional practices, on the other, came with the headlong expansion of 3x, which had disastrous consequences as its slow maturation and planters' early cutting drove down industrial yields and overall production.

Recognizing the expansion of 3x as a mistake, and seeing the broader consequences of that misstep and those associated with it, requires understanding the details of varietal management and parsing the nitty-gritty of payment schemes in the fields. These complex processes were not uniform across the cane zone, so we must also look at intraregional variation and discrepancies in longer-term patterns of land access, food growing, and more. Together, these factors make up the shifting conditions faced by workers. While the attempts to rationalize production processes did result in the expansion of cane, they also had ripple effects in social relations. Besides being increasingly displaced from land, workers experienced a shift from standard customs to the increased pressure of production-based payment systems that tied them to tasks for longer period of time. Workers resisted the increased demands in a bid to preserve flexibility and control over their own time, including through the "strike for lower wages." Running parallel to these events was the particularly interesting case of fire, which saw the transformation of a natural environmental phenomenon into a highly politicized discourse, followed by its instrumentalization into the most basic and routine of processes. Worker resistance and the new political opening associated with President João Goulart and Pernambuco Governor Miguel Arraes led to worker mobilization and a massive, unprecedented strike. This confrontation with the old planter discourse of the laboring landscape, coming as planters were already reeling from the passage of the Estatuto do Trabalhador Rural (ETR, Rural Worker Statute) and the 3x fiasco, created an opening that led to a greatly increased role for the state.

The military took command in 1964 and approached the region's modernization convinced of its ability to manage the challenges that had defeated the planters. This ethic, an effort to harness the region to a vision of transformative control that was yet another discourse of landscape, was on clear display in the agronomist J. Motta Maia's exhortations to allow agricultural science to unleash a revolution. The effort was pursued with a deep conviction of the power of science and planning to overcome all challenges. Antônio Alves succinctly captured the state's belief in growth as progress, what Warren Dean described as the "development imperative."[6] Systematic state intervention largely opened up the planters' domains of power by the 1970s, as workers became state clients as well as wage earners. Workers no longer saw themselves entirely as captives of cane and planter power, in the way that the worker Manoel do Ó had described so poignantly life in the late nineteenth century.

At the same time that the ETR increased labor costs and provided planters with a further reason to restrict workers' access to land, cane fields expanded and occupied some of the space earlier reserved for workers. This conjunction consolidated rural proletarianization. In contrast to the predictable trajectory, this did not take place through a process of labor market privatization. Rather, workers exited the profoundly private realm of the plantation and entered a regime of state regulation and oversight. For many workers, the state stamped its approval on their new status in surprisingly concrete form, awarding them industrial classification on their work cards. Increasingly, women secured their own work cards, or at least worked with a contractor for their own wages rather than as adjuncts to fathers or husbands.

The 1970s boom offers a remarkable parallel to the experience in the 1960s. In both cases, modernization and agricultural rationalization provoked worker responses that culminated in strikes. The agricultural contexts that characterized the periods leading up to the two strikes were not identical, but both mobilizations must be understood in the context of prevailing agricultural circumstances. As with the planters' discourse in the earlier period, the dictatorship's effort to command through technocracy rested on a faulty assumption of the state's capacity to control. Diverted by its illusion of command, the state had helped create the conditions for a bold challenge to its control over labor. Laws outlined the boundaries of behavior and channeled complaints through rural unions and the court system. Workers could bring work complaints to the labor courts with representation by the union, or the federation could make broader appeals to the regional labor delegate. This system reinforced collective organization and encouraged unions to master labor law.

The workers' experiences contributed to the deepening of class identity as they came to see themselves as a clear group with shared challenges and aims. Steve Striffler describes a comparable process that took place on the southern Ecuadorian coast from the 1930s through the 1960s. He argues that peasants in this banana region were shaped into a class through dialogue with the state. They were "organized by and into the state," Striffler writes, becoming increasingly dependent and occupying a space "that could be managed, granted rights, denied resources, disciplined, or simply ignored." Pernambuco's sugarcane workers occupied juridical and political spaces (the unions) and engaged in a dialogue with the state over benefits and services, while battling producers in the state's courts. These engagements transformed the workers just as the Ecuadorian peasants were transformed. The coup had not been a complete defeat for the rural labor movement, though as in Ecuador workers' options were narrowed to paths defined by their interaction with the state.[7]

The unintended consequence for the state was that this new class, recognizing itself as such, proved capable of an admirable level of collective action that challenged not just workers' conditions in the cane fields, but the legitimacy of the state itself. The workers were driven to action through changes in their laboring conditions—increased expulsions from *engenhos*, decreased access to land, and frustration with the environmental damage they witnessed as a product of the cane agro-industry. The strategies they pursued depended on their relationship with the state. As they inhabited roles established by the dictatorship, workers came to recognize themselves as a class, rather than as an atomized group of individuals working on separate plantations.[8] The dire environmental consequences of the state's approach to modernization, ranging from pollution to destructive floods, also gave the lie to the dream of authoritarian efficiency and control. In a final ironic twist, the increased production levels of the period came not through enhanced yields or efficiencies but through the simple expansion of cane agriculture—an old, rather than modern, method if ever there were one. The dictatorship's approach had failed miserably, and the cane zone's stinking rivers and energized unions testified to that fact.

After the Strikes: Rights and Land

In June 1981 a group of twenty-eight families living on Engenho Goitá Grande in the northern *mata* negotiated with the *engenho* owner, Cristina Cavalcante de Luna, to buy their land. The families had lived on the *engenho* between twenty and forty years, and in 1981 most paid Cavalcante rent for the small plots where they grew food crops and cane. In the wake of the 1979 and 1980 strikes, the federal government had established the Programa Especial de Apoio às Populações Pobres das Zonas Canavieiras do Nordeste (PROCANOR, Special Assistance Program for the Poor of the Sugar Zones of the Northeast), with the responsibility of providing credit to family-scale farmers for buying land. It was this program that the Goitá Grande tenants planned to use to finance their purchase. The deal hit a snag, however, when a man named Artur Correia de Oliveira showed up claiming to be the new owner and began forcing the families off the *engenho*. In response to the dispute, the Federação dos Trabalhadores na Agricultura do Estado de Pernambuco (FETAPE, Federation of Workers in Agriculture in the State of Pernambuco) sent urgent letters to the governor, the Instituto Nacional de Colonização e Reforma Agrária (INCRA, National Institute for Settlement and Agrarian Reform), and the police. FETAPE president José Rodrigues da Silva's description of the conflict and his appeals to the authorities show that some aspects of rural relations in the cane zone had changed markedly over the previous decades, while others had remained the same.[9]

Rodrigues da Silva complained that Correia de Oliveira had created a climate of terror and was forcing renters to sign eviction agreements. Though the police had looked into the matter, they had failed to reestablish a secure environment for the families. Correia de Oliveira's "ostentatiously armed" henchmen remained posted around the property even though, as Rodrigues da Silva pointed out, he could not invoke the rights of property since he was not the legitimate owner of the *engenho* but simply sought to intimidate the poor renters. "For the sake of argument," the union leader wrote, "even if [Correia de Oliveira] were owner of the land, he wouldn't be owner of the people, nor endowed with police powers." To the governor, Rodrigues da Silva accused the interloper of "improperly using the name of the Fourth Army [posted in Recife] against the workers, and declaring that he 'commanded' [*manda*] in all of the official spheres of Power." He noted that the Goitá Grande families produced food for the regional market, a clear point of interest for a state that saw inflated food prices as a trigger for rural unrest. He also argued that the families' attempt to buy their plots fit the aims of the military regime's Land Statute and adhered precisely to the contours of the recently established PROCANOR.[10]

The physical repression suffered by the *engenho* residents appeared to fit a pattern that extended deep into the cane zone's history. The heavily armed enforcers were playing a role that Gregório Bezerra remembered the ex-slave Pelegrino playing when Bezerra was a child on the Engenho Brejinho. Though Rodrigues da Silva explicitly countered the idea that Correia de Oliveira could be "owner of the people" even if he were the actual owner of the *engenho*, the FETAPE leader's comments recall Bezerra's description of a new *senhor* arriving at Brejinho and assuming ownership as "lord of the big house, of the mill, and of everything, including the inhabitants."[11] The renters' struggle also fit a secular pattern of attachment to land, stretching from slaves' production of food on *engenho* plots to the coveted *roças* of the *morada* system and to workers' resistance to being expelled onto "the street."

Correia de Oliveira's justification of control over the land through his ability to command also resonated historically, harkening back to a time when *senhores de engenho* measured their clout by the reach of their command. A crucial difference, however, lay in the authority that he invoked; it was not just his own word that carried power; he also claimed that he had pull in all "official spheres of Power," including the Fourth Army. For their part, the renters had recourse to a higher power, too, through their union federation. It was partially due to Bezerra's efforts in the 1960s that such an organization existed and could act effectively in the workers' interests. Though the police appear to have responded to an earlier FETAPE complaint, their files do not contain responses to the July 1 letters. Still, the very fact that a union federation could support rural working families trying to buy their own land, and defend them against aggression, sets this situation apart from earlier periods.

Violence and coercion remained all too common during the 1980s, but transparency and visibility had also increased. For example, one of the Goitá Grande renters, Fortunato dos Santos, gave an interview to Globo television about Correia de Oliveira's actions.[12] Episodes of repression were much more likely to be investigated and adjudicated than sixty or forty or even twenty years earlier.[13] Indeed, repression had increasingly shifted to the state, as opposed to the private sphere. This process had begun earlier, manifest in such episodes as planters' demand in 1960 that Governor Cid Sampaio respond to cane fires. The expanding role of the state brought shifts in the workers' and planters' conditions and social relations that had consequences for their modes of mobilization. The strike of 1979 and those in following years showed workers demonstrating powerful solidarity and collective voice. After that demonstration, what defined the horizon of freedom for workers? How had changes over the previous decades transformed workers' aspirations?

Proálcool continued to support cane ethanol production, and the rising supply of alcohol fueled more and more cars and trucks, just as the program's founders had hoped. Over 85 percent of the cars sold in Brazil between 1984 and 1989 ran on ethanol as their primary fuel. In 1989 more than 92 percent of new cars sold ran on alcohol (or 80 percent of all vehicle sales), and the country had a total of 4.5 million cars running on pure ethanol.[14] At the end of the 1970s, when auto companies responded to ethanol's increased availability and began mass production of engines that ran on pure alcohol, an engineer gloomily predicted that sugarcane workers' fate had been sealed: "Now the wheels of the entire world's cars will turn on the hunger of the Northeast."[15]

The engineer meant that alcohol's predominance would never allow workers access to land, and landownership did remain extremely concentrated in the region. The military regime's Land Law of 1964 had little effect and was followed by no other major land reform legislation. In 1980 properties of fewer than ten hectares accounted for more than three quarters of private landholdings yet occupied only 6 percent of the land. Only 5 percent of properties were larger than one hundred hectares, yet these estates accounted for four-fifths of the land area. Almost precisely this percentage of the *zona da mata*'s land—81 percent—was occupied by cane, while less than a fifth was devoted to food crops such as corn and manioc.[16] In 1985 only about a quarter of workers still had access to land on *engenhos* for subsistence crops;

another 17 percent had access to land elsewhere (probably on the cane zone's periphery).[17]

Some in the labor movement faulted ethanol production for completing workers' alienation from land.[18] But this had been a long-term process, and earlier signs had pointed to changing worker desires. Padre Paulo Crespo's failed two-hectare campaign in 1965 was one such sign. Even though the peasant leagues in the 1950s had been predicated on securing land, and the rural unions incorporated land reform as a central demand, by the end of the 1960s union leaders had acknowledged that their own pursuit of land reform was out of step with the rank and file. When given a hypothetical choice in a 1985 survey between a better wage and a piece of land, the workers were about evenly split. But they overwhelmingly identified wages as contributing most to their subsistence. A study from the time suggests that although a strand of "mythical" thinking about land remained—a deeply held belief in its ability to bring freedom and independence—workers generally saw themselves as wage earners.[19] In light of these observations, the Goitá Grande renters' fight appears more like the last gasp of an impulse for land than a new battle in an active war.

As former union leaders in Vicência, the old workers Paulo and Severino had keen insight into the worker struggles of the 1980s and 1990s. And above all else, they emphasized workers' efforts to protect or increase their wages. Discussing the 1979 strike, Severino said, "So they sat and began to negotiate the wage." He paused and corrected himself. "Not just the wage but the question of the Task Table, which was very important, the question of how much for clearing brush, how much for planting cane, how much for cutting cane."[20] When strikes became annual events after 1979, the negotiations focused more and more on wage levels and less on other benefits or on land access. Discussing these mobilizations, Paulo said simply that they aimed to protect wages. Then he too stopped and corrected himself: "It isn't just the wage, it's the *rights*." A wage could be set at any level, he pointed out, but without an established Task Table a planter could demand from workers any level of work in order to earn their wage.[21]

Severino and Paulo recognized the unions' interest in demanding that all labor rights be respected—the Task Table, the thirteenth wage, holidays, indemnification. But in describing the unions' strategic use of strikes—their primary tool for exerting pressure—the two men treated these benefits almost as afterthoughts. This is also how they treated demands for land. Severino's and Paulo's comments reveal the extent to which workers focused on wages alone in their ongoing confrontations with planters.[22] Neither wage raises nor access to land were meaningful without enforcement of the Task Table, which protected against exploitative task sizes and the abuse of workers' time.[23] But planters consistently violated the table and failed to provide benefits, forcing workers continually to seek justice in the courts. In an open letter to the *Diário de Pernambuco* in 1983, union officials summed up their members' plight and the planters' strategy: "Problems of social order such as the expulsion of *moradores* from their subsistence plots, incompliance with labor law, etc., began long ago, as can be seen through thousands of labor complaints judged in the Labor Court since 1963."[24] Through a war of attrition, workers were worn down until many simply hoped to be paid wages regularly, something that one scholar has called "the end of the peasantry."[25]

The end of the military regime was marked in Pernambuco by a highly symbolic echo of an earlier political moment when Miguel Arraes was reelected governor in 1986. But the return of the workers' idol failed to revive the political moment and objectives of that previous time of struggle and success. Arraes's administration did not bring higher wages or systematic land redistribution. Severino actually felt that conditions worsened with the return of civilian government and said he preferred the dictatorship because then he knew who the enemy was: the bosses. Afterward, he felt corruption was everywhere, not just among exploitative planters.[26] This period also brought the brutal economic realities of neoliberalism. When the cane industry lost subsidies and price supports, eighteen *usinas* and distilleries closed in Pernambuco, leaving only twenty-three in the region.[27]

Some efforts to claim land emerged to take advantage of the political and economic moment. The Movimento Sem-Terra (MST, Landless Workers' Movement) and FETAPE began organizing in the region and leading squatter occupations of unproductive land.[28] Wendy Wolford suggests that the MST had to overcome sugar workers' conceptions of access to land that placed it wholly within "the hierarchical ordering of plantation space. . . . The rural workers conceived of full employment and paternalistic protection on the plantation as the ideal rather than ownership of a small plot of land."[29] A downward spiral of sugar and ethanol prices during the 1990s helped draw unemployed workers' interest in the occupations, but the fragility of their commitment to small farming on the settlements was demonstrated when many workers left to work for *usinas* again when sugar prices rose in the early 2000s.

Severino and Paulo both complained about the lackluster commitment to land among their younger colleagues. Born at a time when access to land meant a type of freedom, Severino remained convinced of its emancipatory potential. His descriptions of land frequently sounded like slogans: "Land is life with abundance"; "The garden is a prepared meal"; and "Working in your field means having a full stomach."[30] But he was disappointed in his fellow workers, especially those younger than himself, who he thought had "stopped having a feeling for the land."[31] Paulo agreed, saying people no longer knew how to work the land—"There isn't this love [for the land]," he said, that comes from intimate experience.[32] Though he acknowledged that land reform settlements and occupations represented an advance in the fight for land, he said that even those in the camps "only know how to work for wages . . . and wait for the government to send money."[33] Severino said rural workers did not understand the value of land, and he angrily described a failed occupation he led in 1998, when he could not convince his followers to hold out in the face of the landowner's resistance. "The people [here] are weak," he said. "Land is no longer the answer for these people."[34]

It is ironic that the fight for land reform was undermined by a continued cane monoculture, the very same crop that had inspired the extreme concentration of land in the first place. But there are ironies in Severino's and Paulo's comments as well. They expressed their disappointment partly as nostalgia; their longing for a time when people had "a feeling for the land" sounds in many ways like the nostalgic discourse of Freyre and Bello for a time when *senhores* had a love of the land. In 2003 I asked a foreman on an *engenho* in the *mata norte* whether he would prefer to own land or have a secure job. "Ah, everyone would like to have his own little piece of land," he said, "but I never had one nor can I. I have to live my whole life just working other people's land, no?"[35] His pragmatic answer reflected a clear-eyed sense of economic hierarchies. Without any realistic chance of ever owning land, why even discuss the idea? Twenty years after the Goitá Grande struggle, workers' priorities appear to have shifted, though as Wolford has shown new impulses can emerge quickly.[36]

The *Zona da Mata* after Sugar?

A particularly revealing episode at Usina Catende during the 1990s and 2000s offers insight into the *zona da mata*'s present and future. Catende, with its nationally prominent past, is now a workers' cooperative. In chapter 4 I told the story of Tenente Azevedo's stewardship of Catende, when it showcased the greatest changes associated with the rise of *usina* production in the 1930s and 1940s. For a time the largest *usina* in the country, Catende served as a model for others and earned praise from the IAA for its innovative spirit. However, the company approached collapse in the early 1990s and reemerged as a hybrid enterprise partially under worker control. The mill had gradually declined over the decades, failing to weather the end of artificial price supports in 1990 and direct competition with efficient producers in the south (both of which came with the abolition of the IAA). Faced with debts exceeding the property's value, the owners fired twenty-three hundred workers in 1993 and refused to pay back wages or indemnification, then began auctioning trucks, tractors, and other salable pieces of the business. The workers protested, forcing the owners to abandon the mill, and filed a bankruptcy claim for the mill since they owned a great deal of debt in the form of unpaid wages. The owners filed a competing claim and the mill fell under court control in 1997.[37]

Rural union leaders and their allies established a cooperative called the Companhia Agrícola Harmonia, intending it to take over once the mill exited bankruptcy. A decade later, however, Catende still found itself mired in legal wrangling over the transition. The Banco do Brasil, as a major creditor, assigned a former *usina* manager named Mário Borba to run Catende, and he worked with the directors of Harmonia to operate the mill and coordinate production in the fields. Workers shared their opinions with their colleagues and the management team at regular meetings held on *engenhos* and in the town of Catende and its neighboring villages.[38]

In 2003 Catende functioned with 1,350 workers holding professional cards and an additional 1,300 during the harvest (down from more than five thousand workers when the mill was economically healthy).[39] The mill's particular crisis fit within the larger downturn, which had also affected workers. Increased herbicide use and mechanical collectors, among other things, had cut down on labor needs during the 1980s. Using chemicals instead, planters began to dispense with one of the traditional two "cleanings" or weedings of cane fields. They also began to order more cane cut "loose," meaning workers did not bundle the cane after cutting. This alone reduced labor needs for cane cutting by nearly half, since the daily task size for wrapped cane was 1 to 1.2 tons per worker and 2 to 2.4 tons for loose cane. The mechanical collectors that gather cane in the fields and load it onto trucks for transport to the mill also trim labor needs. Combining the labor savings for cutting and transporting cane with the new techniques in the 1980s, planters eliminated an estimated forty-six workers per day per hectare.[40]

The democratically run, ecologically sensitive cooperative of Harmonia was structured as a response to the historical problems of exploitation and underemployment. With union involvement and transparent management, the company provided badly needed jobs and a chance for self-determination. The management team began "Cana de morador," a program to support worker production of cane by distributing plant cane and fertilizer to workers on credit so they can grow cane on their plots of land on the *engenhos*. They can rent more land or negotiate with their neighbors to increase their holdings. More and more workers joined as sugar prices rose, and in 2003 the program's two hundred participants were supplying more than half of the mill's cane.[41] "Cana de morador" carries great symbolic weight and the potential to displace norms of worker subjugation and exploitation in the region. If planters historically disapproved of workers planting fruit trees because it would give them too great a sense of ownership (and potentially self-empowerment), then planting cane—the crop of power and control—surely grants even more of a sense of autonomy.

Despite the apparently revolutionary potential of Harmonia, many workers criticized it. In interviews with a dozen workers, I frequently heard complaints about how the management team exercised its authority. Workers spoke about demands that they do certain types of work, at certain times and in certain places, about the supply of cane or the withholding of pay, and said these were reminiscent of the ways planters traditionally treated workers. A significant minority of workers appeared to feel as though they were being treated as they would by any given boss.[42]

And "Cana de morador" has the troubling effect of reinforcing the hold of a crop that has dominated the region for centuries. The focus on cane, a union leader said, was crippling the Companhia's crop diversification projects.[43] Cane dominates life so completely that workers, like their elite ancestors in the region, cannot imagine a future without sugar; they still believe in it. Juan, Catende's Cuban field manager who is much beloved (if not always understood) by workers, showed a clear passion for cane. "Do you know how many things we make from sugarcane in Cuba?" he asked, before quickly answering, "More than three hundred!" He ticked off products ranging from fuels and animal feeds to cosmetics and cure-alls.[44] Juan hopes that Catende will grow into a model for worker-run enterprise in the region, though his enthusiasm for cane makes him sound like boosters of the crop from traditional planter groups. A representative of the Pernambuco Sugar Suppliers' Association, for example, told me the region's destiny had always depended on cane and remained tied to it. "If it weren't for sugarcane," he claimed, "the entire Northeast would be a desert."[45]

Perhaps he was right. Brazil as a whole has profited from cane. In the first decade of the twenty-first century, a new ethanol boom gripped the country, driving the proliferation of new distilleries and a rapid expansion of cane fields. The worldwide pursuit of renewable energy sources and a growing debate about "green economies" contributed to ethanol's rise. But it was also spurred by the familiar specter of rising oil prices, especially the dramatic spike in 2007–8. In 2006–7 Brazil produced 4.6 billion gallons of ethanol, and over 80 percent of passenger cars sold in the country that year were flex fuel cars that run on ethanol or gasoline or any mixture of the two—an echo of the 1980s.[46] Plans for major, agriculturally based fuel production are afoot the world over as the ethanol industry continues to grow in Brazil.[47] The country's success earned praise and interest from abroad.[48]

Most of the growth in ethanol production has taken place in Brazil's south and center-west. Over 80 percent of the cane agro-industry is now located in the states of Minas Gerais, Rio de Janeiro, and São Paulo; 70 percent of production takes place in São Paulo alone.[49] Production in the south continues to be more efficient and competitive, even drawing migrant labor from traditional cane regions of the Northeast such as Paraíba, Alagoas, and Pernambuco. Pernambuco's proud leadership of the nation's cane industry may be long past, but the state remains committed to cane. Will the decisive efficiencies of other producing regions force Pernambuco to wean itself from cane? Will Catende serve as the model that so many hope, of worker self-determination and crop diversification? Will the view of the Cuban field manager Juan take hold, that sugarcane is the crop of the past *and* the crop of the future—incredibly efficient in its use of sun and rain and destined to provide sustenance, energy, and more to Brazil's workers?

Cícero, a foreman on a Catende *engenho*, spoke skeptically about plans for diversifying the *usina*'s agricultural production, developing orange and coffee trees and potentially farming fish. "They've already tried [diversification] three or four times," he said. He shook his head and said "the security here is always cane."[50] He said of people on the *engenho* where he lived, "You see here people who used to plant big gardens [*roça*]—cassava, manioc—today they don't want to. A mango tree, they took it down; they ripped up banana trees to plant cane." It was causing problems at the market, he said, since prices were higher for food, and fewer people grew their own.[51] Even with access to land and the freedom to plant whatever they wished, some people remained captives of cane. Pernambuco and sugarcane are still united in the twenty-first century. Perhaps the region really is wedded to the crop in the way people have claimed emphatically for centuries—it is the *zona da mata*'s "natural vocation."[52]

Landscape proves its worth as a conceptual tool for narrating history by helping us to understand this durable bond. People's vision of their location—of the possibilities and realities of their lives—are structured by landscape discourses. The concept allows us to think about the material impacts of these discourses, weaving together cultural threads with the environment itself. Even the modernizing state had a version, though it did not have a José Lins do Rego to express the discourse in literature. Its controlling technocratic approach led to the dictatorship's failures. Landscape is "a construction, a composition of [the] world"; it is a *way of seeing* the environment.[53] People see their surroundings in ways that relate to their ideas of identity, power, and social organization. Decisions about changing the landscape thus will always derive from or result in social change. Outlining landscape discourses means attending to how social structures impinge on the material environment. Examining these discourses improves our understanding of the motivations and perspectives that guide people's actions, and helps explain why cane's roots in the *zona da mata* run so deep. Deep enough that planters and workers alike can hardly think of a life without it.

NOTES

Abbreviations

The following abbreviations are used in the notes.

APE	Arquivo Público Estadual Jordão Emerenciano (Pernambuco State Archive), Recife
A-TRT	Arquivo do Tribunal Regional de Trabalho, Vitória de Santo Antão
CEHIBRA	Centro de Documentação e de Estudos da História Brasileira, Fundação Joaquim Nabuco, Recife
DOPS	Departamento de Ordem Político e Social
FGV-CPDOC	Centro de Pesquisa e Documentação História Contemporânea do Brasil, Fundação Getúlio Vargas, Rio de Janeiro
FJN	Fundação Joaquim Nabuco, Recife
JCJ	Junta de Conciliação e Julgamento
NARA-US	U.S. National Archives and Records Administration, Washington, D.C.
RG	Record Group

Introduction

1. Freyre, *Nordeste*, 18.
2. Freyre, *Mansions and Shanties*, 11.
3. Nabuco, *Abolitionism*, 111.
4. The worker-centered novel is Lins do Rego, *Moleque Ricardo*. Early in the book, Ricardo travels to Recife and begins work in a bakery.
5. The sparse literature on the abolition period marks a profound difference between the literature on Pernambuco and the literatures for southern Brazilian states, particularly Rio de Janeiro and São Paulo. Scholars have published extensive work on those state's experiences during the transition in the late nineteenth century from slavery to postslavery. See, e.g., D. A. S. de Moura, *Saindo das sombras*; Mattos de Castro, *Das cores do siléncio*; Chalhoub, *Visões da liberdade*. It is interesting that one of the few studies of this period for Pernambuco was written by an American: Galloway, "Last Years of Slavery."
6. A text capturing the political effervescence of the time partly through first-person reportage is Page, *Revolution That Never Was*.
7. Tad Szulc, "Northeast Brazil Poverty Breeds Threat of a Revolt," *New York Times*, October 31, 1960.
8. Levinson and Onís, *Alliance*, 244–46; "Kennedy Will Begin Brazil Visit July 30," *New York Times*, May 25, 1962.
9. Pernambuco's importance to the national movement of rural workers is demonstrated by the fact that a Pernambucan—José Francisco da Silva—was president of the national

confederation of rural unions (CONTAG) from 1968 through 1989. A. Pereira, *End of the Peasantry*, 6.

10. Marquardt, "Pesticides, Parakeets, and Unions," 33. Biorn Maybury-Lewis's chapter on Pernambuco in *The Politics of the Possible* and Anthony Pereira's *The End of the Peasantry* are two English-language contributions to a well-developed bibliography on the rural labor movement.

11. See the acknowledgements in Sigaud, "Percepção do salário," 317. A year before she died, Lygia Sigaud wrote an analysis of the anthropologists' experience. Sigaud, "Collective Ethnographer." Members of the working group called Estudo Comparativo do Desenvolvimento Regional included Moacir Palmeira, Lygia Sigaud, José Sérgio Leite Lopes, Afrânio Garcia, Beatriz Heredia, Marie-France Garcia, Luiz Maria Gatti, Roberto Ringuelet, Vera Echenique, and Rosilene Alvim. Some of the monographs that came out of the project included Sigaud, *Clandestinos e direitos*; Leite Lopes, *Vapor do diabo*; Heredia, *Morada da vida*; and A. Garcia, *Terra de trabalho*.

12. Scheper-Hughes, *Death without Weeping*. Nancy Scheper-Hughes described her book as "a descent into a Brazilian heart of darkness" (xiii).

13. Dabat, *Moradores de engenho*.

14. "Relatório," June 11, 1960, in SSP 29343: Ligas Camponesas; engenho Malemba (Paudalho), Acervo DOPS, APE.

15. The Pindobal workers' actions also illustrate Ted Steinberg's arguments that power "operates through and across landscapes" in much the same way that people agree it functions along the axes of race, class, and gender. Steinberg, "Down to Earth," 802–3.

16. "Relatório dos acontecimentos delituosos," May 14, 1960, in SSP 29343, Acervo DOPS, APE.

17. Raffles, *In Amazonia*, 61, 8.

18. Cosgrove, "Prospect, Perspective," 50. Olwig, "Recovering the Substantive Nature of Landscape," 634–37. Schama, *Landscape and Memory*, 10.

19. Much scholarship on landscape over the past several decades, within geography but also by historians, has taken place in a tradition pioneered by Carl Ortwin Sauer, a Latin Americanist geographer who taught at Berkeley early in the twentieth century. Under Sauer's influence, landscape became the basic unit of geography as a discipline. He called it "an area made up of a distinct association of forms, both physical and cultural." Landscape was displaced by Richard Hartshorne, who disliked the cultural and even psychological overtones of landscape and centered "region" instead. Sauer, "Morphology of Landscape." Hartshorne, *Nature of Geography*. Harvey, *Condition of Postmodernity*, 203–7. Lefebvre, *Production of Space*, 73–76.

20. Thomas Greider and Lorraine Garkovich define landscape simply as "the symbolic environments created by human acts of conferring meaning to nature and the environment, of giving the environment definition and form from a particular angle of vision and through a special filter of values and beliefs." Greider and Garkovich, "Landscapes," 1. Crumley, "Historical Ecology," 6. Balée, "Historical Ecology," 16. Schama, *Landscape and Memory*, 6–7. Busch et al., *Making Nature*, 4.

21. Sartre, *Philosophy of Jean-Paul Sartre*, 89. Following Martin Heidegger, Sartre's perspective here does not separate living and working into discrete categories. Heidegger wrote: "The manner in which we humans *are* on earth is . . . dwelling. To be a human being means to be on the earth as a mortal. It means to dwell." Heidegger, "Building Dwelling

Thinking," 339. Casey, "How to Get from Space to Place," 39; Basso, "Wisdom Sits in Places," 54.

22. Strohmeier, "Wild West Imagery," 268.

23. Ibid., 258.

24. Tuan, quoted in DuPuis and Vandergeest, *Creating the Countryside*, 18.

25. Cosgrove, *Social Formation*, 19.

26. Pred, *Making Histories*, 126.

27. Williams, *Country and City*, 105.

28. Raffles, *In Amazonia*, 53–68.

29. For the idea of a "labored landscape," see Gidwani, "Labored Landscapes."

30. White, "Are You an Environmentalist?," 173. Sackman, "'Nature's Workshop,'" 29.

31. Worster, *Dust Bowl*; Marquardt, "'Green Havoc,'" 33.

32. This study takes the advice of Donald Worster's well-known article outlining the basic tenets of an "agro-ecological perspective" toward history. Besides paying attention to the physical environment itself—what lies "out there"—and the ways people manipulate it, Worster argued that historians must also pay attention to the ideas guiding people's interactions with nature. Worster, "Transformations of the Earth," 1087–93.

33. Braudel, *On History*, 3, 48–52.

34. Funes Monzote, *From Rainforest to Cane Field*, see, e.g., 5, 86, 90–99, 128.

35. Soluri, "People, Plants, and Pathogens," 485. Soluri, *Banana Cultures*.

36. Roseberry, *Coffee and Capitalism*, 8–10; 82–96; Mintz, *Worker in the Cane*; Mintz, *Sweetness and Power*; Wells, *Strawberry Fields*.

37. Dean, *With Broadax and Firebrand*, 245. See also Dean, "Green Wave of Coffee."

38. Striffler, *In the Shadows of State and Capital*, 60–81.

39. French, "Latin American Labor Studies Boom," 280–81n8.

40. Batalha, Teixeira da Silva, and Fortes, *Culturas de classe*; Fortes, *Nós do Quarto Distrito*; Fontes, *Trabalhadores e cidadãos*. The national history association (ANPUH) has a working group, Worlds of Labor, which embraces comparative work on unfree and free labor.

41. French, *Drowning in Laws*.

42. Negro and Fontes, "Using Police Records," 15–22.

43. Dean, *Brazil and the Struggle for Rubber*; Dean, *With Broadax and Firebrand*; Pádua, *Sopro de destruição*; Funes Monzote, *From Rainforest to Cane Field*; Drummond, *Devastação*. José Augusto Drummond has focused on training a cadre of environmental historians. The four conferences of the Sociedad Latinoamericana de Historia Ambiental (SOLCHA) have been held in Santiago de Chile, Havana, Seville, and Belo Horizonte.

44. A recent collection of essays on Latin American environmental history addresses three themes: human occupation and conflict over territory, environmental impacts of particular commodity production, and conflict between specific environmental knowledges. In their introduction, the editors acknowledge omissions, several of which I will study in this book. Brannstrom and Gallini, "Introduction," 12–20.

45. Rodney, *History of the Guyanese Working People*.

46. Steinberg, "Down to Earth," 802. See also Soluri, "People, Plants, and Pathogens," 467. Cynthia Radding has also made important innovations with her use of political ecology as a framework for understanding environmental and social dynamics of Spanish colonial frontier areas. Radding, *Landscapes*, 5. Radding, *Wandering Peoples*.

47. Peck, "Nature of Labor." Richard White writes compellingly at the beginning of his

study of the Columbia River, "We no longer understand the world through labor. . . . once the energy of the Columbia River was felt in human bones and sinews; human beings knew the river through the work the river demanded of them." David Igler's history of large-scale ranching in the early twentieth century in California's San Joaquin Valley brings close attention to labor relations and conditions together with an analysis of the ranches' environmental impact. White, *Organic Machine*, 4. Igler, *Industrial Cowboys*.

48. "Relatório dos acontecimentos delituosos," May 14, 1960, in SSP 29343, Acervo DOPS, APE.

49. For earlier examples of this line of thought and an analysis of Nabuco's abolitionist-environmentalism, see Pádua, *Sopro de destruição*, 272–80.

50. The reference to *habitus* is an obvious cue that Pierre Bourdieu's ideas form the basis for some of the third chapter's arguments.

51. Albuquerque, "Weaving Tradition."

52. Julião, *Cambão*, 92.

Chapter 1

1. Tollenare, *Notas dominicais*, 102, 62.

2. This chapter owes its approach in part to Fernand Braudel's idea of nested conceptions of time. He felt that historians had to move back and forth between deep time—the *longue durée*—and shorter treatments to produce "a history of gentle rhythms, of groups and groupings." Braudel, *On History*, 3.

3. G. S. de Souza, *Tratado descritivo*, 47. According to the eminent nineteenth-century Pernambucan historian Pereira da Costa, the first person to catch sight of this cape was the Spanish sailor Vicente Yañez Pinzon, one of Columbus's crew on the voyage of 1492. Eight years later, while trying to return to the Caribbean, he was blown southward and saw a cape that he called Santa Maria de la Consolación. Souto Maior and Dantas Silva, *Paisagem pernambucana*, xii. L. Teixeira, "Roteiro," 11.

4. Dean, *With Broadax and Firebrand*, 6. According to the authors of *Dossiê Mata Atlântica*, the forest reached all the way to Rio Grande do Norte and even had small patches in Ceará. They estimate that the forest at one point covered 17 percent of the current territory of Brazil. *Dossiê Mata Atlântica*, 6–7. Vasconcelos Sobrinho, *Regiões naturais*, 62. Francis Dov Por suggests that the Atlantic Forest should actually be considered one of the Amazon's "parent floras," since it predates the Amazon by millennia. Por, *Sooretama*, 15. Lima, *Estudos fitogeográficos*, 21.

5. Brandão, *Diálogos*, 34, 118. Alfred Crosby has remarked that "people of the sixteenth century were not statistically minded, so their estimates of the numbers killed by epidemic disease may be a more accurate measurement of their emotions than of the numbers who really died." The same observation might be made for descriptions of landscape. Crosby, *Columbian Exchange*, 43.

6. Grant, *History of Brazil*, 8; Cardim, *Tratados da terra*, 43–44.

7. Tollenare, *Notas dominicais*, 104, 61.

8. Cardim, *Tratados da terra*, 53, Tollenare, *Notas dominicais*, 103. The three trees used for these examples are, respectively, *pau d'arco*, *acicapugá*, and *tatajuba*. The last is also called *pau amarelo*.

9. Cardim, *Tratados da terra*, 52, 50, 40.

10. Por, *Sooretama*, 69.

11. Dean, *With Broadax and Firebrand*, 29; Schwartz, "Indian Labor," 46.

12. Andrade and Lins, *Pirapama*, 84.

13. *Zoneamento agroecológico do Nordeste*, 1:31. G. O. de Andrade, "Teoria."

14. G. O. de Andrade, "Teoria,'" 17–19; *Zoneamento agroecológico do Nordeste*, 1:31.

15. *Zoneamento agroecológico do Nordeste*, 2:106–7. *Várzea* is sometimes translated simply as "flood plain," but I retain the Portuguese here in order to preserve the richness of the local vocabulary for the sugar landscape.

16. Antonil, *Cultura e opulência*, 105. Perruci, *República das usinas*, 100. C. A. Taunay, *Manual do agricultor*, 45.

17. Freyre, *Nordeste*, 42. Shawn Miller found that certain types of forests and specific species of trees were thought to correlate to the presence of *massapê*. "Some species were so widely accepted as indicators of superior soils that they were denominated *árvore padrão* (standard tree) and were often left standing in the fields like billboards, advertisements of soil quality for possible future buyers, or simply as status symbols." S. W. Miller, *Fruitless Trees*, 34. Warren Dean documents the same phenomenon in south-central Brazilian coffee country. "A reconnaissance was carried out to find tree species that were regarded as *padrões*—indicators—of superior locations for coffee groves," he writes, after which the entire forest was destroyed to prepare for planting, save a few garlic trees. "These were left in place, because they were accepted as the surest of all *padrões* and thus might be displayed to an estate's potential purchaser as proof of the productivity of the groves." Dean, *With Broadax and Firebrand*, 181–82.

18. Quoted in Prado, *Colonial Background*, 31.

19. The characteristic feature of the *mata seca* is the *tabuleiro*, which is more technically called a "low coastal plateau" and translates roughly as "tableland" (as a plural). *Mesorregião da Mata pernambucana*, 36.

20. M. C. de Andrade, "Caracterização geo-econômica," 33.

21. *Mesorregião da Mata pernambucana*, 36.

22. *Programa de ação*, 38.

23. Andrade and Lins, *Pirapama*, 30–31; Koster, *Travels in Brazil*, 1:225.

24. Caminha, "Letter," 33; Cardim, *Tratados da terra*, 31.

25. Shawn Miller emphasizes the importance of the Portuguese experience, coming from a "near-barren metropolis, a world away from Brazil's unimaginably large timbers and vast woodlands," in structuring the colonists' reaction to Brazil's forests, as well as the management policies they instituted. S. W. Miller, *Fruitless Trees*, 46.

26. Grove, *Green Imperialism*, 29–31.

27. Nieuhof, *Memorável viagem*, 68. Moacir Vasconcelos's translation of Nieuhof's observation is more poetic: "Pôsto que situado entre a linha equinocial e o Trópico de Capricórnio, sujeito, portanto, à canícula abrasadora dessas latitudes, o calor é aí consideràvelmente amenizado pelos ventos de leste."

28. Emma Juliana Smith to Sophy Smith, Pernambuco, January 20, 1844, from Emma Juliana and John P. Smith Letterbook, 1843–1845, Rare Book, Manuscript, and Special Collections Library, Duke University.

29. Nieuhof, *Memorável viagem*, 296.

30. Frequently it is not only the sentiment that is repeated but the exact words. Vaz de Caminha is repeatedly invoked as a prophet of Brazil's agricultural glory. See, e.g., A. Fernandes, *Senhor de engenho*, 88.

31. Brandão, *Diálogos*, 11.

32. Nieuhof, *Memorável viagem*, 297.

33. Grant, *History of Brazil*, 6.

34. "Fixar justo preço para o açúcar é dever do govêrno," *Diário de Pernambuco*, May 12, 1963.

35. Souto Maior and Dantas Silva, *Paisagem pernambucana*, xiii.

36. Charles Darwin stopped in Recife with the *Beagle* and was stuck for two weeks by calm weather (much to his chagrin; he did not appreciate the city's charms). He took the opportunity to study the reef structure that formed the city's harbor, a formation he found fascinating. "The Writings of Charles Darwin on the Web," <http://pages.britishlibrary.net/charles.darwin/texts/beagle_voyage/beagle21.html>, accessed October 2, 2003.

37. G. S. de Souza, *Tratado descritivo*, 46.

38. Hans Staden, on board that ship, grumbled that the threatened residents had themselves to blame for their predicament, the Caeté "having been to that point tranquil." Staden, *Primeiros registros*, 40–41.

39. As with the shift from *A Terra de Santa Cruz* to *Brasil*, the name *Nova Lusitânia* fell prey to a locally oriented identifier. The early settlers found an island several leagues north of the cape (Itamaracá), separated from the mainland by a calm canal. The name for the area—Paranambuco—came from the local indigenous words *parana* (sea) and *mbuk* (arm): the canal was the arm of the sea. M. Melo, "Ensaio sobre alguns topônimos," 140. Schwartz, "Indian Labor," 47–48.

40. Galloway, *Sugar Cane Industry*, 50.

41. Varnhagen, *História geral*, 1:181. Dantas Silva, "Açúcar," 16.

42. For advocates of sugar's importance and Pernambuco's centrality to Brazilian history, see Sociedade Auxiliadora, *Trabalhos do Congresso Agrícola*, 69; Milet, *Lavoura da cana*, 17, 20; *Açúcar*; and Freyre, *Casa-grande e senzala*, 34.

43. Galloway, *Sugar Cane Industry*, 11, 62. Until then, it had been called simply *cana*, there being no other varieties to distinguish it from.

44. Vasconcelos Sobrinho, *Regiões naturais*, 252. M. C. de Andrade, "Caracterização geo-econômica," 33.

45. Dean, *With Broadax and Firebrand*, 77.

46. Quoted in ibid.

47. Such naming is not incidental. Robert Mugerauer observes, "Words, including names, are not merely labels, but the evocation of what things are and [of] how they are related to other things in the web of particular lives and places. And more than names and words, there is language itself, which is not any fanciful artifact, but that which has the power to articulate and join humans to plants, animals, and activities in a surrounding world. The entire fabric of a people's meaningful world—the total environment—comes along with the whole of that people's language." Quoted in Cloke, Philo, and Sadler, *Approaching Human Geography*, 90.

48. Galloway, *Sugar Cane Industry*, 72; Schwartz, "Indian Labor," 47–50. J. Miller, "Numbers, Origins, and Destinations of Slaves," 395. Lockhart and Schwartz, *Early Latin America*, 206, 216.

49. S. R. de Moura, *Memórias de um camponês*, 53.

50. C. A. Taunay, *Manual do agricultor*, 108–9; Galloway, *Sugar Cane Industry*, 73; Watts, *West Indies*, 402.

51. Dean, *With Broadax and Firebrand*, 29.

52. M. C. de Andrade, *Rios-do-açúcar*; G. S. de Souza, *Tratado descritivo*, 47.

53. Cardim, *Tratados da terra*, 295. Schwartz, "Indian Labor," 57. An arroba is about twenty-eight pounds. G. S. de Souza, *Tratado descritivo*, 46. Lockhart and Schwartz, *Early Latin America*, 211.

54. Cardim, *Tratados da terra*, 291, 295.

55. Gonsalves de Mello, Introduction, xliii. Lockhart and Schwartz, *Early Latin America*, 202. Dean estimates that the total exports of the Brazilian sugar sector in 1600 amounted to ten thousand tons. Dean, *With Broadax and Firebrand*, 79.

56. Brandão, *Diálogos*, 111–12. Miller finds the complaints about brazilwood scarcity hard to believe. S. W. Miller, *Fruitless Trees*, 35.

57. Brandão, *Diálogos*, 10. One of the most systematic and extended pieces of propaganda extolling the fertility and abundance of Pernambuco, the *Diálogos das grandezas do Brasil* appeared in 1618. Structured in the form of a six-day-long dialogue between Brandônio, a Brazil enthusiast, and his skeptical friend Alviano, the book provides a partisan account of Brazil's potential. Brandônio provides a *capitania* by *capitania* description of the virtues of the colony, moving from north to south. He admits to being focused on the North, having never visited the South, and indeed the dialogues are oriented mostly toward Pernambuco, where the Brandônio character obviously lives. And Alviano, although his opinion of Brazilian soil is such that he "took it for the worst in the world," recognizes the fame Pernambuco had "acquired in the world, for its greatness, richness, and abundance in everything." Brandão, *Diálogos*, 129, 31, 11. In his preface, Leonardo Dantas Silva provides a précis of the debate over the book's authorship. The current consensus among historians is that Ambrósio Fernandes Brandão wrote it after having lived in Pernambuco for twenty-five years, mostly as a landowner in São Lourenço da Mata. José Honório Rodrigues adds that Brandão also lived in Paraíba managing sugar plantations during the first part of the 1610s. Honório Rodrigues, *História da história do Brasil*, 373–74.

58. Andrade and Lins, *Pirapama*, 151.

59. Watts, *West Indies*, 389. Barickman, "'Bit of Land,'" 657, 669.

60. Andrade and Lins, *Pirapama*, 152; Nieuhof, *Memorável viagem*, 289–90. Nieuhof wrote that Nassau's orders to plant one thousand manioc plants every year brought down the price of *farinha* (ground manioc) by several orders of magnitude. The problem of food production had been present before the Dutch and continued long afterward. The Portuguese unsuccessfully attempted to facilitate food production by Indians with land grants to the *aldeias* established for them by Jesuits. Schwartz, "Indian Labor," 54. Cardim wrote about a famine in 1583 caused by a drought. Cardim, *Tratados da terra*, 292.

61. Nieuhof, *Memorável viagem*, 17–19, 82, 294. Nieuhof lived in Brazil from 1640 to 1649.

62. Andrade and Lins, *Pirapama*, 19.

63. Brandão, *Diálogos*, 125; Cardim, *Tratados da terra*, 283; Antonil, *Cultura e opulência*, 130. Shawn Miller has used Cardim and Antonil along with *engenho* records from Bahia to calculate these numbers. S. W. Miller, "Fuelwood in Colonial Brazil," 184.

64. Dean, *With Broadax and Firebrand*, 80.

65. S. W. Miller, "Fuelwood in Colonial Brazil," 184.

66. Antonil, *Cultura e opulência*, 170.
67. Saint-Hilaire, quoted in Varnhagen, *História geral*, 1:92–93.
68. Nieuhof, *Memorável viagem*, 297–98.
69. S. W. Miller, *Fruitless Trees*, 9.
70. Ibid. Viewed in eighteenth-century terms, in which forest utilization trumped forest conservation and certainly preservation, Miller argues, the tragedy of the Atlantic Forest is that it was not simply deforested but destroyed, without any benefit to the Brazilians. Ibid., 6, 9.
71. Miller also notes the continued abundance of forests deep into the nineteenth century. Ibid., 35.
72. Brandão, *Diálogos*, 123.
73. Nieuhof, *Memorável viagem*, 297.
74. Antonil, *Cultura e opulência*, 126–27.
75. Antonil, actually a pseudonym for Andreoni, followed other well-informed Jesuits in describing the sugar industry. We have already seen some of Padre Cardim's sixteenth-century observations. The Jesuits were far from detached and disinterested with regard to the industry. They established sugar plantations of their own in colonial Brazil, including some of the colony's largest. Antonil's book was based in part on his experiences observing the workings of the Jesuits' *engenho* in the Bahian Recôncavo. *Cultura e opulência do Brasil por suas drogas e minas* was a catalog of Brazil's major export commodities, where they were to be found or how they could be extracted or grown, prevailing prices, and the details of management. By far the largest section is on sugar, but Antonil also covered tobacco, precious metals and diamonds, and cattle ranching. Though he probably finished the book before the turn of the century, it only cleared the complex, multitiered censorship process for publication in 1711. Ironically, after all of that it was immediately recalled by the Church. During such a tenuous time, when the security of the colony was by no mean guaranteed, it was seen as inconvenient to have a book in circulation that detailed how to build wealth in Brazil. Honório Rodrigues, *História da história do Brasil*, 394; A. de E. Taunay, "Antonil e sua obra," 20–21.
76. Antonil, *Cultura e opulência*, 68. As cited above, Shawn Miller covers the issue of deforestation associated with *engenho* consumption for Bahia. S. W. Miller, "Fuelwood in Colonial Brazil."
77. Antonil, *Cultura e opulência*, 71 68, 106, 71–72, 94, 105, 68, 130.
78. Barickman, "'Bit of Land,'" 657.
79. Watts, *West Indies*, 393–405.
80. Andrade and Lins, *Pirapama*, 62–63.
81. Galloway, "Agricultural Reform," 767.
82. The watery metaphor is purposeful, echoing what is in some ways the standard discourse about the crop. Leonardo Dantas Silva, in his essay for a 2002 exhibition on sugar in Pernambuco, writes, "The sugar field, as if it were a river overflowing its banks, expanded through the floodplains, consumed the small hills, spread up the slopes, filled the horizon with cane green, replacing the green of the tropical forest." Dantas Silva, "Açúcar," 16.
83. Dean, *With Broadax and Firebrand*, 79. Richards, *Unending Frontier*, 392.
84. Cardim, *Tratados da terra*, 295; Nieuhof, *Memorável viagem*, 17; Antonil, *Cultura e opulência*, 170; Galloway, *Sugar Cane Industry*, 77.
85. Eisenberg, *Sugar Industry in Pernambuco*, 4.

86. Dean, *With Broadax and Firebrand*, 80. Of his total estimate, Dean guesses that one thousand square kilometers were accounted for by cane fields, while the rest represented forest cleared for firewood and building purposes. Of the cane fields he writes, "This was modest depredation; indeed, it represented less than half the area of the present-day municipality of Rio de Janeiro." Richards, *Unending Frontier*, 393.

87. Watts, *West Indies*, 393–400; Funes Monzote, *From Rainforest to Cane Field*, 127–28.

88. Tollenare, *Notas dominicais*, 71.

89. Ibid., 84.

90. Cited in Andrade and Lins, *Pirapama*, 157.

91. Koster, *Travels in Brazil*, 1:256; Henderson, *History of Brazil*, 366.

92. Galloway, *Sugar Cane Industry*, 73.

93. G. S. de Souza, *Tratado descritivo*, 45–46. In 1584, Cardim estimated a population of two thousand "vizinhos." He guessed there was an equal number of slaves. Cardim, *Tratados da terra*, 294. Eisenberg, *Sugar Industry in Pernambuco*, 147. Kidder estimated Recife had a population of sixty thousand. Kidder, *Sketches*, 119. Tollenare claimed it impossible to calculate the population of the *capitania*, "not being in a country of lights." Tollenare, *Notas dominicais*, 84.

94. Fletcher and Kidder, *Brazil and the Brazilians*, 341.

95. Tollenare, *Notas dominicais*, 83.

96. Gardner, *Travels in the Interior*, 73.

97. The vocabulary for describing density of weed, shrub, and tree growth is extensive and varies from region to region as well as over time. When a tract was burned, cleared, and cultivated for a while before being abandoned, what grew back was called *capoeira*. When allowed to grow for a decade or more, it became *capoeirão*. When *capoeira* was cleared and planted, then abandoned again, the growth that returned was called *capoeirinha*, "a scrubby dense growth that indicated to all that the ground was no longer good for cultivation of any kind." S. W. Miller, *Fruitless Trees*, 30.

98. Funes Monzote, *From Rainforest to Cane Field*, 83–85.

99. Grant, *History of Brazil*, 284–85. Grant also correctly predicted that on the return of "this imbecile court" to Portugal after Napoleon's defeat, the colony would separate from the mother country—an observation thirteen years ahead of its time (293).

100. Koster, *Travels in Brazil*, 1:49, 21.

101. Fletcher and Kidder, *Brazil and the Brazilians*, vi, 525.

102. See Antonil's description of huge ranching concerns already by 1692. Also, in 1701 the Crown prohibited ranching within ten leagues of the sea, a measure intended to protect plantations from cattle. S. W. Miller, "Stilt-Root Subsistence," 238.

103. Tollenare, *Notas dominicais*, 83.

104. Henderson, *History of Brazil*, 370–71.

105. Tollenare, *Notas dominicais*, 84, 88. Holloway, *Immigrants on the Land*, 18, 66.

106. Gardner, *Travels in the Interior*, 69–70; Henderson, *History of Brazil*, 387.

107. Henderson, *History of Brazil*, 388.

108. Gardner, *Travels in the Interior*, 68–69.

109. João Correia de Andrade, interview with author, handwritten notes, Engenho Jundiá, Vicência, February 18, 2003.

110. Vasconcelos Sobrinho, *Regiões naturais*, 252.

111. Pickersgill, Barrett, and Andrade-Lima, "Wild Cotton," 51.

112. Barman, *Citizen Emperor*, 139, 215.

113. Sociedade Auxiliadora, *Trabalhos do Congresso Agrícola*, 9, 12.

114. Greenfield, *Realities of Images*; Blake, "Invention of the Nordestino," 6. Gadiel Perruci says that some observers blamed the drought for the disappearance of 300,000–500,000 people, of whom 150,000 died of hunger. Perruci, "Canto do cisne," xiv.

115. Deutsch, "Bridging the Archipelago," 190; Leff, "Economic Development in Brazil," 35.

116. Perruci, "Canto do cisne," xvi.

117. Ibid., xxi.

118. Sociedade Auxiliadora, *Trabalhos do congresso agrícola*, 31.

119. Ibid., 10. The *quebra-kilos* was a smaller regional revolt that had none of the republican zeal that made the elites so proud of the 1817, 1824, and 1848 uprisings. It was apparently in response to the 1872 application of an 1862 law by which Brazil adopted the metric system. Where I translated "refugees," Carneiro da Cunha wrote "retirantes," the term for the *sertanejos* forced to leave their homes and travel coastward because of drought.

120. Eisenberg, *Sugar Industry in Pernambuco*, 158, 180.

121. Sociedade Auxiliadora, *Trabalhos do Congresso agrícola*, 13–15.

122. C. A. Taunay, *Manual do agricultor*, 106–7.

123. Sociedade Auxiliadora, *Trabalhos do Congresso Agrícola*, 114. See also Milet, *Lavoura da cana*, 16–17.

124. Perruci, *República das usinas*, 112.

125. Pernambuco produced an average of 116,379 tons between 1876 and 1880, while Cuba produced 533,000 tons in 1878, and Demerara produced 88,728 tons on average between 1877 and 1886. Eisenberg, *Sugar Industry in Pernambuco*, 16; Deerr, *History of Sugar*, 1:131, 2:377.

126. Sociedade Auxiliadora, *Trabalhos do Congresso Agrícola*, 114; 10–20.

127. Eisenberg, *Sugar Industry in Pernambuco*, 52.

128. Sociedade Auxiliadora, *Trabalhos do Congresso Agrícola*, 69.

129. Eisenberg, *Sugar Industry in Pernambuco*, 52–55.

130. Mill owners found themselves at the mercy of fluctuating cane production. Without reliable cane supplies, the mills failed to achieve a predictable rhythm. The *usinas*, producing around half of their cane from their own fields, gave far greater control. M. C. de Andrade, *Modernização e pobreza*, 37. Dé Carli, *Processo histórico*, 10.

131. Eisenberg, *Sugar Industry in Pernambuco*, 179–80.

132. E. A. Andrade, *CPRH 25 anos*, 7–9.

133. Costa Lima, "Reserva da biosfera," 12.

134. Andrade and Lins, *Pirapama*, 178, 64. It should be pointed out that strong supporters of sugarcane argue for its benefits in terms of conservation. Nélson Coutinho claims that, "notwithstanding not having a perfected agriculture from the technological point of view, there is no erosion in the zona da mata of Pernambuco." Nélson Coutinho, interview with Célia Maria Leite Costa and Dulce Pandolfi, tape recording and transcript, Rio de Janeiro, August 23, 1979 (FGV-CPDOC), 3.

135. Camilo, "Seca de 1877," 44. Josemir Camilo notes that the landscape was "constantly modified" during the 1870s as a result of new technology, led by the railroads.

136. Ó, *100 anos de suor e sangue*, 21.

Chapter 2

1. Lins do Rego, *Meus verdes anos*, 55–56.
2. Freyre, *Nordeste*, 75.
3. Nabuco, *Discursos parlamentares*, 18–19. We should take note of Nabuco's use of Africa as the comparative referent. He meant the continent to represent the quintessential site of backwardness, which carried on in the complete absence of modernity. And he might have been playing on the anxieties of his colleagues, who knew very well that the better part of their country's inhabitants were African or African-descended. Although slavery could be seen as the problem of the epoch, Nabuco's northeastern peers were more concerned then about the Great Drought that beset the region from 1877 to 1879.
4. Dean, "Latifundia and Land Policy," 619–21.
5. Viotti da Costa, *Brazilian Empire*, 85, 91–92; Dean, *With Broadax and Firebrand*, 201.
6. Dean, "Latifundia and Land Policy," 622.
7. Dean wrote that he intended his book to be "a memoir of the destruction" of the Atlantic Forest. Dean, *With Broadax and Firebrand*, 6, 108, 151–52.
8. Nabuco, *Abolitionism*, 10.
9. Viotti da Costa, *Brazilian Empire*, 79.
10. J. M. de Carvalho, "Modernização frustrada," 47, 51.
11. Galloway, "Last Years of Slavery," 596–97.
12. J. M. de Carvalho, "Modernização frustrada," 40; Viotti da Costa, *Brazilian Empire*, 82.
13. Worster, "Transformations of the Earth," 1099–100.
14. Nabuco, *Abolitionism*, 20, 27, 84.
15. Pádua, "'Cultura esgotadora,'" 160.
16. José Pádua shows that nineteenth-century thought remained in an earlier mode, which viewed environmental degradation as the price of backwardness. It was only in the twentieth century that a new idea came to prevail: that environmental degradation was a price of progress. Pádua, "'Cultura esgotadora,'" 158; Duarte, "Por um pensamento ambiental histórico," 151. "It is impossible to understand the political ecology of seventeenth- and eighteenth-century Brazil," Pádua writes, "without recognizing the identification between technological modernization and the overcoming of ecological disaster" (158).
17. Nabuco, *Abolitionism*, 105, 117, 111 (my emphasis).
18. This view of property fits well within the theoretical structure of Allen Abramson, who elaborates a definition of property that is cultural rather than strictly legal. There is "a jural element of formal legal codes," he writes, "but also as a localized notion" arising out of cultural practice. Abramson, "Mythical Land," 13.
19. Dean, *With Broadax and Firebrand*, 148.
20. French, "História," 14–15.
21. J. M. de Carvalho, "Modernização frustrada," 51.
22. Freyre, "Joaquim Nabuco e as reformas sociais," 15.
23. Nabuco, *Minha formação*, 129.
24. Leff, "Economic Development in Brazil," 35–36.
25. Singer, "Economic Evolution," 63.
26. Eisenberg, *Sugar Industry in Pernambuco*, 52–55; 108–11.

27. Nabuco, *Minha formação*, 132. See also the slightly different translation in Eisenberg, *Sugar Industry in Pernambuco*, 176.

28. Nabuco, *Abolitionism*, 95.

29. Nabuco, *Minha formação*, 132.

30. Dantas Silva, "Nota do editor," vii.

31. Bello, *Memórias*, 85.

32. Ibid., 185. Bello was partly referring to the chaos following the 1930 revolution, about which, like Freyre, he was quite displeased. See also ibid., 182.

33. Ibid., 162.

34. Ibid., 59, 162, 186; 228, 162; 179.

35. Sérgio Milliet would call this "the peaceful satisfactions of *mandonismo* and wealth." Milliet, "Obra de José Lins do Rego," xx.

36. Nabuco, *Minha formação*, 132.

37. Bello, *Memórias*, 217.

38. A substantial literature from geography, anthropology, and history addresses the category of "place." See, e.g., Tuan, "Place," 151–65; and Casey, *Fate of Place*.

39. Nabuco, *Minha formação*, 131. This quotation inspired Caetano Veloso, who includes the words as the lyrics of a song on his 2001 album *Noites do Norte* (Nonesuch). In a biography of her father, Carolina Nabuco referred to these passages as being "pregnant with all the poetry . . . of the institution which [Nabuco] was destined to combat." C. Nabuco, *Life of Joaquim Nabuco*, 8. The wistful tone of the passage might be explained by the time of its writing, when Nabuco lived in London in semiexile from the republic toward which he felt ambivalent.

40. In his introduction to a collection of Freyre's essays published as *Região e tradição*, Lins do Rego recounts being summoned by Freyre to go over his (Lins do Rego's) newspaper articles. Freyre saw "something that interested him" in the pieces and so, Lins do Rego continues, "my education with the master of my age began without my noticing the lessons." He began imitating Freyre's style before slowly developing his own. Lins do Rego, "Notas sobre Gilberto Freyre," 10. Freyre himself was well aware of his influence over Lins do Rego. A late 1925 entry in his diary reads: "J. L. do R. [José Lins do Rego], as well as O. M. [Olívio Montenegro?], and A. F. [Alfredo Freyre?] have been imitating me—they, among various others of lesser stature—my style, my form, even my punctuation. I know that I have a style or a form and a rhythm that is defined partly by punctuation." Freyre, *Tempo morto*, 176.

41. Bello, *Memórias*, xiv.

42. Eisenberg, *Sugar Industry in Pernambuco*, 123.

43. José Lins do Rego, Preface, xix. Freyre obviously was an exceptional case of deep intellectual curiosity and exposition among Pernambuco's elite. Underlining Lins do Rego's point, it is important to keep in mind that these texts betray a much higher degree of reflection, reflexivity, and self-consciousness than should be considered "normal" for a *senhor* of the time. I believe that my argument—that these texts articulate a discourse of land and labor common to a class—still holds true.

44. Some people include *Fogo morto* in the Sugar Cycle. José Aderaldo Castello, who calls it the "synthesis work" of the series, claims that Lins do Rego himself included *Fogo morto*, despite having announced in the preface to *Usina* that the series was over. Castello, "Origens e significado," 30.

45. Lins do Rego, *Usina*, 7.

46. Lins do Rego, *Meus verdes anos*, 6. Not only are the descriptions of the *engenho* and the stories he remembers the same as those experienced by Carlos de Mello, but even the names of the people in his autobiography are the same as those he used in his series of novels.

47. Christine Dabat writes that Lins do Rego and Freyre characterized "a banal episode of any economic history as a fundamental social rupture, a period of crisis, or decadence," because of their attachment to the idea of a glorious past. Dabat, *Moradores de engenho*, 186.

48. Lins do Rego, *Plantation Boy*, 527 (from the book *Banguê*).

49. Of Carlos de Mello's grandfather, Lins do Rego wrote that he prevented any incursions into his forest, because "that would be as if someone asked for a piece of flesh from his own body." Lins do Rego, *Plantation Boy*, 34.

50. Lins do Rego, *Fogo morto*, 220–23. See also the similarity in dynamics between this description and the case discussed by Peter Beattie in "Slave Silvestre's Disputed Sale," 41–65.

51. Lins do Rego, *Meus verdes anos*, 313.

52. Ibid., 331–32, 335–42.

53. Rogers, "José Lins do Rego," 26–27.

54. Lins do Rego, *Plantation Boy*, 220 (from the book *Doidinho*).

55. Lins do Rego, *Pureza*, 125.

56. He writes "me habituava a ver" the peasants in their misery, which Emmi Baum translates as "I accepted their state as something natural." The next phrase, "I thought it quite natural," is "achei muito natural" in the original. Lins do Rego, *Menino de engenho*, 134; Lins do Rego, *Plantation Boy*, 80.

57. Lins do Rego, *Menino de engenho*, 134.

58. Bello, *Memórias*, 217.

59. Yet another iteration of Lins do Rego and Bello's idea appears in an interview that Caio Lima Cavalcanti gave in 1966. From the family that owned the Usina Pedrosa, and brother to Carlos Cavalcanti who governed the state in the 1930s, Caio told the U.S. economist Robert Alexander that he did not "understand the attitude of Americans" toward race since he did not "think that any damage [was] done to him when he [shook] hands with a Negro." In fact, he said, "if [Negroes] are treated right they are exceptionally good people. . . . they are as loyal as his dog [was] to him." Caio Lima Cavalcanti, interview with Robert Alexander, Rio de Janeiro, June 14, 1966 (Robert Alexander Archive, Rutgers University).

60. Trigo, *Engenho e memória*, 218.

61. Bello, *Memórias*, 179.

62. Trigo, *Engenho e memória*, 218–19.

63. Francisco Julião Arruda de Paula, interview with Eliane Moury Fernandes, tape recording and transcript, Recife, 1982 (CEHIBRA), 1.

64. Mota, *Ideologia da cultura brasileira*, 54.

65. Freyre wrote in his diary in 1929, "Everyone knows that the Governor (Estácio Coimbra) now has more confidence in me than in any of his old aides: which gives me a decisive 'influence' in certain affairs." Freyre, *Tempo morto*, 317, 324. He narrates the revolution, which included the invasion and burning of his family home, on 336–37.

66. Mota, *Ideologia da cultura brasileira*, 29–30.

67. Skidmore, *Black into White*, 191.

68. Viotti da Costa, *Brazilian Empire*, 244; Mota makes a similar point in *Ideologia da cultura brasileira*, 55.

69. One biographer notes that Freyre was an urban child, though he spent a vacation on a sugar plantation as a nine-year-old. Chacon, *Gilberto Freyre*, 37.

70. Lins do Rego, "Notas sobre Gilberto Freyre," 15.

71. Freyre, *Masters and Slaves*, xi.

72. Mota, *Ideologia da cultura brasileira*, 56, 58.

73. Freyre, *Nordeste*, 74, 65, 73.

74. Ibid., 26, 162.

75. Nabuco, *Abolitionism*, 113. Another reverberation of Nabuco's reference to scars on the land and society appears in Freyre's *Mansions and Shanties*, in which he refers to "the two running sores of monoculture and slavery, two wide-open mouths that clamored for money and for blacks." Freyre, *Mansions and Shanties*, 11.

76. Freyre, *Nordeste*, 162.

77. Freyre's interchangeable use of *landscape* and *region* perhaps derives from Sauer's observation that they were synonymous. Sauer chose to use *landscape*, and to propose its use as the basic unit of geography, because it lacked the troublesome connotations with which both *region* and *area* were laden. Sauer, "Morphology of Landscape," 321. Freyre says he met Sauer at a 1939 conference at the University of Michigan. Freyre, *Nordeste*, 26–27.

78. C. Buarque, "The Northeast," 272. Also, Durval Muniz de Albuquerque Jr., Joedna Meneses, and Stanley Earl Blake highlight the 1920s as the moment the Northeast emerged as a coherent and recognized unit. Albuquerque, *Invenção do Nordeste*; Blake, "Invention of the Nordestino"; Meneses, "Indústria do atraso ou o atraso da indústria?"

79. Freyre, *Nordeste*, 41–42.

80. Ibid., 17. Freyre wrote, "The word 'northeast' is today disfigured by the expression 'works for the Northeast,' which means: 'works against the droughts.'" Freyre, *Nordeste*, 41.

81. Freyre, "Introdução à 9a edição," 11.

82. Skidmore, "Raízes de Gilberto Freyre," 3. It is characteristic of the scholarship on Freyre that after mentioning these two factors Skidmore says nothing more about climatic or environmental determinism, focusing instead on the issue of race.

83. Freyre, *Masters and Slaves*, 60. Freyre also argued that this view urgently needed to be overturned specifically in the case of Brazil, "for it is the example of which alarmists make so much use in crying about the mixture of races and the malignity of the tropics in support of their thesis that man's degeneration is the effect of climate or of miscegenation." Freyre was far from the first person to challenge these determinisms, a point various scholars have made. Sidney Chalhoub, for instance, discusses Brazilian physicians in the mid-nineteenth century who worked against racial and climatic determinism. Chalhoub, "Politics of Disease Control," 455.

84. Freyre, *Masters and Slaves*, 45.

85. Ibid., 48, 240, 401.

86. Da Cunha, *Rebellion in the Backlands*, 112.

87. Making clear his intellectual debts, in an introduction to Nabuco's autobiography, Freyre lauded both Nabuco and da Cunha for their perspicacious recognition of the connection between man and the land. But while the two men's writing similarly brought the landscape to the page, Freyre argued that Nabuco had even more "terrestrial vigor" than did da Cunha. Freyre, "Introdução à 10a edição," 20.

88. Freyre's citations of these scholars are far too numerous to mention, but for a fairly sustained engagement with some of them, see *Masters and Slaves*, 15, 17, 297, and 330.

Semple and Sauer taught geography in the well-known program at the University of California at Berkeley. Thomas Griffith Taylor wrote *Environment and Race*. And Freyre studied with the eminent anthropologist Franz Boas at Columbia University.

89. Freyre, *Masters and Slaves*, 330. Huntington, *Civilization and Climate*. The book was reprinted and revised ten times between its initial appearance in 1915 and 1948.

90. Semple, *Influences of Geographic Environment*.

91. Freyre, *Masters and Slaves*, 18–19, 81, 179.

92. Ibid., 32, 140, 141.

93. Ibid., 381, 311.

94. Nabuco, *Minha formação*, 131.

95. Freyre, *Nordeste*, 57, 46, 148. Freyre used the term "civilization of sugar" in his introduction to Nabuco's autobiography. Freyre, "Introdução à 9a edição," 10.

96. Freyre, *Masters and Slaves*, 251, 376, 212. Oton Bezerra de Melo, in a 1940 article from the *Folha da manhã*, wrote of "sete mil homens que mourejam de sol a sol." Reprinted in *Homem e terra*.

97. Freyre, *Masters and Slaves*, 247, 235. Freyre pointed to patterns in Iberian culture as the roots of the horror that manual toil inspired in the Portuguese. The tendency was reinforced by the use of African slave labor.

98. See Cosgrove, "Prospect, Perspective," 55–56.

99. Nabuco, *Minha formação*, 132; Freyre, *Mansions and Shanties*, 131.

100. This notion of conflating the human and natural fuel comes from Nabuco, *Minha formação*, 134.

101. Nabuco, *Abolicionismo*, 60, 59; translation from Nabuco, *Abolitionism*, 10–11. Nabuco's geological metaphor nicely evokes the conflation of cultural and "natural" issues that is being pinpointed here.

102. Nabuco, *Abolicionismo*, 60, 65; Nabuco, *Abolitionism*, 11, 17.

103. D. A. S. de Moura, *Saindo das sombras*, 100.

104. R. J. Scott, *Slave Emancipation in Cuba*, 281.

105. D. A. S. de Moura, *Saindo das sombras*, 141.

106. Freyre, *Região e tradição*, 175–76.

107. Mota, *Ideologia da cultura brasileira*, 55.

108. Milliet, "Obra de José Lins do Rego," xx.

109. Bezerra, *Memórias*, 1:34–35, 46, 51.

110. Bezerra was born in 1900 (the same year as Freyre) and Lins do Rego in 1901. The particular experience Bezerra relates took place in 1907. Bezerra, *Memórias*, 1:51.

111. In his fascinating history of the large ranching concern Miller and Lux, David Igler writes that the company's "power ultimately derived from the ability to tap both human and natural energy for its own ends." Of course, this is arguably the case for any agribusiness. Pernambucan planters did not recognize a difference between these types of energy. Igler, *Industrial Cowboys*, 124.

112. Raiz, "Trabalhador negro," 191–94.

113. He also describes it as a "durably installed generative principle of regulated improvisations." Bourdieu, *Outline of a Theory of Practice*, 76, 17, 18, 78.

114. Ibid., 76–79.

115. Ibid., 78, 82, 81, 15 (my emphasis).

116. Nabuco, *Minha formação*, 129. In a similar passage in a collection of his early writings,

Freyre notices the presence of crucial themes that he continued to pursue over his career, including ecological ones. Freyre, *Tempo de aprendiz*, 33.

117. Freyre, "Introdução à 9a edição," 8.

118. Lins do Rego, *Meus verdes anos*, 5.

119. Interestingly, this is a point that Durval Muniz de Albuquerque Jr. recently made (after I had worked out these problems along similar lines). He writes, "The sociological writings of Gilberto Freyre and the writings of the so-called 1930s novelists make memory the principal material of their work. These novelists endeavor to build the Northeast out of their memories of their childhoods, in which the forms of social relationships that are now under threat were dominant. They make their own narratives a manifestation of traditional and popular culture threatened by the modern world and use them as expressions of the regional." Albuquerque, "Weaving Tradition," 55. As I noted above, this common thread tying the authors together relates partly to the type of texts under analysis. Though this argument draws on sociology, history, and political tract, there are also three memoirs represented.

120. Ted Steinberg offers clear illustrations of how power and the environment interact in historical change. Steinberg, "Down to Earth," 802.

121. Bello, *Memórias*, 39. Much like Nabuco's exclamation about the pardon by the "sainted blacks" of the whites, Bello suggested that "the tragic sentiments of hate and rancor did not separate, nor should they separate, the former slaves from the white man who exploited them for three centuries. In the immense sensitivity of their soul, all of this hate falls and submerges completely in charity, drowning itself" (ibid.).

122. Albuquerque writes, "Nostalgia is what we feel when we perceive ourselves as losing a cherished part of our very being—of spaces we have occupied as our own." Albuquerque, "Weaving Tradition," 42.

123. Schama, *Landscape and Memory*, 7.

124. Allen Abramson argues that the eye cast over a landscape "bathes in the nostalgia of human phenomena, lost. This is why iconic landscapes in their respective national cultures have always tended to inspire wistful feelings as well as feelings of exhilaration or national pride." Landscapes bring forth memories, he writes, "not so much because their earthy tangibility helps trigger memories which we have culturally archived there, but because landscape embeds in symbols mainly that which has been lost and, consequently, that which can only be retrieved precisely as recollection and memory." Abramson, "Mythical Land," 4–5.

125. Milet, *Lavoura da cana*, 85.

Chapter 3

1. Ó, *100 anos de suor e sangue*, 21, 17, 29. The phrase "green horizons" appears five times in the first thirty pages of the book. Manoel's escape mirrors that of Ricardo, the protagonist of José Lins do Rego's 1935 novel about the experiences of a young black boy from an *engenho*. Ricardo plots his flight, watching daily as the trains come and go at the local station. He sees the passengers, with hats on their heads and holding umbrellas, going to Recife, Paraíba, Campina Grande, "people speaking of the market, of cities, of lands that weren't *engenhos*." Lins do Rego, *Moleque Ricardo*, 7.

2. White, "Are You an Environmentalist?," 172.

3. Paulo, interview with the author, tape recording, Murupé, Vicência, March 27, 2003.

4. Tollenare, *Notas dominicais*, 85.

5. Ibid., 95, 97.

6. Galloway, "Last Years of Slavery," 601. Maria de Nazareth Baudel Wanderley also dates the modern system of *morada* to the transition away from slave labor, in *Capital e propriedade fundiária*, 37.

7. Dabat, *Moradores de engenho*, 80–81. For Jamaica's experience after emancipation, see Holt, *Problem of Freedom*.

8. Bello, *Memórias*, 39.

9. Lins do Rego, *Plantation Boy*, 49.

10. Galloway, "Sugar Industry of Pernambuco," 298.

11. Quotation from Galloway, "Last Years of Slavery," 602; Eisenberg, *Sugar Industry in Pernambuco*, 179–80.

12. For a broad discussion of continuities between slavery and postslavery with observations of recent historiography, see French and Rogers, "Slavery as a 'Sinister Principle' of Authority."

13. M. J. M. de Carvalho, *Liberdade*, 213.

14. Reis, "'Revolution of the *Ganhadores*,'" 359.

15. Ó, *100 anos de suor e sangue*, 30, 21.

16. Galloway, *Sugar Cane Industry*, 159. Eisenberg, *Sugar Industry in Pernambuco*, 218.

17. This postemancipation development, especially of the industrial aspect of sugar production, is explained by the planter drive for control. John Rodrigue observes of postbellum Louisiana planters, "Because they were unable to establish control over labor in the fields, planters focused their efforts at improving efficiency in the processing phase of sugar production—that is, at the sugar mill." Rodrigue, *Reconstruction in the Cane Fields*, 116.

18. Galloway, "Sugar Industry of Pernambuco," 300–302. By 1920, 1,023 *engenhos* had ceased milling.

19. "O President Getúlio Vargas salvou Pernambuco da ruina," *Brasil açucareiro* 4, no. 4 (December 1934): 20.

20. Raiz, "Trabalhador negro," 191.

21. Eisenberg, *Sugar Industry in Pernambuco*, 5–20, 186, 239–40.

22. Bello, *Memórias*, 133.

23. Holloway, *Immigrants on the Land*, 111–38.

24. Galloway, "Last Years of Slavery," 602.

25. Watts, *West Indies*, 386. Trevor Burnard goes so far as to argue that slaves in eighteenth-century Jamaica were "proto-capitalist" because of their engagement with the market system through sale of surplus from garden plots. He suggests that this tradition tied slaves to their particular plantations (and plots) and made them less likely to rebel against their situations. Burnard, *Mastery, Tyranny, and Desire*, 153. For more on provision grounds, see Barickman, "'Bit of Land'"; and Schwartz, *Slaves, Peasants, and Rebels*.

26. Palmeira, "Casa e trabalho,", 314. Julião, *Cambão*.

27. Systematic study of *morada* began in the late 1960s and early 1970s, propelled by a group of innovative and energetic scholars from the Universidade Federal do Rio de

Janeiro's graduate program in anthropology, housed at the National Museum and united in a working group called "Projeto estudo comparativo do desenvolvimento regional." See the note on these scholars in the introduction.

28. Cabral, "Processo de proletarianização," 161.

29. Paulo, interview with the author, tape recording, Murupé, Vicência, March 21 and 27, 2003.

30. Diégues, *População e açúcar*, 127.

31. Bezerra, *Memórias*, 1:34–35, 46, 51.

32. Sigaud, "Armadilhas da honra," 138; Palmeira, "Casa e trabalho," 306.

33. S. R. de Moura, *Memórias de um camponês*, 142–44. The Jesuit João Antonil wrote in his early-eighteenth-century text that transporting cane during rainy and muddy times effectively meant killing oxen. A. de E. Taunay, *André João Antonil*, 115.

34. S. R. de Moura, *Memórias de um camponês*, 142–44.

35. Harvests have been ending earlier and earlier in the cane zone, as transportation and processing of cane has gotten more efficient.

36. Diégues, *População e açúcar*, 137–38.

37. S. R. de Moura, *Memórias de um camponês*, 168.

38. Paulo, interview with the author, March 21, 2003. Severino, interview with the author, tape recording, Vicência, April 1, 2003. Andrade and Andrade, *Cana-de-açúcar*, 56.

39. Bastos, *Ligas camponesas*, 30.

40. Workers refer to the need to "roça mato," when this clearing is required. Many different tiers of difficulty are distinguished, depending on the density of growth—*mato, mato grosso, capoeira, capoeirão, mata*, etc. On the detailed vocabulary for various stages of growth, see S. W. Miller, *Fruitless Trees*, 30.

41. Diégues, *População e açúcar*, 134–35. The first method is called "rêgo," the second "mergulho." In commercial cane agriculture, cane is almost never planted from seed. Rather, the plants' reproduction is asexual. J. C. de Andrade, *Escorço histórico*, 33.

42. Diégues, *População e açúcar*, 136. Diégues, writing in 1954, mentioned that the number of cleanings had recently been increasing (137).

43. Bezerra, *Memórias*, 1:34.

44. S. R. de Moura, *Memórias de um camponês*, 209–10. Dantas, "Recuperação da lavoura," 33.

45. J. C. de Andrade, *Escorço histórico*, 93–98.

46. Ribeiro, *Folclore do açúcar*, 177.

47. Bezerra, *Memórias*, 1:35–36, 38. The archive of Recife's Museu do Açúcar holds a fascinating exchange from the early 1940s between an old *senhor de engenho* in Pernambuco and someone in Rio de Janeiro researching a book on ox carts. The scholar sent a detailed set of questions about ox carts to the man about the construction, use, and folklore of ox carts in the cane zone. "Questionário sobre carro de boi (1942)," pasta Engenho Amaragí, Arquivo Museu do Açúcar, FJN.

48. Freire, *Pedagogy of Hope*, 19–20.

49. Freyre, *Mansions and Shanties*, 302; Vieira Filho, "Bumba-me-Boi do Maranhão," *Brasil açucareiro*, 72, no. 2 (1968): 102. As Richard White puts it, "For millennia humans have known animals largely through work. Work gave the people who trained and worked with animals a particular knowledge of them. . . . 'the horses themselves became . . . part of the man that drove them.'" White, "Are You an Environmentalist?," 177.

50. Pandeiro, "Boi tungão."
51. Bello, *Memórias*, 56.
52. See Barros, *Década 20*, 48.
53. Real, *O folclore no carnaval*, 73–74. This edition is a reprint of the original 1967 edition; Vicente, *Maracatu rural*, 26.
54. Adrião Caminha Filho, "A agricultura brasileira e o Estado Novo," *Brasil açucareiro* 19, no. 1 (January 1942): 5.
55. Sigaud, "Percepção do salário," 320–21.
56. Severino, interview with the author, April 1, 2003. Two translation notes: "I have a degree" from "me formei" and "hard ground and thick brush" from "terra dura e mato grosso."
57. Ronivaldo, interview with the author, tape recording, Engenho Humaitá, Palmares, January 25, 2003.
58. Palmeira, "Casa e trabalho," 310–11. Workers appeared to use a common shorthand to indicate that they did everything involved with *engenho* work: "I clear brush and cut cane [*roço mato e corto cana*]." Sidney Mintz describes his informant don Taso's narrative of maturing and becoming a good worker. By a certain point, he "mastered all the jobs in the cane (except, perhaps significantly, cane cutting)." Mintz, *Worker in the Cane*, 263.
59. Ronivaldo, interview with the author, January 25, 2003.
60. Striffler, *In the Shadows of State and Capital*, 43, 47.
61. Freyre outlines the abortive study in *Nordeste*, 162. His collaborators were to be Pernambucano, Olívio Montenegro, and Sílvio Rabelo. Pernambucano's article was coauthored with Arnaldo di Lascio, Jarbas Pernambucano, and Almir Guimaraes and appeared as "Alguns dados anthropologicos da população do Recife," in Freyre et al., *Novos estudos afro-brasileiros*.
62. Fontes, *Nordeste em São Paulo*, 17–18, 41–54.
63. Gaspari, *Ditadura derrotada*, 33.
64. Freyre, *Nordeste*, 179.
65. Leite Lopes, *Tecelagem dos conflitos*, 165.
66. See Barros, *Decada 20*.
67. Freyre, *Mansions and Shanties*, 110.
68. Lins do Rego, *Moleque Ricardo*, 15; Rogers, "José Lins do Rego," 26–27.
69. Leite Lopes, *Tecelagem dos conflitos*, 16, 165. The factory site took its name from the fact that the land had been granted to the Paulista adventurer Navarro as a reward for his role in defeating the runaway slave community of Palmares at the end of the seventeenth century (145). The overtones of subjugation, played out in the exploitation of a generation of workers descended from those runaway slaves of nearly 250 years earlier, are interesting to consider.
70. Leite Lopes, *Tecelagem dos conflitos*, 117. See also Leite Lopes's recent film, with Celso Brandão and Rosilene Alvim, *Tecido memória*.
71. Castro's best-known work—*The Geography of Hunger*, which in editions after 1953 became *The Geopolitics of Hunger*—was first published in 1946. A powerful articulation of the ideas he had sustained for over a decade, rooted in empirical findings from his work in Recife but reaching far beyond to make sweeping arguments about the pervasiveness of poverty and its debilitating effects, *The Geography of Hunger* catapulted Castro to prominence. He would go on to write important studies of hunger in other Latin American countries and to

lead the antihunger campaign of the U.N. Food and Agriculture Organization. Castro, "The Geography of Hunger," <http://www.josuedecastro.com.br/port/index.html>, accessed September 8, 2004; Andrade et al., *Josué de Castro*.

72. Castro, *Documentário do Nordeste*, 59, 64–66.

73. Ibid., 8. Olívio Montenegro, introducing the same collection, wrote memorably about the continuity of thought throughout a man's life. "Rarely is the man not perceived in the child," he wrote, "and what many people call precocity, should with more exactitude be called conformity: the child awakening to a life of harmony with the man he will be tomorrow." Ibid., 4. Montenegro wished to praise the quality of Castro's early efforts and link it to his more famous later work, such as *The Geography of Hunger*. But his observations also point to the importance of the young Castro's experiences in shaping his later interests and intellectual perspective.

74. The *palafita* dwellers took center stage in Castro's novel *Of Men and Crabs*. The book focuses on the slum dwellers' reliance on crabs, and Lins do Rego's Ricardo also recognizes their centrality to Recife workers' diets; *Moleque Ricardo*, 57.

75. Castro, *Alimentação e raça*, 95, 104. Leite Lopes, *Tecelagem dos conflitos*, 57.

76. Skidmore, "Raízes de Gilberto Freyre," 13. Castro used Monteiro Lobato's rural everyman—Jeca Tatú—as an illustrative example in *Geography of Hunger*. A foreigner shocked at the malnutrition asks Jeca, "Don't beans, rice, and fruit grow here?" Jeca says no. "But have you planted some of these things, to see if they'll grow?" the foreigner persists. "Of course they'll grow—if you plant them!" Jeca responds. Castro, *Geography of Hunger*, 77–78. After characterizing Jeca Tatú as inherently inferior, in 1919 Lobato portrayed him "as the victim of neglect who had failed to benefit from such controllable factors as correct nutrition and proper medical attention." Skidmore, "Raízes de Gilberto Freyre," 15. Castro, *Alimentação e raça*, 89.

77. Rosa e Silva Neto, "Sugestões para uma nova orientação," 4–7.

78. E. Pinto, *Problema agrário*, 7. Freyre, *Nordeste*, 178–79.

79. White, *Organic Machine*, 7.

80. S. R. de Moura, *Memórias de um camponês*, 144. Worker health became a topic of interest for the IAA in the early 1950s because a federal law directed sugar producers to make arrangements for workers to access medical care. A 1947 study of forty-eight Pernambucan *usinas* (almost all of them) found that only four had doctors in residence. José Leite, "A assistência médico-hospitalar, nas usinas de Pernambuco, à luz de um inquérito," *Brasil açucareiro* 35, no. 1 (January 1950): 91.

81. Ingold, "Temporality of Landscape," 152.

82. Sílvio Pélilco Leitão, "Problemas médico-sociais da lavoura canavieira," *Brasil açucareiro* 28, no. 3 (1946): 80–89.

83. Vasconcelos Torres, "A habitação nas zonas canavieiras do Brasil," *Brasil açucareiro* 23, no. 4 (1944): 42–44. The pictures Torres includes from the Northeast portray homes much as those remaining on *engenhos* continue to look.

84. Albuquerque has characterized the patriarchal perspective as the view "from the porch of the big house." Albuquerque, "Weaving Tradition," 57. Christine Rufino Dabat articulated a similar view in the title of a conference paper, "Da calçada da casa-grande," presented before Albuquerque's article appeared.

85. On overcharging at the *barracão*, see Severino, interview with the author, tape recording, Vicência, April 16, 2003. In a 1962 police report, the officer compared a *barracão*

to the "Esquina do Lafaiete," a legendary spot on the Rua do Imperador in Recife where, as he wrote, "one knows to go to get news first-hand." The corner is memorialized in Paraíso, *Esquina do Lafayette*. "Relatório ao Delegado Auxiliar," January 22, 1962, p. 2 in SSP 29201: Liga Camponesa de São Lourenço da Mata, Acervo DOPS, APE.

86. Freyre delved into the archetectonic layout of the sugar region in "Arquitetura de casas-grandes e sobrados e civilização brasileira do açúcar," in Freyre, *Presença do açúcar*.

87. Eisenberg estimates that mid-nineteenth-century planters used between 16 and 21 percent of cultivable land. Dé Carli claims that the cane for the 1929–30 harvest took up 13 percent of the total *zona da mata*. Using Eisenberg's estimate of 3.4 million hectares for the region, Dé Carli's figure for 1929 would be 450,812 hectares, up from 121,443 seven or so decades earlier. Eisenberg, *Sugar Industry in Pernambuco*, 127; Dé Carli, *Processo histórico*.

88. Usina Pedrosa belonged to one arm of the powerful Cavalcanti family. Rui de Lima Cavalcanti was the owner of the *usina* and his brother Carlos de Barros Cavalcanti managed it, but other family members also had interests in the business. S. R. de Moura, *Memórias de um camponês*, 45. Interestingly, Tenente's innovative head chemist, José Britto Pinheiros Passos, began his career at Pedrosa. Passos, *Cronistórias de minhas atividades*, 51.

89. S. R. de Moura, *Memórias de um camponês*, 38, 43, 44–46. The promotion to *apontador* came only two years after Moura arrived at the *usina*.

90. Ibid. Gilberto Freyre paid close attention to the big house and to the space of home life in general in his famous 1933 book, *The Masters and the Slaves*.

91. Palmeira, "Casa e trabalho," 306–7.

92. Bastos, *Ligas camponesas*, 39.

93. Benedito, interview with the author, tape recording, Engenho Humaitá, Palmares, January 25, 2003.

94. Massey, "Double Articulation," 111–19.

95. In his 1829 "A Letter from Sydney," Edward Wakefield argued that land in Australia should be sold at an artificially high price to ensure that potential laborers did not simply disperse (unproductively) throughout the land-rich colony. Wakefield's ideas became a popular model for other liberal policymakers, including in Brazil. Kittrell, "Wakefield's Scheme," 87–111. Viotti da Costa, *Brazilian Empire*, 82.

96. Sigaud, "Morte do caboclo," 8, 13.

97. Câmara Cascudo, *Sociologia do açúcar*, 248.

98. S. R. de Moura, *Memórias de um camponês*, 198.

99. Basso, "Wisdom Sits in Places," 55.

100. Ribeiro, *Folclore do açúcar*, 28–29.

101. Câmara Cascudo, *Sociologia do açúcar*, 253, 251, 250. There was also a more generalized fear of night in the cane zone. One wonders whether this has anything to do with the night blindness (mentioned in Ribeiro, *Folclore do açúcar*, 148) that can accompany poor nutrition—especially a lack of vitamin A.

102. Severino, interview with the author, April 1, 2003.

103. Rosa e Silva Neto, "Sugestões para uma nova orientação," 7.

104. Ribeiro, *Folclore do açúcar*, 29–33. Ribeiro cites the research of folklorist Ademar Vital from decades earlier. He himself links the legend to totemic traditions among Amerindians and snake cults from Africa. Later in his book (201), Ribeiro asserts that all folklore is "subordinate to ecology."

105. Ronivaldo, interview with the author, January 25, 2003. Through long-term

residence and labor, workers developed a substantial knowledge of place. A secret police investigator describing the difficulty of uncovering suspected Peasant League activity wrote in his report that cane workers "are profoundly knowledgeable of the terrain of the property where they live." Severino Oliveira, Relatório to Delegado, June 15, 1961 in SSP 29796: Ligas Camponesas (1950), Acervo DOPS, APE.

106. "Questionário sobre o carro de boi," FJN.

107. Ronivaldo, interview with the author, January 25, 2003.

108. Ibid. The same phenomenon comes through clearly in Elide Bastos's account of the Galiléia Engenho peasants' reactions in 1963 when the state government tried to relocate them to newly expropriated land. The workers protested, expressing their desire to stay on their own plots even if they were "weak" and eroded. Bastos, *Ligas camponesas*, 21.

109. Ingold, "Temporality of Landscape," 154. "It is from this relational context of people's engagement with the world, in the business of dwelling, that each place draws its unique significance. Thus whereas with space, meanings are *attached to* the world, with the landscape they are *gathered from* it."

110. Julian Thomas argues that distinct relational experiences result in "more than a juxtaposition of representations"; groups can "actually inhabit *different landscapes*, which intersect in various ways." Thomas, "Comments on Part I," 181. For more on landscapes as relational and produced through social interaction, see Thrift, "Steps to an Ecology of Place," 304.

Chapter 4

1. Cid Feijó Sampaio, interview with Aspásia de Alcântara Camargo, Dulce Pandolfi, and Eduardo Raposo, tape recording and transcript, Recife, April and October 1979 (FGV-CPDOC), 56.

2. "Polícia de Catende ausentou-se, mas camponeses ordeiros frustraram as agitações em 'Roçadinho,'" *Diário de Pernambuco*, July 12, 1963.

3. Bello, *Memórias*, 56, 134.

4. With the name "creole cane" (*cana creoula*), this variety took on the aura of a homegrown product, and all later arrivals were by comparison foreign. Of course, the "creole" cane was itself simply an earlier migrant from the east, by way of the Atlantic islands. Much of the first half of this chapter appeared in different form in Rogers, "Geneticistas." I thank *Clio*, and especially Christine Rufino Dabat, for allowing that material to appear here.

5. J. C. de Andrade, *Escorço histórico*, 55–63.

6. Gonsalves de Mello, "Notas acêrca da introdução de vegetais," 42, 52. The botanical garden had fallen into complete disrepair and abandonment by 1844.

7. The occasional Brazilian planter did refer to it as *Taití, Otaheiti*, or *Bourbón*. Watts, *West Indies*, 433. José Clóvis de Andrade suggests that Cayenne cane may have come to Pará's public gardens via the first director, a Frenchman named Gremoullier who emigrated from French Guiana. J. C. de Andrade, *Escorço histórico*, 20. Eisenberg, *Sugar Industry in Pernambuco*, 32. For the linguistic variants, see C. A. Taunay, *Manual do agricultor*, 111.

8. I. M. G. de Almeida, "Gomose da cana de açúcar," 33–39.

9. The new varieties included Bois Rouge, Cavangierie, Ubá, and Salangor. Bento Dantas and Mário Lacerda de Melo, "As áreas cultivadas com as atuais variedades na zona

canavieira de Pernambuco," *Boletim canavieiro* 5, no. 54 (September 1959): 3; J. C. de Andrade, *Escorço histórico*, 63.

10. J. C. de Andrade prints Salgado's April 26, 1892, circular in its entirety; *Escorço histórico*, 63–66. Salgado's communication and Cavalcanti's experimentation demonstrate the degree to which Pernambucan planters saw themselves as lodged within a larger sugar-producing world.

11. Ibid., 75–83; Artur de Siqueira Cavalcanti Júnior, "Reminiscencia sobre a cultura da cana de assucar em Pernambuco," undated manuscript in the Artur de Siqueira Cavalcanti Júnior Collection, FJN. Artur de Siqueira, carried away by his grandfather's achievement, claims that this was the first time *worldwide* that seeds had been purposefully fertilized.

12. Bello, *Memórias*, 58, 59, 133, 162.

13. M. C. de Andrade, *História das usinas*, 78–81, 91–93; Eisenberg, *Sugar Industry in Pernambuco*, 124; Vasconcelos Sobrinho, *Regiões naturais*, 171; Barros, *Década 20*, 49.

14. Mintz, *Worker in the Cane*, 255.

15. González, "Law, Hegemony, and Politics," 79.

16. Ibid.

17. In the alphabet soup of cane variety prefixes, CO designates strains developed at the Sugarcane Breeding Institute in Coimbatore, India, and POJ designates Indonesian strains—from the Proofstation Oost Java.

18. See, e.g., Aguirre, "Creação de novas variedades."

19. Lins, "Natureza e limites do conhecimento geográfico," 272.

20. Aguirre, "Creação de novas variedades," 7.

21. Topik, *Political Economy*, 60–85.

22. Vargas's creation of numerous autarkies followed a pattern of systematic interventionism. González, "Law, Hegemony, and Politics," 171. Singer, "Economic Evolution," 69–70.

23. The station moved to the suburb of Curado and was known by that name. See "Estação experimental da Várzea do Curado," *Brasil açucareiro* 2, no. 1 (1934): 59; "Estação experimental de Curado," *Brasil açucareiro* 24, no. 2 (1944): 88. José Clóvis de Andrade, the author of the study of historic cane varieties in Brazil, directed the research station at Curado between 1947 and 1950. J. C. de Andrade, *Escorço histórico*, 154. He was active on the national sugar scene, presenting a paper at the 1950 Primeiro Congresso Açucareiro Nacional in Rio de Janeiro, "Variedades de cana recomendadas para a zona Norte de Pernambuco."

24. See, e.g., Alan Pickles, "Experiências com insecticida no combate à cigarrinha dos canaviais," *Brasil açucareiro* 30, no. 6 (1947): 59–60; P. E. Turner, "Para dispor da folhagem seca antes de plantar," *Brasil açucareiro* 29, no. 2 (1947): 64.

25. Caminha Filho published early and often in *Brasil açucareiro*. For an early report on his efforts at breeding hybrids, see Caminha Filho, "A cultura da canna de açucar," *Brasil açucareiro* 2, no. 1 (1934): 9.

26. J. C. de Andrade, *Escorço histórico*, 183; *Brasil açucareiro* 5, no. 2 (1935): 61–67.

27. Dantas, *Suscetibilidade de algumas variedades de cana*, 7. Várzea, *Geografía de açúcar*, 261.

28. González, "Law, Hegemony, and Politics," 180–98.

29. Barros, *Década 20*, 45.

30. Beattie, *Human Tradition*, 121–23.

31. Torres, *Condições de vida*. Vasconcelos Torres, "A mobilidade do trabalhador nas zonas canavieiras do Brasil," *Brasil açucareiro* 24, no. 2 (1944): 81–82. Vasconcelos Torres, "O salário do trabalhador na agro-indústria do açúcar," *Brasil açucareiro* 24, no. 3 (1944): 78–79. Vasconcelos Torres also wrote a biography of Oliveira Vianna: Torres, *Oliveira Vianna*.

32. *Arquivos do Instituto de Pesquisas Agronômicas*, i–iv.

33. The idea of "nature's economy," the need for wise conservation in the interest of economic progress, and an explicit, indeed assumed, link between the stewardship of the environment and the reaping of economic rewards has deep roots in Western culture. For the formulation above, see Anker, *Imperial Ecology*, 1.

34. Strauss, "Cana de açúcar e fertilidade," 323.

35. E. A. Andrade, *CPRH 25 anos*, 7–9. Franscisco Dario Mendes da Rocha, "A propósito de um problema: A poluição dos Rios do Açúcar," *Brasil açucareiro* 69, no. 1 (January 1967): 26.

36. Gilberto Freyre attested to the influence of Vasconcelos Sobrinho's study of "man" in his "ecological niche" on other natural and social scientists in the state. Vasconcelos Sobrinho, *Regiões naturais*, 8, 171–73. Vasconcelos Sobrinho headed IPA's Botanical Section. See *Arquivos do Instituto de Pesquisas Agronômicas*, i.

37. Dalmiro Almeida, "O problema do reflorestamento e as usinas," *Brasil açucareiro* 27, no. 1 (1946): 66.

38. A. Fernandes, *Senhor de engenho*, 27–87. *Homem e terra*, 116, 61. Oton Bezerra refers to seven thousand workers, an administration chart from the early 1940s mentions fifty-five hundred. Andrade and Andrade, *Cana-de-açúcar*, 80. The Andrades write that Catende encompassed eighty *engenhos*.

39. *Homem e terra*, ii, 10, 44. Barbosa Lima Sobrinho, "Prefácio-testemunho," 28. When IAA President Sílvio Bastos Tavares visited Pernambuco early in 1951 and toured Catende, he said he had modified "certain impressions formed at a distance, according to which the northeastern worker was a helpless creature abandoned by all." "Visita do presidente do I.A.A. ao estado de Pernambuco," *Brasil açucareiro* 37, no. 4 (April 1951): 66.

40. See Sales, *Ministério da Agricultura*.

41. *Homem e terra*, 6, 147. Hawaii boasted yields of up to 150 tons per hectare. A. Fernandes, *Senhor de engenho*, 90. Sales, *Ministério da Agricultura*, 61.

42. Sales, *Hawaii açucareiro*.

43. *Homem e terra*, 80–120.

44. Octávio Valsechi, "Aspectos da economia açucareira do Nordeste brasileiro," *Brasil açucareiro* 29, no. 4 (1947): 58.

45. Dantas, "Recuperação da lavoura," 43–46.

46. *Homem e terra*, 88, 89, 116, 61; M. C. de Andrade, *História das usinas*, 91, 99, 109.

47. Adrião Caminha Filho, "A cigarrinha dos canaviais em Sergipe," *Brasil açucareiro* 25, no. 1 (1945): 92–96. In his survey of Sergipe's sugar industry, Caminha found that most Sergipano planters relied on the same two varieties of cane that most Brazilian planters used heavily—POJ 2878 and CO 290. The balance tipped heavily toward POJ 2878 in the fields of less advanced mill owners, without newer cooking machinery capable of eking higher yields out of CO 290 juice. The same year, another *Brasil açucareiro* writer lamented the "antiquated forms of planting and treatment of cane fields." Adelino Deícola dos Santos, "Modernização da lavoura canavieira," *Brasil açucareiro* 25, no. 6 (1945): 60.

48. Octávio Valsechi, "Aspectos da economia açucareira do Nordeste brasileiro," *Brasil açucareiro* 29, no. 4 (1947): 60.

49. "Laudo da Avaliação da Safra Pendente à corte, relativa ao anno agrícola 1940–41," in SSP 921: Usina Frei Caneca, Acervo DOPS, APE.

50. "Congresso açucareiro do Nordeste," *Brasil açucareiro* 29, no. 5 (1947): 53.

51. Strauss, "Cana de açúcar e fertilidade," 325.

52. Mário Lacerda de Melo, "Problemas agrícolas e industriais do açúcar em Pernambuco," *Brasil açucareiro* 28, no. 6 (1946): 52. The Paulistas went from 3 million sacks of sugar to 11.5 million. Gileno Dé Carli, "Direito de sobrevivência de Pernambuco," *Brasil açucareiro* 42, no. 6 (December 1953): 57.

53. "Industrial Antônio Ferreira da Costa Azevedo," *Brasil açucareiro* 35, no. 4 (1950): 30.

54. M. B. de Carvalho, "Comentário."

55. "Congresso açucareiro do Nordeste," *Brasil açucareiro* 29, no. 5 (1947): 53.

56. "Um milhão de cruzeiros para a defesa dos canaviais nordestinos," *Brasil açucareiro* 30, no. 1 (1947): 74.

57. Gonsalves de Mello, "Notas acêrca da introdução de vegetais," 33–64; Lima, *Estudos fitogeográficos*.

58. João da Costa Carvalho Lima and Fernando Basto Lima, *Fomento de práticas fitossanitárias na lavoura canavieira*, Comissão de Combate às Pragas da Cana de Açúcar no Estado de Pernambuco, publication no. 13 (1959).

59. McCook, "Promoting the 'Practical,'" 53–54.

60. Ayala, *American Sugar Kingdom*, 235–40.

61. Rosa e Silva Neto, *Contribuição ao estudo da zona da mata*, 131.

62. Mário Coelho, *Instruções para a instalação de pluviômetros e o registro das suas observações*, Comissão de Combate às Pragas da Cana de Açúcar no Estado de Pernambuco (Secção Técnica), publication no. 1 (October 1954).

63. As mentioned earlier, POJ denotes canes from Java and CO canes from Coimbatore. CP designates canes from the Canal Pointe research center in Florida. POJ 2878 was so attractive in part because of its very high sucrose concentrations, often 40 to 50 percent greater than varieties grown earlier. Sampaio, *Problema açucareiro*, 5.

64. Dantas et al., "Novas variedades," 5–6. The research team that completed this study focused on 3X. The disease to which CO 290 was susceptible was "red rot," 5.

65. In the early 1950s, São Paulo researchers experimented with 3X and ultimately rejected it because of its susceptibility to a fungus not prevalent in Pernambuco. Almeida et al., "Florescimento na variedade," 157–81.

66. Dantas and Santos e Silva, *Subsídios*, 15.

67. Sugarcane Breeding Institute, "Breeding," <http://sugarcane-breeding.tn.nic.in/breeding.htm>.

68. Antônio Lazzarini Segalla, "Novas variedades de cana de açúcar," *Brasil açucareiro* 35, no. 3 (1950): 46; "Variedades de cana mais recomendáveis para Pernambuco," *Brasil açucareiro* 35, no. 5 (1950): 39. IAA, *10 congresso*, 134–35.

69. J. C. de Andrade, *Escorço histórico*, 148–50.

70. Ibid., 150.

71. Ibid. IANE stood for Instituto Agronómico do Nordeste (Northeastern Agricultural Institute).

72. Dantas, "Recuperação da lavoura," 13.

73. Ayres Silveira Sousa, "Considerações sôbre algumas variedades de cana," *Jornal do comércio*, February 3, 1956.

74. Dantas, "Recuperação da lavoura," 15. Research in 1958 also showed that 3X was one of the most resistant varieties to another prevalent cane pest. Dantas, *Suscetibilidade de algumas variedades de cana*, 9; Rosa e Silva Neto, *Contribuição ao estudo da zona da mata*, 1.

75. Junqueira and Dantas, "Cana-de-açúcar," 39.

76. In Portuguese, ratoons are called *socas* or *folhas*.

77. Agronomists felt that a harvest should be comprised of equal parts first-growth cane, ratoon cane, and second ratoons. A 1965 study revealed, however, that in 1964, 48 percent of harvest cane was first growth, 42 percent was ratoon, and 10 percent came from second ratoon. Dantas, "Recuperação da lavoura," 13.

78. Dantas et al., "Novas variedades," 6. In a later publication, Dantas wrote that 3X can withstand "near abandonment." Dantas, *Agroindústria canavieira*, 68.

79. Wanderley, *Capital e propriedade fundiária*, 63. Dantas, "Recuperação da lavoura," 16; Dantas et al., "Novas variedades," 6.

80. Dantas and Santos e Silva, *Subsídios*, 14.

81. Bento Dantas and Mário Lacerda de Melo, "As áreas cultivadas com as atuais variedades na zona canavieira de Pernambuco," *Boletim canavieiro* 5, no. 54 (September 1959): 4.

82. Renato Farias, "Ainda a desnecessidade da criação de nova estação experimental de cana," *Boletim canavieiro* 6, no. 65 (1960): 7. This journal was the official organ of Pernambuco's Association of Cane Suppliers.

83. Bento Dantas includes a chart in a 1971 publication confirming that even when mature, 3X provides a lower concentration of sucrose than other varieties at their moment of maturity. Dantas, *Agroindústria canavieira*, 68. "Por um melhor rendimento dos canaviais de Pernambuco," *Boletim canavieiro* 6, no. 69 (1960): 1. J. C. de Andrade incorrectly dates Veiga's visit to 1961, but he attributes to the event an increased attention among planters to questions of agricultural technology. J. C. de Andrade, *Escorço histórico*, 157. Veiga was respected by his Pernambucan colleagues for having bred some successful varieties at the Campos experimental station; Junqueira and Dantas, "Cana-de-açúcar," 35. Before Veiga's visit, the director of Pernambuco's Usina Cooperative had asked the IAA for help in securing new cane varieties well-adapted to the state's conditions. "Novas variedades de cana para Pernambuco," *Brasil açucareiro* 58, nos. 1–2 (1960): 7.

84. Rosa e Silva Neto, *Contribuição ao estudo da zona da mata*, 81.

85. "Renovação dos canaviais melhorará o rendimento: Recomendam técnicos do IAA," *Diário de Pernambuco*, May 12, 1963.

86. E. de S. L. Pinto, "Considerações agronômicas," 13–14.

87. Quoted in Rosa e Silva Neto, *Contribuição ao estudo da zona da mata*, 81, 76.

88. "As cigarrinhas nos canaviais do Brasil," *Brasil açucareiro* 68, no. 3 (September 1966): 9. Gilberto Miller Azzi, "Medidas de precaução recomendadas para a introdução de novas variedades de canas em uma zona produtora," *Brasil açucareiro* 69, no. 3 (March 1967): 55.

89. Cabral, "Processo de proletarianização," 170.

90. "Esgotamento das terras da zona cannavieira de Pernambuco," *Brasil açucareiro* 7, no. 6 (1936): 452. Diégues, *População e açúcar*, 200; Junqueira and Dantas, "Cana-de-açúcar," 38.

91. Rosa e Silva Neto, *Contribuição ao estudo da zona da mata*, 81. M. L. de Melo, *Paisagens do Nordeste*, 102.

92. José Leite, "A assistência médico-hospitalar, nas usinas de Pernambuco, à luz de um inquérito," *Brasil açucareiro* 35, no. 1 (January 1950): 92.

93. Bello, *Memórias*, 133–34.

94. Bastos, *Ligas camponesas*, 58.

95. Celso Filho, "O estatuto da lavoura canavieira e a política demográfica," *Brasil açucareiro* 24, no. 3 (1944): 86. The fear of worker mobility was a chronic feature of long-term postemancipation history in tropical commodities.

96. Barbosa Lima Sobrinho, *Problemas econômicas e sociais*, 216–17. Barbosa Lima's distinction between mill workers and field workers is a false one, since a single worker frequently occupied both categories at different points in her or his career. Vasconcelos Torres, "A mobilidade do trabalhador nas zonas canavieiras do Brasil," *Brasil açucareiro* 24, no. 2 (1944): 81.

97. Eudes de Souza Leão Pinto, interview with Eliane Moury Fernandes, tape recording and transcript, February 23, 1983 (CEHIBRA), 93. Pinto also claimed the federal minister of agriculture was involved in the effort. That was not covered in the only other mention of this effort I have found: "Problemas açucareiros do Nordeste numa entrevista do presidente do I.A.A.," *Brasil açucareiro* 41, no. 4 (April 1953): 47. This article highlights the efforts of the archbishop of Olinda and Recife, Dom Antônio de Almeida Morais Júnior.

98. José Sérgio Leite Lopes covers the various routes workers took to mill employment for a slightly later period (1960s–70s). He makes it clear that patronage had a great deal to do with securing the prized jobs in the *usina*. Leite Lopes, *Vapor do diabo*, 152–53.

99. "Escandalosa frasa a generosidade de 'Tenente,'" *Folha da manhã*, August 18, 1945; "Esclarecimentos da Usina Catende, S.A.," *Jornal do comércio*, August 26, 1945.

100. Cabral, "Processo de proletarianização," 170. True to Cabral's claim, Catende's production expanded from 304,000 sixty-kilogram sacks of sugar in 1934 to 676,000 in 1954 and 736,000 in 1964. M. C. de Andrade, *História das usinas*, 91, 99, 109.

101. Sidney Mintz's thirty-year-old essay on rural proletarians is still a useful tool for thinking through the puzzles of agricultural rationalization and the consequences for workers. Mintz, "Rural Proletariat," 34–51.

102. Palmeira, "Casa e trabalho," 309.

103. Contrary to those who argue that daily wages ended abruptly in the 1930s, displaced by payment according to task work, this was a long, uneven process. For the former argument, see, e.g., Cabral, "Processo de proletarianização," 171.

104. Dean, *Rio Claro*, 174–80. Montrie, *Making a Living*, 45–46.

105. Mintz, *Worker in the Cane*, 70.

106. Várzea, *Geografía de açúcar*, 261–63. Várzea cited *tarefa* sizes of 3,025, 4,400, and 3,630 square meters and others that varied randomly from property to property.

107. Galloway, *Sugar Cane Industry*, 71.

108. For a comparatively recent account of a beating with a *vara*, see, e.g., "Apenas 10 usinas estão funcionando," *Diário de Pernambuco*, September 30, 1980. The article describes a confrontation during the strike of that year, when a union delegate happened on a planter distributing tasks to scab workers. After a short, bitter exchange, the planter picked up a *vara* and used it to beat the delegate.

109. Palmeira, "Aftermath of Peasant Mobilization," 82.

110. Mário Lacerda de Melo, "Problemas agrícolas e industriais do açúcar em

Pernambuco," *Brasil açucareiro* 28, no. 6 (1946): 51. *Brasil açucareiro* reprinted in 1939 an article from the Argentine sugar journal that catalogued wage levels in the British West Indies, Fiji, India, Mauritius, Sri Lanka, Nigeria, Uganda, Singapore, and the United States. "Salarios de trabalhadores rurais," *Brasil açucareiro* 13, no. 6 (August 1939): 69–70. The wage level at the beginning of the 1960s was 450 cruzeiros, but most planters paid 150–250 cruzeiros to men and half that to women and children performing the same work. Bezerra, *Memórias*, 2:175.

111. Mário Lacerda de Melo, "Problemas agrícolas e industriais do açúcar em Pernambuco," *Brasil açucareiro* 28, no. 6 (1946): 51. Melo's argument fits into a tradition of liberal reformers seeking to yank "backward" populations into modern, liberal, capitalistic society. For a remarkable parallel of his analysis, see the discourse of colonial agents working to manage the transition from slavery to free labor in the British West Indies, analyzed by Thomas Holt in *Problem of Freedom*, chap. 2.

112. Diégues, *População e açúcar*, 128. Diégues also noted that the *conta* was known in some areas of the Northeast as the "ticuca." Sigaud, "Percepção do salário," 321. Sigaud offers the following quick chronology of work modes: First, the *eito*, then the *tarefa* in the 1930s (625 *braças*), then the *conta* (ten *braças*). Sigaud, *Clandestinos e direitos*, 136–37.

113. Paulo, interview with the author, tape recording, Murupé, Vicência, March 27, 2003.

114. Severino, interview with the author, tape recording, Vicência, April 1, 2004.

115. J. C. Scott, *Seeing Like a State*, 22–25.

116. R. F. de Carvalho, *Melhoramento e mecanizagem*, 34.

117. Dantas, "Recuperação da lavoura," 13.

118. Mintz, *Worker in the Cane*, 136–37.

119. Rottenberg, "Negotiated Wage Payments," 403.

120. "Ligas camponesas, agindo com outra denominação, ameaçam a tranquilidade, em Goiana," *Jornal do comércio*, April 21, 1959, in SSP 7956: Liga Camponesa de Goiana, Acervo DOPS, APE.

121. "Relatório das ocorrências observadas neste município, do dia 30 do mês findo, até a presente data," November 13, 1959, in SSP 7956, Acervo DOPS, APE.

122. Dantas, "Recuperação da lavoura," 39–40.

123. In his study of mill workers, José Sérgio Leite Lopes also heard plenty of testimony from rural workers about the manipulation of task size and remuneration. Leite Lopes, *Vapor do diabo*, 165.

124. After a worker completed a task, the *engenho* "fiscal," or inspector, would study the work done, almost always reducing the worker's wages because of allegedly poor work. Sigaud, "Idealization of the Past," 181. Sigaud, "Percepção do salário," 318–20.

125. David Montgomery offers the most eloquent discussion of these issues for the early-twentieth-century United States in *Workers' Control in America*; especially relevant are chaps. 1–3.

126. Relatório, June 11, 1960, in SSP 29343: Ligas Camponesas; engenho Malemba (Paudalho), Acervo DOPS, APE. The fact that they described a *braça* as 2.1 meters rather than 2.2 points to the fact that many of these norms were still being negotiated at the time.

127. Benedito, interview with the author, tape recording, Engenho Humaitá, Palmares, January 25, 2003.

128. These cases are valuable precisely because they include testimony from people

describing in some detail their work relationships and responsibilities. Because the bureaucracy of the labor courts, of which the *juntas de conciliação e julgamento* are the local bodies, was only set up for rural areas in the wake of 1963's Estatuto do Trabalhador Rural, there are no records for earlier periods. However, my argument here is that the language and practices of work organization were changing at precisely this time; in part, I also hold, as a result of the new mediating institutions of the labor courts.

129. JCJ Palmares, processo 541/64: *João Rodrigues do Nascimento vs. Engenho Camorim Grande (Cirilo Pereira de Souza)*, hearing March 6, 1964, A-TRT. Another plaintiff also spoke of getting paid for each *conta* completed while the opposing lawyer said he worked by the task. JCJ Nazaré da Mata, processo 86/63: *Manoel Cícero Joaquim de Santana vs. Engenho Caraúba*, hearing July 18, 1963, A-TRT. JCJ Palmares, processo 392/64: *Maria Firmino da Silva vs. Dr. Antonio Siqueira (Engenho São Gregorio)*, hearing October 15, 1963, A-TRT. JCJ Nazaré da Mata, processo 88/63: *José Pedro do Nascimento vs. Engenho Pindoba*, hearing June 26, 1963, A-TRT.

130. E.g., JCJ Nazaré da Mata, processo 88/63, hearing June 26, 1963.

131. "Polícia de Catende ausentou-se, mas camponeses ordeiros frustraram as agitações em 'Roçadinho'," *Diário de Pernambuco*, July 12, 1963. The author of the article actually referred to the *conta* as ten *meters* square, rather than ten *braças*. This is surely a product of the reporter's ignorance (ironic considering the condescending and sarcastic tone of the article).

132. Sampaio, *Problema açucareiro*, 10.

133. Sampaio, interview with Aspásia de Alcântara Camargo, Dulce Pandolfi, and Eduardo Raposo, April and October 1979, 65.

134. Sampaio's views echoed the frustrated comments of a Mississippi Freedmen's Bureau officer who in 1867 complained that many freedpeople were "inexcusably idle and slothful," managing to get by just fishing and hunting. Chad Montrie points out that through hunting and fishing, those freedmen likely "sought and found supplements to family subsistence, freedom from oversight and physical abuse, as well as opportunities for asserting their masculinity." The analysis here is similar, that the "strike for lower wages" was a bid for greater freedom, from the repressive control of field work. Montrie, *Making a Living*, 35–36.

135. Junqueira and Dantas, "Cana-de-açúcar," 38.

136. Maciel, *Nível de vida do trabalhador*, 36–38. *Levantamento sócio-econômico*, 47.

137. Cabral, "Processo de proletarianização," 170–71.

138. The planters' concern parallels the early-twentieth-century case of business and government interests associated with California agriculture, who tried to achieve a "social 'fixing' [of labor], in order to protect the productive landscape." In that case, it was also in planters' interest to assure that "labor remained *geographically* mobile," so that it would be perpetually available. Mitchell, *Lie of the Land*, 80. This same desire to fix rural workers was the ideological rationale for later government- and planter-initiated efforts at agrarian reform.

Chapter 5

1. Investigator Oliveira came from the Delegacía de Ordem Política e Social, or DOPS, an agency created in 1924 and closed in 1982 that operated in every state and had links to

federal police programs. The DOPS "was designed to control civil society, which it sought to do through a system of vigilant spying, intimidation, arrests, torture, and seizure of personal property." Negro and Fontes, "Using Police Records," 16–17.

2. "Relatório ao delegado auxiliar," January 22, 1962, p. 2 in SSP 29201: Liga Camponesa de São Lourenço da Mata, Acervo DOPS, APE. The *Diário de Pernambuco* covered the incident on January 17, and Oliveira mentions national attention.

3. Castro, *Zona explosiva*. The English version was published as *Death in the Northeast*.

4. Montenegro, "Labirintos do medo," 223–24. Montenegro independently found the Oliveira case that I had discovered in 2002, and he mentions it in his article.

5. Andrade and Andrade, *Cana-de-açúcar*, 12.

6. Mello, "Mata norte e mata sul," 84–91.

7. Wanderley, *Capital e propriedade fundiária*, 69. The "transition from *engenho* to *usina* production" has become a reified category in scholarship about the *zona da mata*, and we must always keep in mind that this was not a moment, or a swift and decisive process. It was a long, gradual, and uneven movement from one internally differentiated regime to another. Writing in 1975, Mário Lacerda de Melo underlines the heterogeneity of the process and the fact that production levels were *still* quite irregular from place to place in the region. M. L. de Melo, *Açúcar e homem*, 45.

8. Mello, "Mata norte e mata sul," 84; Eisenberg, *Sugar Industry in Pernambuco*, 242–43; Rosa e Silva Neto, *Contribuição ao estudo da zona da mata*, 28–30.

9. M. L. de Melo, *Paisagens do Nordeste*, 102.

10. *Anuário estatístico de Pernambuco, 1959*, 21. Fundação Joaquim Nabuco, CEHIBRA, Arquivo Museu do Açúcar, folder: "Estatísticas," "Produção de cana-de-açúcar, segundo os municípios."

11. *Anuário estatístico de Pernambuco, 1959*, 69. The numbers were similar for surrounding years (i.e., this was not a one-year fluke). Maciel, *Nível de vida do trabalhador*, 42.

12. Bento Dantas and Mário Lacerda de Melo, "As áreas cultivadas com as atuais variedades na zona canavieira de Pernambuco," *Boletim canavieiro* 5, no. 54 (September 1959): 4. Bento Dantas e José Lacerda de Melo, *Proporção entre canaviais de várzea e de ladeira, em Pernambuco* and *A participação das diversas fôlhas na cana de moagem*, Comissão de Combate às Pragas da Cana de Açúcar no Estado de Pernambuco, publications nos. 10 and 11 (Recife: April 1959), 3.

13. Maciel, *Nível de vida do trabalhador*, 38.

14. Paulo, interview with the author, tape recording, Murupé, Vicência, March 21, 2003.

15. Sigaud, "Idealization of the Past," 180.

16. JCJ Nazaré da Mata, 213/63.

17. This was one aspect of life that Bello found so comforting, especially when former slaves remained on his *engenho* following abolition. Bello, *Memórias*, 216–22.

18. Maciel, *Nível de vida do trabalhador*, 18–22.

19. Ibid., 25–32; 18–21.

20. Castro, *Death in the Northeast*, 3.

21. Bezerra, *Memórias*, 1:139–47. Skidmore, *Politics in Brazil*, 20–23.

22. Chilcote, *Brazilian Communist Party*, 156–57. See SSP 7601: Liga Camponesa de Beberibe; SSP 7581: Liga Camponesa de Ibura; SSP 29187: Liga Camponesa de Moreno; and SSP 7947: Liga Camponesa de Iputinga, all from Acervo DOPS, APE. There were also periodic attempts to make more significant inroads. In 1945, Gregório Bezerra reports his

involvement in a regional organizing push in the *agreste*, immediately to the west of the *zona da mata*. Bezerra, *Memórias*, 1:329.

23. Welch, *Seed Was Planted*, 215, 92.

24. Letter from Diniz de Souza Leão, 15 October 1946, in SSP 7581, Acervo DOPS, APE.

25. Police documentation on this early cohort of peasant leagues is scarce. One such notice, from 1950, concerns the establishment of a league in Escada. See SSP 7602: Liga Camponesa de Escada, Acervo DOPS, APE. For the boilerplate text, see SSP 29796: Ligas Camponesas (1950), Acervo DOPS, APE. A document dated 1947 reads "Estatutos da Liga Camponeza de ———."

26. Wilkie, "Report on Rural Syndicates." Decree Law no. 7,038 was passed on the seventh anniversary of the Estado Novo, on November 10, 1944. Brayner, *Partido Comunista*, 71.

27. See the string of correspondence from state and federal authorities trying to verify with local *cartórios* the legal registration of the union in Carpina and São Lourenço: SSP 1606: Sindicato dos Trabalhadores Rurais de São Lourenço, Acervo DOPS, APE. A 1956 letter from the minister of labor to the secretary of public security in Pernambuco confirmed that "*the* rural union of this state, recognized by this Ministry, with an official letter on January 3, 1956, is of the rural workers of Barreiros, Rio Formoso, and Serinhaem." SSP 1634: Sindicato dos Trabalhadores Rurais de Rio Formoso, Acervo DOPS, APE.

28. *Jornal do comércio*, January 1, 1958, and January 21, 1959, and Letter to Ministro de Trabalho, Indústria e Comércio, September 13, 1956, all in SSP 28857: Sindicato dos Trabalhadores Rurais de Barreiros, Acervo DOPS, APE.

29. Martins, *Camponeses e política*, 81. *Fôlha do povo*, September 21, 1954, clipping in SSP 28702: Associação dos Trabalhadores Agrícolas e Camponeses de Pernambuco, Acervo DOPS, APE.

30. Welch provides a rich account of the national conference; Welch, *Seed Was Planted*, 189–96.

31. Brayner, *Partido Comunista*, 72, 63–64. For surveillance reports on suspected militants, see SSP 28856: Sindicato dos Trabalhadores Rurais de Carpina, Acervo DOPS, APE; SSP 31215: Sindicato dos Trabalhadores Rurais de Paudalho, Carpina, e São Lourenço, Acervo DOPS, APE.

32. Julião, *Cambão*, 96. The peasants' modest aims contrast sharply with Julião's characterization of the leagues' history: "The *latifundio* is a wasteland in the North-East as it is everywhere else. And on that wasteland a flower was born big and beautiful as the enormous water lily of the Amazon: the League. It developed fast, like flames through dry straw, spread by winds blowing from Cuba."

33. Page, *Revolution That Never Was*, 42.

34. Soares, *Frente do Recife*, 15–18; Bezerra, *Memórias*, 2:169.

35. Sampaio, *Mensagem apresentada*, 6.

36. Bastos, *Ligas camponesas*, 20.

37. In an interview from 1984, Cid Sampaio said he convened discussion meetings with other northeastern governors in the mid-1950s that led to the establishment of SUDENE. Cid Feijó Sampaio, interview with Eliane Moury Fernandes, tape recording and transcript, Recife, April 1984 (CEHIBRA), 5. This chapter draws on a series of interviews held in CEHIBRA as part of a project on the memory of 1964 carried out by the unit of the Fundação Joaquim Nabuco. Several years after I used them at the Fundação, some in transcribed and some in audio form, a catalog of the interviews was published with the transcripts of

several. Andrade and Fernandes, *Vencedores e vencidos*; and E. M. Fernandes, *Movimento político-militar*. F. de Oliveira, *Elegia*.

38. F. de Oliveira, *Elegia*, 115–16.

39. Besides Callado's books, see Barreto, *Julião*; Horowitz, *Revolution in Brazil*; E. Pinto, *Problema agrário*; Price, "Brazilian Land Reform Statute"; Castro, *Death in the Northeast*; Rosa e Silva Neto, *Contribuição ao estudo da zona da mata*; Hewitt, "Brazil"; Page, *Revolution That Never Was*; Leeds, "Brazil"; Oliveira, *Elegia*; Mallon, "Peasants and Rural Laborers"; Palmeira, "Aftermath of Peasant Mobilization"; Azevedo, *Ligas camponesas*; Bastos, *Ligas camponesas*; and F. de A. L. de Souza, *Nordeste*.

40. Callado, *Industriais da sêca*. That book was followed by *Tempo de Arraes*. The coup took place between the reporting for *Tempo de Arraes* and its publication in book form.

41. Skidmore, *Politics in Brazil*, 218.

42. Callado, *Tempo de Arraes*, 31, 87, 101.

43. Ibid., 83. "Brazil is such a timid country that until now it hasn't had a history," 85.

44. Maciel, *Nível de vida do trabalhador*, 9, 40–47. Gonçalves, "Condições de vida do trabalhador," 117.

45. Pearson, "Small Farmer and Rural Worker Pressure Groups," 162; Martins, *Camponeses e política*, 81.

46. De Kadt, *Catholic Radicals*, 102. Paulo Freire's educational theories played a prominent (though brief) role in the popular literacy movement of the early 1960s. On Freire, see Kirkendall, "Entering History," 168–89.

47. Pearson, "Small Farmer and Rural Worker Pressure Groups," 150–51. It must be noted that Catholic radicals did not form a homogenous group. There were deep divisions between AP activists, for instance, and their more moderate colleagues. The AP advocated radical modifications to the country's social and political structure.

48. M. C. de Andrade, *Terra e homem*, 13.

49. Tad Szulc, "Northeast Brazil Poverty Breeds Threat of a Revolt," *New York Times*, October 31, 1960. Readers with the Cuban Revolution fresh in their minds would have to wait only until the third paragraph before Castro appeared the article: according to Szulc, Castro was presented to students and peasants as a hero. Szulc, "Marxists Are Organizing Peasants in Brazil," *New York Times*, November 1, 1960. Stefan Robock emphasized the importance of Szulc's articles, along with an ABC documentary called *The Troubled Land*, to the awareness among North Americans of the Brazilian Northeast. Robock, *Brazil's Developing Northeast*, 5.

50. Tad Szulc, "Northeast Brazil Poverty Breeds Threat of a Revolt," *New York Times*, October 31, 1960. The *Times* continued to cover Pernambuco, which was also the subject of several popular and academic books. Ongoing events in the state are the heart of Irving Horowitz's *Revolution in Brazil* (1964).

51. In Ecuador, for example, peasant organizations and rural unions spread in the 1940s and 1950s. Striffler, *In the Shadows of State and Capital*, 52.

52. "Kennedy Will Begin Brazil Visit July 30," *New York Times*, May 25, 1962. Telegram, Rio to Recife, July 27, 1961, and Rio to Recife, October 24, 1961, Classified General Records, 1956–63, RG 84, NARA-US. In the latter telegram, U.S. Ambassador Lincoln Gordon and Secretary of State Dean Rusk advise the Recife consul, "Presentation [of the generator] should, of course, make suitable references to fact gift results from Edward Kennedy's recent visit and illustrative President's personal interest plight Brazilian farmers."

53. Ernest S. Guaderrama, "Operations Memorandum: Request for Additional Personnel," quoted in Kirkendall, "Entering History," 168. Skidmore, *Politics in Brazil*, 217. Levinson and Onís, *Alliance*, 244–46.

54. Welch, *Seed Was Planted*, 260–61, 273–74. Padre Paulo Crespo, interview with Eliane Moury Fernandes and Jorge Zaverucha, tape recording and transcript, Recife, July 10, 1985 (CEHIBRA), 15.

55. "Mensagem apresentada pelo exmo. Sr. Governador Miguel Arraes de Alencar à Assembléia Legislativa do Estado em 15 de março de 1963,", Dep. III, Prat. D, Vol. 281, Acervo Governo do Estado, APE.

56. Sigaud, "Idealization of the Past," 169. Moura also reports the importance of Arraes's order. S. R. de Moura, *Memórias de um camponês*, 157. Dorany Sampaio's praise for Arraes gives an indication of how the police had behaved in the past: "He never used force, or the hoof of a horse, or dogs against the peasants." Dorany Sampaio, interview with Eliane Moury Fernandes, tape recording and transcript, Recife, July 22, 1982 (CEHIBRA), 25–26.

57. Francisco Julião, "Carta de Alforria do Camponês," *A liga* 6 (1963): 4.

58. Paulo, interview with the author, March 21, 2003. Severino, interview with the author, tape recording, Vicência, April 1, 2003.

59. Callado, *Tempo de Arraes*, 92. Bezerra, *Memórias*, 2:173–74.

60. Paulo, interview with the author, March 21, 2003. Severino, interview with the author, April 1, 2003.

61. "Unions Compete in Rural Brazil," *New York Times*, April 30, 1963.

62. Wilkie, "Report on Rural Syndicates," 1–16.

63. Hewitt, "Brazil," 391; Pearson, "Small Farmer and Rural Worker Pressure Groups," 162–64.

64. "Iniciada terceira greve em engenho da Usina Bulhões," *Diário de Pernambuco*, May 10, 1963. "Novas greves de camponeses e incidentes com proprietários rurais na Zona da Mata," *Diário de Pernambuco*, May 7, 1963. The article stated: "The aim of the stoppage was to cut the foreman's *braça*."

65. *Diário de Pernambuco*, May 22, 24, 25; June 7; July 12, 26; August 3, 11, 13; September 4, 12, 21, 22, 24, 25, 27; October 1, 2, 3, 4, 6, 8, 10, 13, 22, 23, 25, 26; November 7, 1963.

66. "Querem os sindicatos rurais sob orientação do Padre Crespo," *Diário de Pernambuco*, May 18, 1963; "Engenhos pararam em Jaboatão: Luta de sindicatos rurais," *Diário de Pernambuco*, May 22, 1963.

67. "Camponeses dos sindicatos rurais furaram o cêrco policial e protestaram contra os pelegos vermelhos," *Diário de Pernambuco*, September 3, 1963. Pearson discusses the Arraes government's efforts to "destroy SORPE's influence in the Pernambuco sindicatos" and the battle for Jaboatão. Pearson, "Small Farmer and Rural Worker Pressure Groups," 166, 170–72.

68. Wilkie, "Report on Rural Syndicates," 1–16.

69. Bezerra, *Memórias*, 2:175–79.

70. "Brazil Strike Brings Violence," *New York Times*, November 19, 1963.

71. "Paro totalmente atividade dos trabalhadores da cana," *Diário de Pernambuco*, November 20, 1963. Planters signed the agreement ending the strike with the understanding that the IAA would raise sugar prices, foreshadowing a pattern of planter strategy during strikes that would emerge in 1979 and the early 1980s. Hewitt, "Brazil," 397.

72. Quoted in L. Pereira, *Trabalho e desenvolvimento*, 121.

73. Sigaud, "Idealization of the Past."

74. Benedito, interview with the author, tape recording, Engenho Humaitá, Palmares, January 25, 2003.

75. Severino brought this up twice in an interview with the author, April 1, 2003. Bezerra also notes the importance of having a bed with a mattress. Bezerra, *Memórias*, 2:179. Of more than four thousand people surveyed by Maciel, only 1,294 reported owning beds. Maciel, *Nível de vida do trabalhador*, 25–32.

76. "Dr. Miguel Arrais" [sic], *Brasil açucareiro* 26, no. 5 (1945): 79.

77. Hélio Mariano, interview with Eliane Moury Fernandes, tape recording and transcript, Recife, October 1983 (CEHIBRA), 23 (my emphasis).

78. Dorany Sampaio, interview with Eliane Moury Fernandes, 25–26.

79. Bonham C. Richardson writes, "Caribbean fire carries with it a strong social connotation, a distinction created by the region's human history of remarkable and persistent social oppression exerted by local planters and other officials, oppression that has in turn been met by the resistance and creativity of the region's working peoples. Slavery, insurrections, threats, and planned destruction in the region all have been accented by either fire or threats of fire." Richardson, *Igniting the Caribbean's Past*, 5–6, 99–125.

80. Antonil, in his 1711 guide to sugar agriculture, warns planters that careless slaves might inadvertently start fires with their pipes. Antonil, *Cultura e opulência*, 114, 84.

81. See, e.g., JCJ Catende, processo 28/71: *João Isidio da Silva vs. Catende*, A-TRT.

82. Joaquim Correia Xavier de Andrade, interview with the author, handwritten notes, Recife, January 29, 2003. Eudes de Souza Leão Pinto, interview with Eliane Moury Fernandes, tape recording and transcript, Recife, August 1983 (CEHIBRA), 92.

83. Letter to Secretario de Segurança Pública, January 22, 1947, in SSP 776: Usina Central Olho d'Agua, Acervo DOPS, APE. The *usina* also complained that an area of cane field had been destroyed by animals.

84. Report of February 6, 1947, in SSP 29179: Liga Camponesa de També, Acervo DOPS, APE.

85. Report of Aurino Xavier de Oliveira, Euclides de Souza Arruda, and Aluizio Carneiro, December 23, 1959, in SSP 29182: Liga Camponesa de Nazaré da Mata, Acervo DOPS, APE.

86. Workers also used fire to settle disputes among themselves. A worker told the author in a 2003 interview that sometimes *engenho* residents angry with a neighbor will resort to fire. Ronivaldo, interview with the author, tape recording, Engenho Humaitá, Palmares, January 25, 2003. (And this phenomenon was observed by the author.)

87. Hélio Mariano, interview with Eliane Moury Fernandes, 35.

88. "Crime contra a economia canavieira," *Boletim canavieiro* 1, no. 11 (February 1956): 1.

89. Letter to Secretário de Segurança Pública, February 4, 1956, in SSP 470: Companhia Uzina Tiúma, Acervo DOPS, APE.

90. Letter to Secretário de Segurança Pública, November 28, 1957, in SSP 470: Companhia Uzina Tiúma, Acervo DOPS, APE.

91. Letter to Secretário de Segurança Pública, February 4, 1956, in SSP 470: Companhia Uzina Tiúma, Acervo DOPS, APE.

92. Report, February 14, 1944, in SSP 806A: Usina Catende, Acervo DOPS, APE.

93. See, e.g., reports on investigations (November 18, 1956, and December 19, 1957) and another letter from a planter in SSP 7577: Liga Camponesa de Carpina, Acervo DOPS, APE. A report of an arrest in 1944 appears in SSP 810: Usina Salgado, Acervo DOPS, APE.

94. "Sem trégua a volúpia dos incendiários: Ofícios trocados entre o governador do Estado e o presidente da Associação dos Fornecedores de Cana de Pernambuco," *Boletim canavieiro* 5, no. 60 (March 1960): 1.

95. It appears that Sampaio's patrol may have been prepared before its formal announcement, since a report covering the period December 1959 through February 1960 covering cane fires appears in SSP 29796: Ligas Camponesas (1950), Acervo DOPS, APE. See "Relatório do serviço prestado pela Patrulha Volante da 3a Zona." In the same file, "Relação de 30 Sgt. Severino Batista" covers six *usinas* in the North, noting twelve fires. Another report, from December 5, 1960, covers *usinas* in the South.

96. "Revoltante desfaçatez," *Jornal do comércio*, January 17, 1960, in SSP 29184: Liga Camponesa de Maraial, Acervo DOPS, APE.

97. "Agricultura pode parar em Vicência: Incendiários agem," *Diário de Pernambuco*, August 3, 1963.

98. A fascinating and encyclopedic U.S. diplomatic report on the shape of Cid Sampaio's political machine documents his influence and control over both the *Jornal do comércio* and the *Diário de Pernambuco*. Edward Rowell, "The Institutional and Economic Support of the Sampaio Political Machine," April 27, 1966, pp. 10–11, in RG 59, box 1927, NARA-US. Thanks to John D. French for alerting me to this document.

99. Amaro Valentim do Nascimento, interview with Eliane Moury Fernandes, tape recording and transcript, Recife, May 1986 (CEHIBRA), 9; Padre Paulo Crespo, interview with Eliane Moury Fernandes, 28; Padre Wanderley Simões, interview with Eliane Moury Fernandes, tape recording and transcript, Recife, October 1985 (CEHIBRA), 6. Artur de Lima Cavalcanti, interview with Eliane Moury Fernandes, tape recording and transcript, Recife, May 1983 (CEHIBRA), 23. It is interesting how prominent a place fire occupies in memories of this period. Both Mariano and Sampaio responded to Eliane Fernandes's questions that mentioned "agitation, strikes, fires in the fields," so they were prompted somewhat to consider fire. But other people interviewed in the 1964 project brought up the subject of fires without suggestion. Joaquim Correia Xavier de Andrade, interview with the author, January 29, 2003.

100. Relatório ao Delegado Auxiliar, January 22, 1962, 2, in SSP 29201: Liga Camponesa de São Lourenço da Mata, Acervo DOPS, APE. The *Diário de Pernambuco*'s January 17 story printed a picture of the boy who claimed he saw the plane and the dropped object.

101. Relatório ao Delegado Auxiliar, January 22, 1962, 4, 5, 7, 10, in SSP 29201: Liga Camponesa de São Lourenço da Mata, Acervo DOPS, APE.

102. Report of Aurino Xavier de Oliveira, Euclides de Souza Arruda, and Aluizio Carneiro, December 23, 1959, in SSP 29182: Liga Camponesa de Nazaré da Mata, Acervo DOPS, APE.

103. "Cortador de cana prêso e espancado pela polícia: ateou fôgo aos canavias," *Jornal do comércio*, January 16, 1960, in SSP 29184: Liga Camponesa de Maraial, Acervo DOPS, APE.

104. Ibid., 14.

105. AMCON RECIFE to Dept. of State, January 23, 1962, in RG 59, box 1577, file 732.0011-262, NARA-US.

106. Report of Aurino Xavier de Oliveira, Euclides de Souza Arruda, and Aluizio Carneiro, December 23, 1959, in SSP 29182: Liga Camponesa de Nazaré da Mata, Acervo DOPS, APE.

107. Eudes de Souza Leão Pinto, interview with Eliane Moury Fernandes, 93.

108. Armando Samico, interview with Eliane Moury Fernandes, tape recording and transcript, Recife, May 1983 (CEHIBRA), 28–29.

109. Report to the Delegado Auxiliar, September 20, 1955, in SSP: 7887: Liga Camponesa de Jaboatão, Acervo DOPS, APE.

110. Eudes de Souza Leão Pinto, interview with Eliane Moury Fernandes, 93.

111. Interrogation of Severino José da Silva, January 8, 1960, in SSP 29,177: Liga Camponesa de Ribeirão, Acervo DOPS, APE.

112. S. R. de Moura, *Memórias de um camponês*, 160. "They are absolutely right, since they are all workers in the same industry, one that pays [mill workers] more," Moura wrote. Indeed, not only did they earn more than field workers, the mill workers won an additional raise in June. "Concedido aumento dos trabalhadores na indústria do açúcar," *Diário de Pernambuco*, June 19, 1963.

113. Gregório Bezerra, interview with Eliane Moury Fernandes, tape recording and transcript, Recife, May 1982 (CEHIBRA), 111.

114. Richardson, *Igniting the Caribbean's Past*, 116.

115. *Anuário estatístico de Pernambuco, 1959*, 165.

116. Naíde Regueira Teodósio, interview with Eliane Moury Fernandes, tape recording and transcript, February 1984 (CEHIBRA), 14.

117. Paulo Crespo, interview with Eliane Moury Fernandes, 28.

118. "Revoltante desfaçatez," *Jornal do comércio*, January 17, 1960.

119. Adelino Deícola dos Santos, "Modernização da lavoura canavieira," *Brasil açucareiro* 25, no. 6 (1945): 61.

120. Report of Aurino Xavier de Oliveira, Euclides de Souza Arruda, and Aluizio Carneiro, December 23, 1959, in SSP 29182: Liga Camponesa de Nazaré da Mata, Acervo DOPS, APE.

121. Jaime Rocha de Almeida, "Canas queimadas," *Brasil açucareiro* 27, no. 1 (1946): 51.

122. Mintz, *Worker in the Cane*, 202.

123. Ibid., 52–57.

124. "Talão de pesagem e cana queimada," *Boletim canavieiro* 4, no. 41 (August 1958): 8.

125. "Corte de canas queimadas," *Boletim canavieiro* 6, no. 68 (November 1960): 1.

126. Dantas, "Recuperação da lavoura," 39–40.

127. A. C. Barnes, "Incendios nos canaviais e seu controle," *Brasil açucareiro* 66, no. 2 (August 1965): 31.

128. Cláudia Hamasaki dates the cutting of burned cane to the 1960s. Hamasaki, "Setor sucroalcooleiro," 57. Severino, interview with the author, April 1, 2003.

129. Rufino de Araújo, *Trator e "burro sem rabo,"* 115.

130. *Levantamento das condições ambientais*, 57–58.

131. Andrade Neto, "Questão ambiental," 12. During the harvest, twenty thousand cubic meters of wash water are released by *usinas* every day (19).

132. A Pernambuco state planning document indicates that pre-harvest fires lower sucrose by 8 percent, while Joaquim Andrade claims a 30 percent loss. *Avaliação e perspectivas da agroindústria*, 14; Andrade Neto, "Questão ambiental," 4. Resende et al., "Long-Term Effects of Pre-harvest Burning," 339–51.

133. *Agenda 21*, 45. Cançado et al., "Impact of Sugar Cane-Burning Emissions," 725–29.

134. Strauss, "Cana de açúcar e fertilidade," 323.

Chapter 6

1. Bezerra, *Memórias*, 2:186, 193–98. There is a well-known picture of Bezerra (reproduced in his *Memórias*, 2:196) sitting in his underwear in the parking lot of the motor pool at the Casa Forte barracks in Recife. Three armed soldiers stand around him and on the wall behind him a painted exhortation to the troops is visible: "Fulfill your responsibility [*Cumpre o teu dever*]."

2. J. C. Scott, *Seeing Like a State*, chap. 1.

3. The establishment of SUDENE, with its scale and authorization to craft development plans for the whole region, obviously predated the military government and represented the ambitions of a central planning model.

4. Gonçalves surveyed 477 families, and a total of 2,573 people. Maciel, *Nível de vida do trabalhador*, 25, 22, 41, 35–38, 50; Gonçalves, "Condições de vida do trabalhador," 129, 127, 142. For the disappearance of the *barracões*—the *engenho* stores—see "Informação no 218/77/ASI/DRT/PE," in SSP 772: Usina São José, Acervo DOPS, APE.

5. Paulo, interview with the author, tape recording and transcript, Murupé, Vicência, March 21, 2003. Paulo's statement echoes many other workers' about the pre-union period. The Communist union organizer Gregório Bezerra uses a similar language, saying that the victims of the rural bourgeoisie "had no one to appeal to." Bezerra, *Memórias*, 1:176.

6. Lygia Sigaud wrote several remarkable essays on this topic, including "Idealization of the Past," "Morte do caboclo," and "Percepção do salário."

7. Koury, "Movimentos sociais," 19–46, 32–33; "DRT considera ilegal greve do Cabo e adverte que decretará intervenção," *Diário da noite*, September 17, 1968. The Cabo union felt safe organizing a strike because of its relationship to Padre Antônio Melo, who had cultivated connections in the military regime.

8. "Usineiros e trabalhadores debatem uniformização de tarefas no campo," *Diário de Pernambuco*, June 29–30, 1963.

9. The newspaper printed the final agreement. It appears in "Proposta conjunta das tabelas das tarefas do campo," in SSP 28688: Federação dos Trabalhadores Rurais de Pernambuco, Acervo DOPS, APE. Originally, the table also allowed workers to choose between getting a set daily wage or working by the task (being a "diarista" or a "tarefeiro"). Bezerra, *Memórias*, 2:178.

10. Paulo, interview with the author, tape recording, Murupé, Vicência, March 27, 2003.

11. Bezerra, *Memórias*, 2:178. M. C. de Andrade, *Area do sistema canavieiro*, 224.

12. "Porque os trabalhadores rurais não acreditam em tabelas," October 28, 1968, in SSP 1352: CONTAG, Acervo DOPS, APE. "Comissão para o trabalho," *Brasil açucareiro* 66, no. 5 (November 1965): 14. The issue was not resolved with the military's table. Shortly after the 1964 table was finished, planters demanded yet another reformulation, and by 1965 another table was promulgated that would remain in effect through the 1970s (having satisfied planters' criteria). The union federation, for its part, claimed that the 1965 table was clearly skewed against the workers, setting standards impossible to meet in one day. The union repeatedly requested to return to the "Table of the Revolution" that had been "democratically established" in 1964. "Porque os trabalhadores rurais não acreditam em tabelas," and "Sindicato de Escada está sob intervenção," *Jornal do comércio*, May 27, 1969.

13. "Proposta conjunta das tabelas das tarefas do campo," in SSP 28688, Acervo DOPS, APE.

14. Ronivaldo, interview with the author, tape recording, Engenho Humaitá, Palmares, January 25, 2003. Ronivaldo had to think for a minute before describing "cana boa," since it seemed an intuitively obvious category to him.

15. John D. French argues for the importance of imperfect but tangible tools of struggle more systematically and on the broader scale of Brazilian labor law in general. In a footnote, French cites Francisco Julião's comments on labor law. Although he acknowledged that labor law "does not and never has functioned fully," Julião held that the rural wage earners' struggle must be carried out according to those laws, since they hold "a promise of liberation." French, *Drowning in Laws*, 175.

16. *Levantamento sócio-econômico*, 19. Almost 85 percent of rural workers had *cartas profissionais* (professional cards), the documents that proved employment and without which workers had no access to benefits. Many scholars agree that after a period of harsh repression, the military government deliberately allowed the union movement to operate. See versions of this argument in A. Pereira, *End of the Peasantry*; Maybury-Lewis, *Politics of the Possible*; and Houtzager, "State and Unions," 106–8.

17. Padre Melo, a colleague of Padre Crespo's in the unionization effort, even trolled the police stations in the aftermath of the coup indicating which detained rural activists should be set free. Padre Antônio Melo, interview with Aspásia Camargo and Dulce Pandolfi, tape recording and transcript, Cabo, Pernambuco, April 1978 (FGV-CPDOC), 68–69. The fact that enough union members had been imprisoned for Melo to make his trips to the jails indicates that there was a significant repression after the coup. I do not underestimate that reality.

18. "Exmo. Sr. Ministro do Trabalho e da Previdência Social," in SSP 31496: Federação dos Trabalhadores na Agricultura de Pernambuco (VI), Acervo DOPS, APE.

19. William P. Kelly, "Labor Leader Pessimistic re Current Situation in Sugar Belt," April 28, 1966, p. 2, in RG 59, box 1927, NARA-US.

20. "Informe do comando do 3o distrito naval, 07/13/67," in SSP 806A: Usina Catende, Acervo DOPS, APE. Workers in Escada also testified that it was impossible to cut one hundred square *braças* in one day, in SSP 28876: Sindicato Rural do Município de Escada, Acervo DOPS, APE.

21. *Levantamento sócio-econômico*, 44. Edward Rowell, "The Institutional and Economic Support of the Sampaio Political Machine," April 27, 1966, p. 8, in RG 59, box 1927, NARA-US.

22. Koury, "Movimentos sociais," 21.

23. Maybury-Lewis, *Politics of the Possible*, 39. FUNRURAL was later buttressed by a program called PRORURAL. In 1977 FUNRURAL was incorporated into the national welfare system, which slowed the efficiency of payments to rural workers.

24. J. C. Scott, *Seeing Like a State*, 2–3, 32, 78.

25. The Northeast's market share was dropping even as the federal government was pursuing a "Plan for the Expansion of the National Sugar Industry." M. C. de Andrade, *Modernização e pobreza*, 173. Dantas, *Agroindústria canavieira*, 36–38.

26. Dantas, *Agroindústria canavieira*, 40, 42. Dantas and Santos e Silva, *Subsídios*, 17. Dantas and Santos e Silva offered as an example of Pernambuco's stunning inefficiency a comparison of statistics from the 1950s: Hawaiian *usinas* got 34 tons of sugar per hectare, while Pernambucan *usinas* got 3.4 tons per hectare.

27. Dantas and Santos e Silva, *Subsídios*, 53. R. F. de Carvalho, *Melhoramento e mecanizagem*, 4.

28. Price, "Brazilian Land Reform Statute," 1, 49–54. Skidmore, *Politics of Military Rule*, 298–99.

29. Price, "Brazilian Land Reform Statute," 49–50. *Brasil açucareiro* 6, no. 2 (August 1965): 12.

30. GERAN also drew the cautious approval of worker representatives. Moacir Palmeira argues that both the Northeast Development Agency (SUDENE) and GERAN attempted to reconcile *usineiro* and *fornecedor* interests with the needs of rural workers. Palmeira, "Aftermath of Peasant Mobilization," 91.

31. Eudes de Souza Leão Pinto, interview with Eliane Moury Fernandes, tape recording and transcript, August 1983 (CEHIBRA), 5–6. He also cowrote the Land Statute of 1964, discussed below.

32. Sigaud, "Armadilhas da honra," 131–32.

33. J. C. Scott, *Seeing Like a State*, 48. Scott writes that similar tools of state oversight and regulation "threatened to destroy a great deal of local power and autonomy. It is no wonder, then, that they should have been so vigorously resisted."

34. *Levantamento sócio-econômico*, 43–44.

35. Ibid., 80, 74. Forty-four percent of respondents to the union survey said the lawyers coerced them into making a deal. Amaral, "Aspectos das relações de trabalho," 185. "Processos ajuizados em 1968," in SSP 1607: Sindicato de Trabalhadores Rurais de Vitória de Santo Antão, Acervo DOPS, APE.

36. Cícero, interview with the author, tape recording, Jaqueira, March 22, 2003.

37. JCJ Palmares, processo 569/64: *José Bezerra da Silva vs. Catende, Termo de reclamação*, February 17, 1964, A-TRT.

38. JCJ Palmares, processo 153/64: *Victor Inácio da Silva vs. Engenho Galo (Agua Preta)*, hearing November 25, 1963, A-TRT (my emphasis).

39. JCJ Nazaré da Mata, processo 86/63: *Manoel Cícero Joaquim de Santana vs. Engenho Caraúba (José Bandeira de Oliveira)*, hearing July 18, 1963, A-TRT. José Bandeira may have felt that acknowledging that some workers on his land did not answer directly to him would have been persuasive. After all, if a *senhor de engenho* admitted to a lack of control over land or labor, even indirectly, it cut against the notions of command that drove this class's ethic.

40. *Anuário estatístico de Pernambuco*, 21.

41. There were plenty of rescissions in 1964 as well. Of the 421 files that survive, 211 were rescissions (processo numbers run through 809, so presumably there were at least 809 total cases). JCJ Palmares, processos 1964–1966, A-TRT.

42. See JCJ Nazaré da Mata, processo 86/63: *Termo de reclamação*, May 27, 1963, for an example of contractor-planter agreements, as spelled out in court testimony.

43. A 1972 document produced by FETAPE argues that temporary labor exceeded permanent, contracted labor by a wide margin, but FETAPE President Almeida do Nascimento only gave numbers for the country as a whole. At that scale, temporary workers outnumber contracted workers, 6.8 million to 1.2 million. "Relatório no. 01/76, Recife," February 9, 1976, in SSP 31496: FETAPE VII, Acervo DOPS, APE. Just as contract work grew in Brazil, the entire world experienced a sharp rise in nontraditional or "informal" labor relations in the 1970s, a feature of the crisis in world capitalism and the growing flexibilization of the labor force. See, e.g., Kalleberg, "Nonstandard Employment Relations," 341–65.

44. Already in 1963, a *senhor de engenho* (José Bandeira, mentioned above) testified in

the JCJ that he had tenants who worked for him and others without *fichas*. JCJ Nazaré da Mata, processo 86/63, hearing July 18, 1963. Sigaud, *Clandestinos e direitos*, 126–27. Espedito Rufino de Araújo notes that particular mills had reputations for being especially legalistic and always hiring *fichado* workers, while others were known for having large proportions of *clandestinos*. Rufino de Araújo, *Trator e "burro sem rabo,"* 132–33.

45. "Ministério do Exército Informação," May 5, 19792, in SSP 774: Usina Santa Tereza, Acervo DOPS, APE.

46. "Testemunha," July 19, 1970, in SSP 1664: Sindicato dos Trabalhadores Rurais de Moreno, Acervo DOPS, APE.

47. *Levantamento sócio-econômico*, 65.

48. Linhart, *Açúcar e fome*, 23, 46.

49. Maciel, *Nível de vida do trabalhador*, 25, 21.

50. Of the thirty-two complete records of rural workers' disputes with planters from the first six months of 1963 in Nazaré da Mata, eighteen involved workers who had been associated with the specific *engenho* for fewer than five years. There are many fewer records for the same time in Palmares—eight, of which four are cases involving work relationships less than a decade old. Considering case volumes for the next several years, it seems clear that many cases from these early months have been lost. JCJ Nazaré da Mata, processos 1963, A-TRT.

51. *Levantamento sócio-econômico*, 21. The study was carried out through questionnaires, circulated in April, May, and June of 1968.

52. Rufino de Araújo, *Trator e "burro sem rabo,"* 126.

53. INJPS, "Trabalhador rural volante," 65.

54. *Programa de ação*, 17. "In the 1960s and 1970s, there proceeded in every cane area of the country, with varying intensity, the process of 'deruralization' of the rural workers, as proprietors promoted a policy of expelling them from their lands." M. C. de Andrade, *Modernização e pobreza*, 210–11.

55. JCJ Nazaré da Mata, processo 49/77: *Severina Rosa da Conceição e outras (4) vs. Engenho Boa Vista (BA)*, hearing March 17, 1977, A-TRT.

56. JCJ Nazaré da Mata, processo 49/77, April 26, 1977, hearing.

57. *Levantamento sócio-econômico*, 58.

58. Houtzager, "State and Unions," 123. It is worth noting that a witness for Emília da Silva testified that her husband sometimes helped her also, when she was working for a contractor. On such occasions, their total production was accounted for in the wages paid to her. JCJ Catende, processo 118/75: *Emília Maria da Silva vs. Usina Catende*, June 12, 1975, hearing, A-TRT.

59. JCJ Nazaré da Mata, processo 49/77, July 13 judgment.

60. Neiman and Quaranta, "Restructuring and Functional Flexibilization," 45–46.

61. JCJ Catende, processo 118/75, June 12, 1975, hearing and July 1, 1975, judgment.

62. Paulo, interview with the author, March 21, 2003. Paulo actually said, "No one wants to see a person with liberty [*Ninguém quer ver a pessoa com liberdade não*]," a much more existential claim. Considering the context of the interview, though, my translation is appropriate. Paulo's ready response to my question about the planters' reasons for expelling workers was that the planters did it "with hate for the worker, because they are paying the salary, holidays, the thirteenth salary." Paulo, interview with the author, March 27, 2003. Paulo's may be a stock, stable explanation. Sigaud noted the regularity with which

workers referred to planters' "hate" for workers after the growth of worker rights. Sigaud, *Clandestinos e direitos*, 12.

63. Bento, interview with the author, tape recording, Murupé, Vicência, April 3, 2003.
64. Sigaud, *Clandestinos e direitos*, 50. *Levantamento sócio-econômico*, 23.
65. Cícero, interview with the author, tape recording, Jaqueira, February 22, 2003.
66. Sigaud, *Clandestinos e direitos*, 136.
67. Linhart, *Açúcar e fome*, 28–29.
68. "Relatório para Sr. Dr. Delegado Regional do Trabalho em Pernambuco," August 25, 1977, in SSP 772: Usina São José, Acervo DOPS, APE.
69. Leite Lopes, *Vapor do diabo*, 166. Sigaud, *Clandestinos e direitos*, 206.
70. Paulo, interview with the author, March 27, 2003.
71. Linhart, *Açúcar e fome*, 45.
72. Castro, *Death in the Northeast*, 83.
73. Page, *Revolution That Never Was*, 42–50.
74. The proceedings were published originally as E. Pinto, *Problema agrário* and reprinted later in Freyre et al., *Cana e reforma agrária*.
75. INJPS, "Trabalhador rural volante," 60. The law was known as a "lei dos sítios" or a "lei de dois hectares."
76. Crespo circulated many flyers and explanations of the law, many of which (including "Need for the Application of Decree no. 57,020/65" and "Social-Economic Situation of the Cane-Growing Zone of Pernambuco") are in SSP 29679: Serviço de Orientação Rural de Pernambuco, Acervo DOPS, APE. In Manuel Andrade's succinct analysis, the two-hectare law "failed because it attempted to resuscitate a situation that the advance of capitalism had already eliminated." M. C. de Andrade, *Modernização e pobreza*, 113.
77. Report from Luiz Martins de Miranda Filho to Delegacia de Segurança Social, March 14, 1967, in SSP 29679: Serviço de Orientação Rural de Pernambuco, Acervo DOPS, APE.
78. "Relatório do I Encontro das Federações dos Trabalhadores na Agricultura do Nordeste," in SSP 1352: CONTAG, Acervo DOPS, APE.
79. *Levantamento sócio-econômico*, 60–61, 87. Tellingly, this report also called for GERAN to cut the sugar workforce by *two-thirds* and to find the excess workers not land but other jobs (13).
80. Cícero, interview with the author, February 22, 2003. The issue of preference—not economic necessity—in land is what Ana Amaral addresses when she writes that "the worker assumes a cultural attitude that leaves him without the aspiration to return to the situation of the farmer." Amaral, "Aspectos das relações de trabalho," 185. Even twenty years after its promulgation, the two-hectare law continued to be the central demand of the national agricultural workers confederation. M. C. de Andrade, *Area do sistema canavieiro*, 225.
81. *Levantamento sócio-econômico*, 82.
82. Severino, interview with the author, tape recording, Vicência, April 16, 2003.
83. Leeds, "Brazil," 194.
84. Sindicato dos Trabalhadores Rurais de Rio Formoso to Secretaria de Segurança Pública, September 10, 1968, in SSP 1634: Sindicato dos Trabalhadores Rurais de Rio Formoso, Acervo DOPS, APE. M. C. de Andradae, *Area do sistema canavieiro*, 219. *Levantamento sócio-econômico*, 47.
85. The dictatorship's stated strategy for the Northeast and the GTIA's plans must be seen

in the context of a conspicuous U.S. presence in the heavily staffed Recife consulate and the promise of huge sums of aid for agrarian reform projects. President Kennedy committed $131 million to the Northeast for 1962 and 1963 alone. "Kennedy Will Begin Brazil Visit July 30," *New York Times*, May 25, 1962. The preoccupation with the U.S. presence continued. A 1970 GERAN plan for educational projects in the sugar zone included the use of Peace Corps volunteers to do educational work. GERAN, *Programa de melhoria*, 9. The experiences of some volunteers in the *zona da mata* can be found on Returned Volunteer discussion boards.

86. Dantas, *Agroindústria canavieira*, 49. This recalls Mitchell's discussion of the "social 'fixing' [of labor], in order to protect the productive landscape." Mitchell, *Lie of the Land*, 80.

87. Anthony Hall writes that "only 12 per cent of PROTERRA funds were devoted toward land redistribution in 1973, decreasing to 4 per cent in 1974." Hall, "Innovation and Social Structure," 144. Barbara Nunberg refers to the widespread belief that local oligarch opposition and subtle antagonism from the Instituto do Açúcar e do Álcool doomed GERAN. Nunberg, "Structural Change," 67.

88. Houtzager, "State and Unions," 106.

89. Greider and Garkovich, "Landscapes," 2.

Chapter 7

1. From his first extended example in *Seeing Like a State*—the emergence of scientific forestry in Germany—James Scott shows that modernist states' aspirations of control crossed seamlessly from the "natural" to the social realms. In both arenas, the same ideology of regimentation, rationalization, and control prevailed and in both cases, the schemes of control actually change the realities they were meant to capture or describe. J. C. Scott, *Seeing Like a State*, 11–25. Scott describes high modernist ideology as a "musclebound version of the self-confidence about scientific and technical progress, the expansion of production, the growing satisfaction of human needs, the mastery of nature (including human nature), and, above all, the rational design of social order commensurate with the scientific understanding of natural laws" (4).

2. Dean, *With Broadax and Firebrand*, 266. Scott argues that the temporal emphasis of high modernist states is on the future. Planning, then, becomes a scientific process endowed with great prestige and importance. J. C. Scott, *Seeing Like a State*, 95–96.

3. Singer, "Economic Evolution," 81–82. Skidmore, *Brazil*, 169–71.

4. J. Motta Maia, "A revolução dos agrónomos," *Brasil açucareiro* 68, no. 4 (October 1966): 12. Maia was head of the IAA's Department of Production Assistance.

5. Gerson Quirino Bastos, interview with the author, handwritten notes, Recife, March 25, 2003. In the late 1980s, Manuel Correia de Andrade reported, 3X still occupied about 20 percent of Pernambuco's cane fields. M. C. de Andrade, *Area do sistema canavieiro*, 390–91.

6. Dantas and Santos e Silva, *Subsídios*, 72. M. C. de Andrade, *Modernização e pobreza*, 42, 173. Under PLANALSUCAR, new cane varieties were developed or imported from elsewhere with the intention of raising overall average sucrose concentrations of cane arriving at mills. The rationalization and support programs were aimed at improving agricultural conditions, creating new experimental stations, and modernizing the defenses against sugarcane pests. Proálcool was created to foment cane alcohol production.

7. Gerson Quirino Bastos, interview with the author, handwritten notes, Recife, March 25, 2003.

8. Dantas and Santos e Silva, *Subsídios*, 4.

9. IJNPS, "Trabalhador rural volante," 22, 25.

10. Figueiredo, "Campo histórico," 33.

11. Malavolta, "Tendências no uso de fertilizantes," 663.

12. "Plant a Second Green Revolution," *Christian Science Monitor*, June 3, 2008, <http://www.csmonitor.com/2008/0603/p08s01-comv.html>, accessed June 26, 2008.

13. M. C. de Andrade, *Modernização e pobreza*, 35. IJNPS, "Trabalhador rural volante," 22, 25. Dantas and Santos e Silva, *Subsídios*, 4, 10. One study argues that the "principal instrument used by the Brazilian government to stimulate the growth in sugarcane production and the capacity for its industrial transformation into alcohol, starting in 1975, was the credit subsidy conceded to approved projects." Bray, Ferreira, and Ruas, *Políticas da agroindustria canavieira*, 57. Figueiredo, "Campo histórico," 33.

14. May, "Modern Tragedy," 24.

15. IJNPS, "Trabalhador rural volante," 32–34.

16. Malavolta and Rocha, "Recent Brazilian Experience," 297. Dantas, "Recuperação da lavoura," 43–46. IJNPS, "Trabalhador rural volante," 32–34.

17. Malavolta and Rocha, "Recent Brazilian Experience," 297.

18. According to a 1979 article in *A lavoura*, the magazine of the Sociedade Nacional de Agricultura (SNA, National Agricultural Society), the cane research stations were experiencing "a period of well-funded hope." Amaro Cavalcanti, "Breve rememoração histórica da cana-de-açúcar," *A lavoura* 82 (September–October 1979): 42. Still, a number of people in the field complained that there were too few agronomists, since the salaries in the profession remained poor.

19. Rental agreement included in JCJ Catende, processo 44/71: *Manoel Francisco dos Santos e outros (34) vs. Engenho Flor do Bosque*, A-TRT.

20. IJNPS, "Trabalhador rural volante," 29. Rufino de Araújo, *Trator e "burro sem rabo,"* 69, 26, 72.

21. M. L. de Melo, *Paisagens do Nordeste*, 102. Melo's observation is also quoted by Manuel Correia de Andrade, who noted the slow beginnings of fertilizer use in his 1958 essay on the Rio Sirijí. M. C. de Andrade, "Vale do Sirijí," 72.

22. R. F. de Carvalho, *Melhoramento e mecanizagem*, 5, 7–8.

23. Andrade Neto, "Questão ambiental," 10. M. C. de Andrade, *Modernização e pobreza*, 110. "PRAC—Programa de Ação Coordenadora 1972–75, Governo do Estado," 91, in Acervo Ministerio de Agricultura, APE.

24. The region's average population density was more than two hundred people per square kilometer and more than three hundred people per square kilometer by the 1980s. Dantas, *Agroindústria canavieira*, 50. M. C. de Correia, *Modernização e pobreza*, 113. Rufino de Araújo, *Trator e "burro sem rabo,"* 18.

25. IJNPS, "Trabalhador rural volante," 42. Rufino de Araújo, *Trator e "burro sem rabo,"* 23–25.

26. Dantas, "Recuperação da lavoura," 65. IJNPS, "Trabalhador rural volante," 42, 48.

27. In his 1964 report Ronildo Ferreiro de Carvalho stated: "We know perfectly well that the Pernambucan humid zone does not possess the desired topography for complete mechanization of this [sugar] or any other crop." R. F. de Carvalho, *Melhoramento e mecanizagem*, 3.

28. Peter Houtzager suggests that the military government paid particular attention to

the cane region both in its efforts to remake the union movement and in the larger project of creating a new agricultural model. Houtzager, "State and Unions," 107–8.

29. Scott emphasizes the infatuation with planning and the creation of task-targeted agencies. J. C. Scott, *Seeing Like a State*, 95.

30. Vasconcelos Sobrinho, *Regiões naturais*, 171–73. Rufino de Araújo, *Trator e "burro sem rabo,"* 66. Rufino de Araújo contends that the period of the sugar boom (1972–87) saw rampant defiance of this law.

31. Cícero, interview with the author, tape recording, Jaqueira, February 22, 2003.

32. R. F. de Carvalho, *Melhoramento e mecanizagem*, 5, 7–8.

33. *Sistemas de produção florestal para a zona da mata*, 2. Costa Lima, "Reserva da biosfera." Edson Miranda agrees with Costa Lima: "Even the scraps of Atlantic Forest, once dense and rich with animals, were destroyed by the cane fields, above all in the last twenty-five years." Miranda, *Chapéu de palha*, 14.

34. The first experiments with burning bagasse took place in the middle of the nineteenth century. The practice succeeded in preventing even greater deforestation than that which took place. Andrade and Lins, *Pirapama*, 135; M. C. de Andrade, "Poluição dos cursos d'água," 95–97.

35. Cavalcanti, *Vinhoto*, 2. M. C. de Andrade, "Poluição dos cursos d'água," 95. Bueno, *Pró-álcool*, 40. Rufino de Araújo, *Trator e "burro sem rabo,"* 80, 106.

36. E. A. Andrade, *CPRH 25 anos*, 7–9. Franscisco Dario Mendes da Rocha, "A propósito de um problema: A poluição dos Rios do Açúcar," *Brasil açucareiro* 69, no. 1 (January 1967): 26.

37. Rabello, "Pobre Rio Siriji," 245–46. M. C. de Andrade, "Poluição dos cursos d'água," 95. Andrade had also written about the Siriji for an essay he submitted when applying to teach at the Colégio Estadual de Pernambuco in 1958. M. C. de Andrade, "Vale do Siriji," 55. M. C. de Andrade, *Modernização e pobreza*, 113.

38. Cavalcanti, *Vinhoto*, 3. Lins, "Efeitos sociais," 199. M. C. de Andrade, "Homem e natureza," 15. A long-time IAA employee interviewed in 1979 drew attention to the major polluting effects of *vinhoto*. Nelson Coutinho, interview with Célia Maria Leite Costa and Dulce Pandolfi, tape recording and transcript, Rio de Janeiro, August 8, 1979 (FGV-CPDOC), 33. E. A. Andrade, *CPRH 25 anos*, 9. "Regimento Interno da Associação Pernambucana de Defesa da Natureza," in SSP 4428: Associação Pernambucana de Defesa da Natureza (ASPAN), Acervo DOPS, APE. ASPAN and its leaders were monitored by the DOPS. Ricardo Braga, the first president, went on to head the Pernambuco Ecological Society.

39. Ferti-irrigation finally became common practice in the 1980s. Rufino de Araújo, *Trator e "burro sem rabo,"* 78. Andrade Neto, "Questão ambiental," 19. Not surprisingly, Catende had been an innovator in developing this technique. Tenente's chemist, José Brito Pinheiros Passos, conducted experiments evaporating the water from mill waste and eventually simply pumping it directly onto fields. M. C. de Andrade, "Poluição nos cursos d'água," 100.

40. "Região de contrastes," *Diário de Pernambuco*, 3 July 1962.

41. F. de M. Freyre, *Simpósio sobre as enchentes*, 9–22.

42. S. R. de Moura, *Memórias de um camponês*, 145–46.

43. F. de M. Freyre, *Simpósio sobre as enchentes*, 19–44. Andrade and Andrade, *Cana-de-açúcar*, 22.

44. André de Oliveira Cavalcanti, "Planejamento urbano e ecologia aplicada: Abordagem ao caso da mata pernambucana," *Revista pernambucana de desenvolvimento* 3, no. 2 (1976): 250.

45. Gonçalves, "Condições de vida do trabalhador," 130, 132. Maciel, *Nível de vida do trabalhador*, 30.

46. Bezerra, *Memórias*, 1:49.

47. Leite Lopes, *Vapor do diabo*, 157–58.

48. Benedito, interview with the author, tape recording, Engenho Humaitá, Palmares, January 25, 2003. Another worker, when asked whether he fished the stream running past his house on an *engenho*, said the fish were gone. Cícero, interview with the author, handwritten notes, Jaqueira, October 14, 2002.

49. Leite Lopes, *Vapor do diabo*, 157–58.

50. Freyre, *Nordeste*, 75.

51. Ibid., 166, 167, 171.

52. Koury, "Trabalhador da cana," 33.

53. Ronivaldo, interview with the author, tape recording, Engenho Humaitá, Palmares, March 25, 2003. Through long-term residence and labor, workers developed intimate local knowledge. A secret police investigator describing the difficulty of uncovering suspected Peasant League activity wrote in his report that cane workers "are profoundly knowledgeable of the terrain of the property where they live." Severino Oliveira, Relatório to Delegado, June 15, 1961 in SSP 29796: Ligas Camponesas (1950), Acervo DOPS, APE.

54. J. Motta Maia, "A revolução dos agrónomos," *Brasil açucareiro* 68, no. 4 (October 1966): 12.

55. The documentation regarding both of Emília da Silva's cases is filed in JCJ Catende, processo 25/77: *Emília Maria da Silva vs. Usina Catende*, A-TRT. This chapter draws on research in the A-TRT for two sample JCJs: Palmares/Catende in the *mata sul* and Nazaré da Mata in the *mata norte*, because of a generalized use of the Catende and Vicência municipalities throughout the text as examples of their respective subregions. In 1963 the Palmares JCJ served Catende; the municipality's own junta was established in 1971. Since Vicência was under the jurisdiction of the Nazaré da Mata JCJ for this entire period, I examined the Nazaré junta's files. I chose sample years from the foundation of the courts (1963–64) through 1982, including 1969, 1971, and 1977.

56. JCJ Catende, processo 25/77, hearing March 10, 1977.

57. There were at least seventy-six requests for reclassification in the surviving records, including workers employed by mills other than Catende. Lima began bringing the cases in groups of six or more plaintiffs. JCJ Catende, processo 186/76: *María José da Conceição vs. Usina Catende*, A-TRT.

58. Súmula 196, <http://www.trt02.gov.br/geral/tribunal2/Trib_Sup/STF/SUM_STF.html#196>, accessed December 1, 2006.

59. Súmulas do TST, Enunciado no. 57, <http://www.Soleis.adv.br/sumulastst.htm>, accessed June 10, 2005. The measure was canceled in 1993.

60. JCJ Catende, processo 25/77, letter from Catende to TRT, January 2, 1978. See the TRT's gloss of the mill lawyers' arguments in JCJ Catende, processo 1075/77: *Maria da Conceição vs. Usina Catende*, judgment, October 11, 1977, filed with JCJ Catende, processo 47/77: *Maria da Conceição vs. Usina Catende*, A-TRT.

61. Complementary Laws 11 and 16, from 1971 and 1973 respectively, created the Program of Assistance to the Rural Worker (PRORURAL) and restated rural workers' support by both PRORURAL and FUNRURAL. And Law 5889, also from 1973, defined a rural worker and mandated the category's oversight by the CLT. For all three, see the Sistema de Informações

do Congresso Nacional: <http://www6.senado.gov.br/sicon/PreparaPesquisaLegislacao. action>, accessed March 15, 2009.

62. This argument appeared in JCJ Catende, processo 48/77: *Alaíde Maria vs. Usina Catende*, letter from Catende, March 27, 1977, A-TRT.

63. JCJ Catende, processo 25/77, parecer, June 29, 1978.

64. The Willys jeep is appraised in JCJ Catende, processo 1075/77.

65. According to the Convenção Coletiva de Trabalho between the sugar producers and their mill workers (a document Floriano de Lima included in court documents), the 42 percent wage increase the workers received in 1975 gave them a monthly wage of 477.12 cruzeiros. Processo 25/77: "Convenção Coletiva de Trabalho."

66. Houtzager, "State and Unions," 107.

67. See JCJ Catende, processos 1–50 and 250–422, 1977.

68. "DRT considera ilegal greve do Cabo e adverte que decretará intervenção," *Diário da noite*, September 17, 1968.

69. "Sindicatos reúnem-se no Interior," *Diário de Pernambuco*, August 23, 1979.

70. "Greves ilegais serão reprimidas com rigor," *Diário de Pernambuco*, August 16, 1979.

71. Law 4,330. The metaphor is from Koury, "Movimentos sociais," 32. Amaral, "Aspectos das relações de trabalho," 186.

72. Sigaud, *Greve nos engenhos*, 13–17.

73. Amaral, "Aspectos das relações de trabalho," 189.

74. Cícero, interview with the author, tape recording, Jaqueira, March 22, 2003.

75. "Mobilização de rurícolas gera preocupações" and "Arraes chega hoje ao Recife," *Diário de Pernambuco*, September 16, 1979.

76. Sigaud, *Greve nos engenhos*, 30–41.

77. Amaral, "Aspectos das relações de trabalho," 190–91.

78. Sigaud, *Greve nos engenhos*, 29. SSP 26993: Movimento da Zona Canavieira, Acervo DOPS, APE.

79. Mapoteca 1, gaveta D, file 31858: Centro de Informações, Acervo DOPS, APE. Fourteen unions were listed as "favorable," only three of which were in the North—Goiana, Itambé, and Aliança. Thirty unions were "contrary" to the regime, and the union at Ponte dos Carvalhos was listed as "undecided."

80. On Padre Melo's actions to minimize repression, see chap. 6, n. 17. Andrade and Andrade, *Cana-de-açúcar*, 22.

81. Sigaud herself took place in the organizing efforts that led to the strikes of 1979–85.

82. Sigaud, "Luta de classes," 319–24.

83. Sigaud, *Greve nos engenhos*, 42.

84. Movimento dos Trabalhadores Rurais Sem Terra, *Assassinatos no campo*, 553, 578–80.

85. Rufino de Araújo, *Trator e "burro sem rabo,"* 281.

86. Ibid., 324–25. Sigaud defends the strikes' efficacy in "Collective Ethnographer," 12.

87. Rosa, "Novas faces," 476–77.

88. A. Pereira, *End of the Peasantry*, 72.

89. "CPI do açúcar chega ao fim com relatório," unidentified newspaper clip in SSP 31496: FETAPE (II), Acervo DOPS, APE.

90. *Pernambuco, problemática de suas atividades*, viii; IJNPS, "Trabalhador rural volante," 29. Rufino de Araújo, *Trator e "burro sem rabo,"* 69, 26, 72.

91. Sugar production from 1952 to 1973 went from 9.8 million sacks to 18 million. IJNPS, "Trabalhador rural volante," 42.

92. *Avaliação e perspectivas da agroindústria canavieira*, 11–13.

93. M. C. de Andrade, *Modernização e pobreza*, 27–30, 113.

94. Dantas and Santos e Silva, *Subsídios*, 72.

95. José Guilherme Azevedo Quiroz, interview with the author, tape recording, Usina Cruangí, May 17 2003.

96. Lins do Rego, *Usina*, 196.

97. Lins do Rego, *Plantation Boy*, 24.

Conclusion

1. "Proálcool degrada ecologia dos rios," *Diário de Pernambuco*, January 16, 1983. Alves worked for the state environmental and water resources agency (CPRH). Proálcool was also responsible for significant deforestation in São Paulo and water degradation in Campos (Rio de Janeiro state). Dean, *With Broadax and Firebrand*, 294.

2. The spill fueled rising public criticism of the cane industry's pollution, prompting the state legislature to pass a law in November prohibiting new distilleries that lacked plans and equipment for wastewater treatment. A stronger law passed in 1987 with wider-ranging regulation of the cane industry. E. A. Andrade, *CPRH 25 anos*, 14.

3. J. C. Scott, *Seeing Like a State*, 87.

4. "Proálcool degrada ecologia dos rios," *Diário de Pernambuco*, January 16, 1983.

5. The appropriate historical analysis of the region's economic and labor history has been the subject of long-term battles, especially among Marxist historians (should it be seen as feudal? as precapitalist?). Christine Rufino Dabat covers this debate thoroughly. Dabat, *Moradores de engenho*, 273–342.

6. Dean, *With Broadax and Firebrand*, 266.

7. Striffler, *In the Shadows of State and Capital*, 64, 74, 81.

8. Sigaud, *Clandestinos e direitos*, 210. In this book (published before the strike), Sigaud mentions the new forms of sociability that became prevalent among workers after moving off of the *engenhos*. When living on the *engenhos*, workers were isolated from one another, living next to their plots. In the small towns of the region, they were much closer together and could commune more. This social consequence was neither uniform nor inevitable, however. Rebecca Scott notes workers' hidden advantages of better communication as they continued to live on plantations in postemancipation Louisiana. R. J. Scott, *Degrees of Freedom*, 83.

9. PROCANOR was established following the 1979 strike to help combat poverty in the cane zone and increase the production of food crops. Howe and Goodman, *Smallholders and Structural Change*, 125. FETAPE letter to Secretaria de Segurança Pública de Pernambuco, July 1, 1981; FETAPE letter to INCRA, July 1, 1981; and FETAPE letter to governor, June 28, 1981, all in SSP 31496: Federação dos Trabalhadores na Agricultura de Pernambuco, Acervo DOPS, APE.

10. FETAPE letter to Secretaria de Segurança Pública and FETAPE letter to governor in SSP 31496: Federação dos Trabalhadores na Agricultura de Pernambuco, Acervo DOPS, APE.

11. Bezerra, *Memórias*, 1:51.

12. FETAPE letter to Secretaria de Segurança Pública in SSP 31496: Federação dos Trabalhadores na Agricultura de Pernambuco, Acervo DOPS, APE.

13. Planter abuses were so routine as to be easily exposed in the labor courts. In a typical 1982 case from Nazaré da Mata, a work gang contractor with a close relationship to the *senhor de engenho* defendant claimed that all of his workers signed pay slips confirming their wages each day. Hearing this testimony, the justices had the presence of mind to recall the first witness—one of the contractor's workers—who had signed his testimony with a thumbprint. The terse "reinterrogation" reads "the deponent confirms that he is illiterate and that the signatures appearing on payment slips exhibited to the court were signed by another person and not by the deponent." JCJ Nazaré da Mata, processo 83/82: *Manoel Gomes da Silva vs. Engenho Cha Grande*, hearing January 25, 1982.

14. Rask, "Social Costs of Ethanol Production," 628–31.

15. Linhart, *Açúcar e fome*, 41–42.

16. Rufino de Araújo, *Trator e "burro sem rabo,"* 16–24.

17. Ibid., 143.

18. "Trabalhador rural volante," 14.

19. Rufino de Araújo, *Trator e "burro sem rabo,"* 166–67.

20. Severino, interview with the author, tape recording, Vicência, April 16, 2003.

21. Paulo, interview with the author, tape recording, Murupé, Vicência, March 27, 2003.

22. Even earlier in the 1970s, Sigaud mentioned that the wage was seen by workers as the most important right. Sigaud, "Idealization of the Past," 167. In worker discourse, she writes, "the indicator of the good in the present is fundamentally the wage."

23. Sigaud, *Greve nos engenhos*, 44–45.

24. "Sindicato desmente ocorrência de atos violentos no campo," *Diário de Pernambuco*, May 17, 1983.

25. A. Pereira, *End of the Peasantry*.

26. Severino, interview with the author, April 16, 2003.

27. Nascimento, "Do 'beco dos sapos' aos canaviais," 115.

28. Rosa, "Novas faces," 1–22.

29. Wolford, "Of Land and Labor," 147–70.

30. Severino, interview with the author, tape recording, Vicência, April 4, 2003.

31. Severino, interview with the author, April 16, 2003.

32. Paulo, interview with the author, tape recording, Murupé, Vicência, March 21, 2003.

33. Ibid.

34. Severino, interview with the author, April 16, 2003.

35. Jeison, interview with the author, tape recording, Murupé, Vicência, February 27, 2003.

36. Wolford, "Of Land and Labor"; Wolford, *This Land Is Ours*, chap. 5.

37. Marivaldo de Andrade, interview with the author, handwritten notes, Catende, October 14, 2002. Extremely helpful has been a series of interviews, collected by José Francisco de Melo Neto, with mill directors, union leaders, *engenho* association leaders, and others. Melo Neto, *Usina Catende*.

38. Interviews with Inaldo Felix da Silva, Lenivaldo Marques da Silva Lima, and Arnaldo Liberato da Silva, in Melo Neto, *Usina Catende*.

39. Ibid.

40. Rufino de Araújo, *Trator e "burro sem rabo,"* 83, 85, 87.

41. Arnaldo Liberato and Marivaldo de Andrade, interviews with the author, handwritten notes, Catende, October 14, 2002. Also Interviews with Arnaldo Liberato da Silva and Amaro Jovino, in Melo Neto, *Usina Catende*.

42. Ronivaldo, interview with the author, tape recording, Engenho Humaitá, Palmares, January 25, 2003; Cícero, interview with the author, February 22, 2003.

43. Marivaldo interview with the author, October 14, 2002.

44. Juan, interview with the author, tape recording, Catende, May 29, 2003.

45. Expedito Correia de Oliveira Andrade, interview with the author, handwritten notes, Recife, February 20, 2003.

46. Budny, "Global Dynamics of Biofuels," 5.

47. Fabiola Salvador, "Plantio em áreas da Amazônia é possível, diz Stephanes," *O estado de São Paulo*, July 29, 2008, <http://www.estadao.com.br/vidae/not_vid214083,0.htm>, accessed July 31, 2008.

48. "A Survey of Brazil: The Economy of Heat," *Economist*, April 14, 2007, 68.

49. Rask, "Social Costs of Ethanol Production," 629–30.

50. Cícero, interview with the author, tape recording, Jaqueira, February 22, 2003.

51. Cícero, interview with the author, tape recording, Jaqueira, March 22, 2003.

52. Christine Dabat comments insightfully on the secular mythology of Pernambuco's "natural vocation" for cane. Dabat, *Moradores de engenho*, 68–72.

53. Cosgrove, *Social Formation*, 13.

BIBLIOGRAPHY

Archival Sources

Arquivo do Tribunal Regional de Trabalho, Vitória de Santo Antão
    Junta de Conciliação e Julgamento
        Catende: 1968, 1969, 1971, 1977, 1980
        Nazaré da Mata: 1963, 1968, 1969, 1971, 1977, 1980
        Palmares: 1964
Arquivo Público Estadual Jordão Emerenciano, Recife
    Acervo Departamento de Ordem Político e Social
    Acervo Governo do Estado
    Acervo Secretária de Agricultura
    Acervo Secretária do Governo
Biblioteca Pública Estadual, Recife
    Coleção Pernambucana
Duke University, Rare Book, Manuscript, and Special Collections Library
    Emma Juliana and John P. Smith Letterbook, 1843–45
Fundação Getúlio Vargas, Centro de Pesquisa e Documentação História
        Contemporânea do Brasil, Rio de Janeiro
    Recorded and transcribed interviews
Fundação Joaquim Nabuco, Recife
    Centro de Documentação e de Estudos da História Brasileira
      Iconography Section
      Oral History Section
      Manuscript files
        Arquivo José Britto Pinheiro Passos
        Arquivo José Campello
        Arquivo Museu do Açúcar
        Coleção Agueda Cavalcanti Lins
        Coleção Antonio de Barros Carvalho
        Coleção Artur de Siqueira Cavalcanti Junior
        Coleção Aurino José Duarte
        Coleção Paulo Guerra
      Periodicals
        Boletim canavieiro, 1955–60
        Diário de Pernambuco, 1962–83
        A lavoura, 1945–85
      Published materials
Instituto de Planejamento de Pernambuco, Recife
    Published materials and planning documents

Instituto Histórico e Geográfico Brasileiro, Rio de Janeiro
  Published materials
New York Public Library, New York, N.Y.
  Brasil açucareiro, 1935–1979
  Published materials
North Carolina State University, Raleigh
  Comissão de Combate às Pragas da Cana de Açúcar no Estado
    de Pernambuco publications
  Escola Agrícola de Piracicaba publications
  Instituto Pernambucano de Agricultura publications
U.S. National Archives and Records Administration, Washington, D.C.
  State Department Records (Record Groups 59 and 84)

Interviews by the Author

Recorded Interviews

Antônio, Engenho Jaqueira, Usina Catende, Palmares, January 4, 2003; 1.5 hrs.
Antônio, Engenho Jaqueira, Usina Catende, Palmares, February 1, 2003; 30 mins.
Antônio Francisco, president, Assentamento Gregório Bezerra, Vicência, April 3, 2003; 2 hrs.
Benedito, Engenho Humaitá, Usina Catende, Palmares, January 25, 2003; 1 hr.
Benedito, Engenho Humaitá, Usina Catende, Palmares, February 8, 2003; 1 hr.
Bento, Murupé, Vicência, April 3, 2003; 1 hr.
Cícero, foreman, Engenho União, Usina Catende, Jaqueira, February 22, 2003; 1 hr.
Cícero and Ednía, Ramos, Engenho União, Usina Catende, Jaqueira, March 22, 2003; 1.5 hrs.
Dulce, president, Associação do Engenho Jaqueira, Usina Catende, Palmares, February 1, 2003; 45 mins.
Elías, Sebastião, and Severino, Assentamento Barra Nova, Jaqueira, May 23, 2003; 1.5 hrs.
Eramos, Murupé, Vicência, February 28, 2003; 35 mins.
Irací, secretary, STR Vicência, March 7, 2003; 30 mins.
Irací, secretary, STR Vicência, March 27, 2003; 45 mins.
Jeison, Engenho Salão, Vicência, February 27, 2003; 45 mins.
João Correia de Andrade, owner, Engenho Jundiá, Vicência, February 18, 2003; 1.5 hrs.
José Guilherme Azevedo Queiroz, mill owner, Usina Cruangí, Timbaúba, May 17, 2003; 1 hr. 15 mins.
José Joaquim, president, STR Catende, recorded at Usina Catende, May 13, 2003; 50 mins.
José Roberto and Antônio, Engenho Jaqueira, Usina Catende, Palmares, December 21, 2002; 2 hrs.
Juan, director, agricultural operations, Usina Catende, May 29, 2003; 50 mins.
Mané, Engenho Araticuns, Vicência, February 27, 2003; 45 mins.
Marilí, president, Associação do Engenho Humaitá, Usina Catende, Palmares, February 8, 2003; 40 mins.
Minervino Ferreira, Anônio, and Severino Ramos Bernardes, president and secretary, STR Vicência, José Americo, April 4, 2003; 1 hr.
Paulo, former president, STR Vicência, Murupé, Vicência, March 21, 2003; 1.5 hrs.
Paulo, former president, STR Vicência, Murupé, Vicência, March 27, 2003; 1.5 hrs.

Ronivaldo, Engenho Humaitá, Usina Catende, Palmares, January 25, 2003; 1.5 hrs.
Severino, Vicência, April 1, 2003; 2 hrs.
Severino, Vicência, April 16, 2003; 2 hrs.

Interviews with Handwritten Notes

Andrade, Marivaldo de, president, Empresa Harmonia, Usina Catende, October 24, 2002
Barros, Osvaldo Henrique, economist, FJN, March 28, 2003
Braga, Bete, CPRH and ASPAN, May 7, 2003
Cavalcanti, Clóvis, director, Centro de Pesquisas em Ciências Sociais, FJN, December 30, 2002
Cícero, foreman, Engenho União, Usina Catende, May 9, 2003
Correia de Andrade, Joaquim Xavier, director, Masters Program in Geography, UFPE, January 29, 2003
Correia de Oliveira Andrade, Expedito, head, legal department, Associação dos Fornecedores de Cana de Pernambuco, February 20, 2003
das Dores, Maria, director, SNE, March 24, 2003
Eugênio, Geraldo, president, INCRA-PE, November 20, 2002
Farias, João, president, INCRA-PE, May 15, 2003
Ferreira Gomes, Paulo, biologist, CPRH, January 27, 2003
José Maria, et al., workers at Engenho União, Usina Catende, Jaqueira, February 22, 2003
Josefa, retired worker, Murupé, Vicência, March 7, 2003
Liberato, Arnaldo, advisor, Empresa Harmonia, Usina Catende, October 24, 2002
Liberato, Arnaldo, advisor, Empresa Harmonia, Usina Catende, May 29, 2003
Luís, agronomist, Usina Catende (and 1 hr. tape of seminar), October 24, 2002
Maranhão, Leonardo, president, Sindicato Rural de Vicência, February 28, 2003
Oliveira, Francisco de, agronomist, UFRPE, March 24, 2003
Pula, president, Assentamento Paisagem de Areia, Belém de Maria, May 8, 2003
Quirino Bastos, Gerson, soil scientist, UFRPE, March 25, 2003
Technicians, Instituto Brasileiro do Meio Ambiente (IBAMA), January 27, 2003
Wanderly, Múcio (agronomist), and Roberto Moura (president), IPA, January 7, 2003

Published Sources

Abramson, Allen. "Mythical Land, Legal Boundaries: Wondering about Landscape and Other Tracts." In *Land, Law and Environment: Mythical Land, Legal Boundaries*, edited by Allen Abramson and Dimitros Theodossopolous, 1–30. London: Pluto, 2000.
Açúcar: A civilização que a cana criou. Recife: Instituto Cultural Bandepe, 2002.
Agenda 21: Pesquisa sobre meio ambiente, desenvolvimento e qualidade de vida no estado de Pernambuco, versão preliminar. Recife: CONDEPE, 2001.
Agricultura de Pernambuco: Uma visão de futuro. Recife: Governo de Pernambuco, Secretaria de Agricultura, 1998.
Aguirre, J. M., Jr. "Creação de novas variedades de canna no estado de S. Paulo." Instituto Agronomico do Estado em Campinas. *Boletim technico*, no. 34. São Paulo: 1936.
Albuquerque, Durval Muniz de, Jr. "Weaving Tradition: The Invention of the Brazilian Northeast." *Latin American Perspectives* 31, no. 2 (2004): 42–61.

———. *A invenção do Nordeste e outras artes*. Recife: Massangana, 1999.
Almeida, Irene Maria Gatti de. "Gomose da cana de açúcar no Brasil." *Anais do IX Reunião Itinerante de Fitossandidade do Instituto Biológico: Cana de açúcar*, 18–21. Catanduva: RIFIB, 2003.
Almeida, José Américo de. *A bagaceira*. Rio de Janeiro: Castilho, 1928.
Almeida, Jaime Rocha de, et al. "O florescimento na variedade de cana Co-331 (Co-3X)." *Anais da Escola Superior de Agricultura "Luiz de Queiroz"* 9 (1952): 157–81.
Alvarenga, Octavio Mello. "Reforma agrária e justiça especializada." *A lavoura* 87 (1985): 33–34.
Amaral, Ana Elizabeth Perruci do. "Nourishment, Health, Work, and Socio-economic Status of Six Sugar Cane Agricultural Worker Families, in Barreiros, Pernambuco, Brazil." *Ciência e trópico* 16, no. 2 (1988): 151–58.
———. "Aspectos das relações de trabalho e do movimento sindical da zona da mata sul de Pernambuco." *Ciência e trópico* 12, no. 2 (1984): 183–93.
Andrade, Evângela Azevedo. *CPRH 25 anos: Sua vida, sua história*. Recife: CPRH, 2001.
Andrade, Expedito José Correia de Oliveira. "Explanação sobre a economia canavieira de Pernambuco." Recife: Associação dos Fornecedores de Cana de Pernambuco, 1991.
Andrade, Gilberto Osório de. "Uma teoria sobre a evolução do 'mar-de-morros.'" In *Capítulos de geografia do Nordeste: O estado e os desequilíbrios de desenvolvimento regional*, edited by Manoel Correia de Andrade, 17–23. Recife: n.p., 1982.
Andrade, Gilberto Osório de, and Rachel Caldas Lins. *Pirapama: Um estudo geográfico e histórico*. Recife: Massangana, 1984.
Andrade, José Clóvis de. *Escorço histórico de antigas variedades de cana-de-açúcar*. Maceió: Associação dos Plantadores de Cana de Alagoas, 1985.
———. "Variedades de cana recomendadas para a zona Norte de Pernambuco." In *10 Congresso Açucareiro Nacional, Anais*, vol. 1. Rio de Janeiro: IAA, 1950.
Andrade, Manuel Correia de. "Apresentação." In Henrique Augusto Milet, *A lavoura da cana de açúcar*. Recife: Massangana, 1989.
———. *Area do sistema canavieiro*. SUDENE Estudos Regionais no. 18. Recife: SUDENE/PSU/SRE, 1988.
———. "Caracterização geo-econômica da região da mata de Pernambuco." In *Região, formação social e desenvolvimento—suas interrelações: O caso nordestino*. Recife: Edições IJNPS, 1980.
———. *História das usinas de açúcar de Pernambuco*. 2d ed. Recife: Editora Universitária, 2001.
———. "Homem e natureza: Por um política de meio ambiente para o Brasil." *Revista Pernambucana de Desenvolvimento* 14, no. 1–2 (1993): 7–24.
———. *Modernização e pobreza: A expansão da agroindústria canavieira e seu impacto ecológico e social*. São Paulo: Editora da Universidade Estadual de São Paulo, 1994.
———. "A poluição dos cursos d'água da região da mata de Pernambuco, pelo despejo de resíduos e águas servidas pelas indústrias." *Boletim do Instituto Joaquim Nabuco de Pesquisas Sociais* 15 (1966): 63–116.
———. *Os rios-do-açúcar do Nordeste oriental*, vol. 4, *Os rios Coruripe, Jiquiá e São Miguel*. Recife: Instituto Joaquim Nabuco, 1959.
———. *A terra e o homem no Nordeste: Contribuição ao estudo da questão agrária no Nordeste*. 6th ed. Recife: Editora Universitária, 1998 [1963].

———. "O Vale do Siriji." *Revista do Museu do Açúcar* 1, no. 6 (1971): 55–98.

Andrade, Manoel Correia de Oliveira, and Sandra Maria Correia de Andrade. *A cana-de-açúcar na região da mata pernambucana; Reestruturação produtiva na área canavieira de Pernambuco nas décadas de 80 e 90: Impacto ambiental, sócio-econômico e político*. Recife: Editora Universitária UFPE, 2001.

Andrade, Manuel Correia de, and Eliane Moury Fernandes, eds. *Vencedores e vencidos: O movimento de 1964 em Pernambuco*. Recife: Massangana, 2004.

———. *O movimento político-militar de 1964 no Nordeste: Catálogo da história oral*. Recife: Massangana, 2004.

Andrade, Manuel Correia de, et al. *Josué de Castro e o Brasil*. São Paulo: Editora Fundação Perseu Abramo, 2003.

Andrade, Sandra Maria Correia de. "Ação sindical no campo a partir da década de 70: O caso dos trabalhadores canavieiros de Pernambuco." Ph.D. diss., Universidade de São Paulo, 1994.

Andrade Neto, Joaquim Correia Xavier de. "A questão ambiental na zona da mata de Pernambuco." Recife: Instituto Inter-americano de Cooperativa para a Agricultura, 1993.

Anker, Peder. *Imperial Ecology: Environmental Order in the British Empire, 1895–1945*. Cambridge: Harvard University Press, 2001.

Antonil, André João. *Cultura e opulência do Brasil por suas drogas e minas*. São Paulo: Companhia Melhoramentos de São Paulo, 1923 [1711].

Anuário estatístico de Pernambuco, 1959. Recife: Imprensa Oficial, 1960.

Araújo, Ricardo Benzaquen de. *Guerra e paz: "Casa-grande e senzala" e a obra de Gilberto Freyre nos anos 30*. Rio de Janeiro: Editora 34, 1994.

Arquivos do Instituto de Pesquisas Agronômicas, vol. 4. Recife: Secretaria de Agricultura, Indústria e Comércio, 1946.

Athayde, Tristão de. "Zé Lins." In *Menino de engenho*, by José Lins do Rêgo, 36th ed, 23–28. Rio de Janeiro: Nova Fronteira, 1986.

Avaliação e perspectiva da agroindústria canavieira em Pernambuco, vol. 1, *Análise do dirigismo estatal*. Recife: Fundação Instituto Pernambucano, 1991.

Ayala, César J. *American Sugar Kingdom: The Plantation Economy of the Spanish Caribbean, 1898–1934*. Chapel Hill: University of North Carolina Press, 1999.

Azevedo, Fernando. *As ligas camponesas*. Rio de Janeiro: Paz e Terra, 1982.

Balée, William. "Historical Ecology: Premises and Postulates." In *Advances in Historical Ecology*, edited by William Balée, 13–29. New York: Columbia University Press, 1998.

Bandeira, Elcia de Torres. "Os usineiros de Pernambuco e a intervenção do estado na agroindústria canavieira, 1889–1933." M.A. thesis, Universidade Federal de Pernambuco, 1989.

Barbosa Lima Sobrinho, Alexandre. "Prefácio-testemunho dos aspectos sócio-econômicos." In *A década 20 em Pernambuco (Uma interpretação)*, by Souza Barros, 2d ed, 21–31. Recife: Prefeitura da Cidade do Recife, 1985.

———. *Problemas econômicas e sociais da lavoura canavieira*. 2d ed. Rio de Janeiro: Zelio Valverde, 1943.

Barickman, B. J. "'A Bit of Land, Which They Call Roça': Slave Provision Grounds in the Bahian Recôncavo, 1780–1860." *Hispanic American Historical Review* 74, no. 4 (1994): 649–87.

Barman, Roderick. *Citizen Emperor: Pedro II and the Making of Brazil, 1825–1891*. Stanford, Calif.: Stanford University Press, 1999.

Barreto, Lêda. *Julião, Nordeste, revolução*. Rio de Janeiro: Civilização Brasileiro, 1963.

Barros, Henrique de, and Blanca Rubio, eds. *Globalización y desarrollo rural en América Latina*. Recife: Imprensa Universitária UFRPE, 2002.

Barros, Souza. *A década 20 em Pernambuco (Uma interpretação)*, 2d ed. Recife: Prefeitura da Cidade do Recife, 1985.

Basso, Keith H. "Wisdom Sits in Places: Notes on a Western Apache Landscape." In *Senses of Place*, edited by Steven Feld and Keith H. Basso, 53–90. Santa Fe, N.M.: School of American Research Press, 1996.

Bastos, Elide Rugai. *As ligas camponesas*. Petrópolis: Vozes, 1984.

Batalha, Claudio H. M., Fernando Teixeira da Silva, and Alexandre Fortes, eds. *Culturas de classe: Identidade e diversidade na formação do operariado*. Campinas: Editora UNICAMP, 2004.

Beattie, Peter M. "Conflicting Penile Codes: Modern Masculinity and Sodomy in the Brazilian Military, 1860–1916." In *Sex and Sexuality in Latin America*, edited by Donna Guy and Daniel Bouldersten 65–85. New York: New York University Press, 1997.

———. "The Slave Silvestre's Disputed Sale: Corporal Punishment, Mental Health, Sexuality, and 'Vices' in Recife, Brazil, 1869–1879." *Estudios interdisciplinarios de América Latina y el Caribe* 16, no. 1 (2005): 41–65.

———, ed. *The Human Tradition in Modern Brazil*. Wilmington, Del.: SR , 2004.

Bello, Júlio. *Memórias de um senhor de engenho*. 3d ed. Recife: FUNDARPE, 1985 [1935].

Bernardes, Lysia Maria Cavalcanti. *Planície litorânea e zona canavieira do estado do Rio de Janeiro*. Rio de Janeiro: Conselho Nacional de Geografia, 1957.

Bezerra, Gregório. *Memórias*. 2 vols. Rio de Janeiro: Civilização Brasileira, 1980.

Blake, Stanley Earl. "The Invention of the Nordestino: Race, Region, and Identity in Northeastern Brazil, 1889–1945." Ph.D. diss., SUNY–Stony Brook, 2001.

Bourdieu, Pierre. *Outline of a Theory of Practice*. Cambridge: Cambridge University Press, 1977.

Bradley, Sandra Maria Correia. *Açúcar e poder: Análise da evolução política de Vicência, um município da microrregião da mata seca pernambucana*. Recife: CONDEPE/FIAM/CEHM, 1977.

Brandão, Ambrósio Fernandes. *Diálogos das grandezas do Brasil*. Recife: Massangana, 1997 [1618].

Brannstrom, Christian, ed. *Territories, Commodities, Knowledges: Latin American Environmental History in the Nineteenth and Twentieth Centuries*. London: Institute for the Study of the Americas, 2004.

Brannstrom, Christian, and Stefani Gallini. "An Introduction to Latin American Environmental History." In *Territories, Commodities, Knowledges: Latin American Environmental History in the Nineteenth and Twentieth Centuries*, edited by Christian Brannstrom, 1–20. London: Institute for the Study of the Americas, 2004.

Braudel, Fernand. *On History*. Translated by Sarah Matthews. Chicago: University of Chicago Press, 1980.

Bray, Sílvio Carlos, Enéas Rente Ferreira, and Davi Guilherme Gaspar Ruas, eds. *As políticas da agroindústria canavieira e o proálcool no Brasil*. São Paulo: UNESP/Marília Publicações, 2000.

Brayner, Flávio Henrique Albert. *Partido Comunista em Pernambuco*. Recife: Massangana, 1989.

Brubaker, Rogers, and Frederick Cooper. "Beyond 'Identity.'" *Theory and Society* 29 (2000): 1–47.

Buarque, Cristovam. "The Northeast: Five Hundred Years of Discoveries." In *Brazil: A Century of Change*, edited by Ignacy Sachs, Jorge Wilhelm, and Paulo Sérgio Pinheiro; translated by Robert N. Anderson, 271–90. Chapel Hill: University of North Carolina Press, 2009.

Buarque, Sérgio. "Desenvolvimento sustentável da zona da mata de Pernambuco: Relatório técnico." Recife: Instituto Inter-americano de Cooperativa para a Agricultura, 1993.

Budny, Daniel. "The Global Dynamics of Biofuels: Potential Supply and Demand for Ethanol and Biodiesel in the Coming Decade." *Brazil Institute Special Report*, no. 3 (Woodrow Wilson International Center for Scholars, April 2007): 1–7.

Bueno, Ricardo. *Pró-álcool: Rumo ao desastre*. 2d ed. Petrópolis: Vozes, 1980.

Burnard, Trevor. *Mastery, Tyranny, and Desire: Thomas Thistlewood and His Slaves in the Anglo-Jamaican World*. Chapel Hill: University of North Carolina Press, 2004.

Busch, Lawrence, et al. *Making Nature, Shaping Culture: Plant Biodiversity in Global Context*. Lincoln: University of Nebraska Press, 1995.

Cabral, Pedro Eugênio Toledo. "O processo de proletarianização do trabalhador canavieiro de Pernambuco." *Revista pernambucana de desenvolvimento* 11, no. 2/12 (1984–86): 159–75.

Callado, Antônio. *Os industriais da sêca e os 'Galileus' de Pernambuco: Aspectos da luta pela reforma agrária no Brasil*. Rio de Janeiro: Civilização Brasileira, 1960.

———. *Tempo de Arraes: Padres e comunistas na revolução sem violência*. Rio de Janeiro: José Alvaro, 1964.

Câmara Cascudo, Luís da. *Sociologia do açúcar: Pesquisa e dedução*. Rio de Janeiro: IAA, 1971.

Camilo, Josemir. "A seca de 1877 e o problema da mão-de-obra na zona da mata de Pernambuco." *Revista de história municipal* 3, no. 4 (1991): 44.

Caminha, Pedro Vaz de. "Letter of Pedro Vaz de Caminha to King Manuel." In *The Voyage of Pedro Alvares Cabral to Brazil and India*. Translated by William Brooks Greenlee. London: Hakluyt Society, 1938.

Camurça, Silvia Maria Sampaio. "Mulheres, política e cotidiano na zona da mata de Pernambuco." M.A. thesis, Universidade Federal de Pernambuco, 2001.

Cançado, E. D., et al. "The Impact of Sugar Cane–Burning Emissions on the Respiratory System of Children and the Elderly." *Environmental Health Perspectives* 114, no. 5 (May 2006): 725–29.

Candido, Antonio, and José Aderaldo Castello. *Presença da literatura brasileira, história e antologia*, vol. 1, *Das origens ao realismo*. 8th ed. Rio de Janeiro: Bertrand Brasil, 1997.

Cardim, Padre Fernão. *Tratados da terra e gente do Brasil*. São Paulo: Companhia Editora Nacional, 1939 [1584].

Carneiro, Edison. *A insurreição praieira, 1848–49*. Rio de Janeiro: Conquista, 1960.

Carter, Paul. *The Road to Botany Bay: An Exploration of Landscape and History*. New York: Knopf, 1988.

Carvalho, Inaiá Maria Moreira de. *O Nordeste e o regime autoritário*. São Paulo: Hucitec-SUDENE, 1987.

Carvalho, Inaiá Maria Moreira de, and Teresa Maria Frota Haguette, eds. *Trabalho e condições de vida no Nordeste brasileiro*. São Paulo: Hucitec, 1984.

Carvalho, José Murilo de. "Modernização frustrada: A política de terras no império." *Revista brasileira de história* 1 (1981): 53.

Carvalho, Marcus J. M. de. *Liberdade: Rotinas e rupturas do escravismo, Recife, 1822–1850*. Recife: Editora Universitária UFPE, 2001.

Carvalho, Mário B. de. "Comentário em torno de uma nova praga da cana de açúcar." *Boletím da Secretaria da Agricultura, Indústria e Comércio* 3, no. 15 (1948): 4–5.

Carvalho, Ronildo Ferreiro de. *Melhoramento e mecanizagem de cultura canavieira em Pernambuco*. Recife: FAP, GEA, 1964.

Carvalho, Zóia Campos de. *Rosto e máscara do senhor de engenho de Pernambuco (1822–1888)*. Recife: Massangana, 1988.

Casey, Edward S. *The Fate of Place: A Philosophical History*. Berkeley: University of California Press, 1997.

———. "How to Get from Space to Place in a Fairly Short Stretch of Time: Phenomenological Prolegomena." In *Senses of Place*, edited by Steven Feld and Keith H. Basso, 13–52. Santa Fe, N.M.: School of American Research Press, 1996.

Castello, José Aderaldo. "Origens e significado de *Menino de engenho*." In *Menino de engenho*, by José Lins do Rego, 28–35. 36th ed. Rio de Janeiro: Nova Fronteira, 1986.

Castro, Josué de. *Alimentação e raça*. Rio de Janeiro: Civilização Brasileira, 1936.

———. *Death in the Northeast*. New York: Random House, 1966.

———. *Documentário do Nordeste*. São Paulo: Brasiliense, 1957.

———. *Geografia da fome; A dilema brasileiro: Pão ou aço*. 14th ed. Rio de Janeiro: Civilização Brasileira, 2001.

———. *The Geography of Hunger*. Boston: Little, Brown, 1952.

———. *Of Men and Crabs*. Translated by Susan Hertelendy. New York: Vanguard, 1970.

———. *Una zona explosiva de América Latina: El Nordeste brasileño*. Buenos Aires: Solar Hachete, 1965.

Caulfield, Sueann. *In Defense of Honor: Sexual Morality, Modernity, and Nation in Early-Twentieth-Century Brazil*. Durham, N.C.: Duke University Press, 2000.

Cavalcanti, Amaro. "Breve rememoração histórica da cana-de-açúcar." *A lavoura* 82 (1979): 38–43.

Cavalcanti, Arthur Lima. *Vinhoto: Um problema com solução; Substitutivo ao Projeto de Lei nº 209, incorporado à Lei nº 9.377*. Recife: Imprensa Oficial, 1983.

Cavalcanti, Paulo. *O caso eu conto, como o caso foi: Memórias*, vol. 4, *A luta clandestina*. Recife: Guararapes, 1985.

Censo agropecuário, 1995–1996, no. 12, *Pernambuco*. Rio de Janeiro: IBGE, 1998.

Chacon, Vamireh. *Gilberto Freyre: Uma biografia intelectual*. Recife: Massangana, 1993.

Chakrabortty, Aditya. "Secret Report: Biofuel Caused Food Crisis; Internal World Bank Study Delivers Blow to Plant Energy Drive." *Guardian*, July 4, 2008. <http://www.guardian.co.uk/environment/2008/jul/03/biofuels.renewableenergy>. Accessed July 8, 2008.

Chalhoub, Sidney. "The Politics of Disease Control: Yellow Fever and Race in Nineteenth-Century Rio de Janeiro." *Journal of Latin American Studies* 25, no. 3 (1993): 441–63.

———. *Visões da liberdade: Uma história das últimas décadas da escravidão na corte*. São Paulo: Companhia das Letras, 1990.

Chilcote, Richard. *The Brazilian Communist Party: Conflict and Integration, 1922–1972.* New York: Oxford University Press, 1974.

Clark, Kenneth. *Landscape into Art.* New York: Harper and Row, 1976.

Cloke, Paul, Chris Philo, and David Sadler, eds. *Approaching Human Geography: An Introduction to Contemporary Theoretical Debates.* New York: Guilford, 1991.

Conforto, Gilberto. "A política de conservação do meio ambiente." *A lavoura* 87 (1985): 28–30.

CONTAG. *As lutas camponesas no Brasil, 1980.* Rio de Janeiro: 1980.

Cosgrove, Denis. "Prospect, Perspective and the Evolution of the Landscape Idea." *Transactions of the Institute of British Geographers* 10 (1985): 45–62.

———. *Social Formation and Symbolic Landscape.* Totowa, N.J.: Barnes and Noble, 1984.

Costa Lima, Maria Lucia Ferreira da. "A reserva da biosfera da Mata Atlântica em Pernambuco: Situação atual, ações e perspectivas." *Série Cadernos da Reserva da Biosfera* 12 (1998): 1–43.

Coutinho, Odilon Ribeiro. "Gilberto e Nabuco." In *Nabuco e a Federação*, edited by Manuel Correia de Andrade and Tereza Cristina de Sousa Dantas, 55–66. Recife: Massangana, 1992.

Crosby, Alfred. *The Columbian Exchange: Biological and Cultural Consequences of 1492.* Westport, Conn.: Greenwood, 1972.

———. "The Past and Present of Environmental History." *American Historical Review* 100, no. 4 (1995): 1177–89.

Crumley, Carole L. "Historical Ecology: A Multidimensional Ecological Orientation." In *Historical Ecology: Cultural Knowledge and Changing Landscapes*, edited by Carole L. Crumley, 1–13. Santa Fe, N.M.: School of American Research Press, 1994.

Cultura e adubação da cana-de-açúcar. São Paulo: Instituto Brasileiro de Potassa, 1964.

Dabat, Christine Rufino. "Da calçada da casa-grande: A visão da mão-de-obra rural na obra de José Lins do Rego." Paper presented at the Fourth Encontro Estadual de História da ANPUH. Recife, September 2002.

———. *Moradores de engenho: Relações de trabalho e condições de vida dos trabalhadores rurais na zona canavieira de Pernambuco segundo a literatura, a academia e os próprios atores sociais.* Recife: Editora Universitária UFPE, 2008.

———. "Os primórdios da cooperative agrícola de Tiriri." *Clio* 16 (1996): 41–63.

Dabat, Christine Rufino, and Leonardo Guimaraes Neto. "Zona da mata: Emprego, relações de trabalho e condições de vida." Recife: Instituto Inter-americano de Cooperativa para a Agricultura, 1993.

da Cunha, Euclides. *Obra completa*, vol. 2. Edited by Afrânio Coutinho. Rio de Janeiro: José Aguilar, 1966.

———. *Rebellion in the Backlands.* Translated by Samuel Putnam. Chicago: University of Chicago Press, 1944 [1902].

Dantas, Bento. *A agroindústria canavieira de Pernambuco: As raízes históricas dos seus problemas, sua situação atual e suas perspectivas.* Publication 10. Recife: GERAN, 1971.

———. "A recuperação da lavoura canavieira de Pernambuco com base no aumento da produtividade e na intensificação da policultura." *Circular* (Recife: Estação Experimental dos Produtores de Açúcar de Pernambuco), no. 5 (1965): 1–80.

———. *Suscetibilidade de algumas variedades de cana de açúcar à Broca da Diatrea.*

Publication 9. Recife: Comissão de Combate às Pragas da Cana de Açúcar no Estado de Pernambuco, October 1958.

Dantas, Bento, and Lúcio dos Santos e Silva. *Subsídios para o programa de desenvolvimento sustentável da zona da mata*. Recife: PROMATA, 1995.

Dantas, Bento, et al. "Novas variedades de cana para plantio na várzea e início de moagem." *Circular* (Recife: Estação Experimental dos Produtores de Açúcar de Pernambuco), no. 2 (1964): 1–10.

Dantas Silva, Leonardo. "Açúcar: O sal da terra pernambucana." In *Açúcar: A civilização que a cana criou*. Recife: Instituto Cultural Bandepe, 2002.

———. "Nota do editor." In *Memórias de um senhor de engenho*, by Júlio Bello, prefaces by Gilberto Freyre and José Lins do Rego, 3d ed, vii–ix. Recife: FUNDARPE, 1985 [1935].

Dean, Warren. *Brazil and the Struggle for Rubber: A Study in Environmental History*. Cambridge: Cambridge University Press, 1987.

———. "The Green Wave of Coffee: Beginnings of Agricultural Research in Brazil." *Hispanic American Historical Review* 69 (1989): 91–115.

———. "Latifundia and Land Policy in Nineteenth-Century Brazil." *Hispanic American Historical Review* 51, no. 4 (1971): 606–25.

———. *Rio Claro: A Brazilian Plantation System, 1820–1920*. Stanford, Calif.: Stanford University Press, 1976.

———. "The Tasks of Latin American Environmental History." In *Changing Tropical Forests: Historical Perspectives on Today's Challenges in Central and South America*, edited by Harold K. Steen and Richard P. Tucker, 5–15. Durham, N.C.: Forest History Society, 1992.

———. *With Broadax and Firebrand: The Destruction of the Brazilian Atlantic Forest*. Berkeley: University of California Press, 1995.

dé Carli, Gileno. *O processo histórico da usina em Pernambuco*. Rio de Janeiro: Irmãos Pongetti, 1942.

Deerr, Noel. *The History of Sugar*, 2 vols. London: Chapman and Hall, 1949, 1950.

de Kadt, Emanuel. *Catholic Radicals in Brazil*. London: Oxford University Press, 1970.

Deutsch, Ruthanne. "Bridging the Archipelago: Cities and Regional Economies in Brazil, 1870–1920." Ph.D. diss., Yale University, 1994.

Diégues, Manuel, Jr. *População e açúcar no Nordeste do Brasil*. Rio de Janeiro: Comissão Nacional de Alimentação, 1954.

Dossiê Mata Atlântica, 2001. Recife: Projeto Monitoramento Participativo da Mata Atlântica, 2001.

Drummond, José Augusto. *Devastação e preservação ambiental*. Rio de Janeiro: Editora da Universidade Federal Fluminense, 1997.

Duarte, Regina Horta. "Nature and Historiography in Brazil, 1937–1945." *Iberoamericana* 3, no. 10 (2003): 27.

———. "Por um pensamento ambiental histórico: O caso do Brasil." *Luso-Brazilian Review* 41, no. 2 (2005): 144–61.

DuPuis, E. Melanie, and Peter Vandergeest, eds. *Creating the Countryside: The Politics of Rural and Environmental Discourse*. Philadelphia: Temple University Press, 1996.

Eisenberg, Peter. "A mentalidade dos fazendeiros no Congresso Agrícola de 1878." In *Modos de produção e realidade brasileira*, edited by José Roberto do Amaral Lapa, 167–94. Petrópolis: Vozes, 1980.

———. *The Sugar Industry in Pernambuco: Modernization without Change, 1840–1910*. Berkeley: University of California Press, 1974.

Estudos afro-brasileiros, vol. 6, Trabalhos apresentados ao 10 Congresso Afro-Brasileiro realizado no Recife, em 1934. Introduction by José Antônio Gonsalves de Mello. Recife: Massangana, 1988.

Estudos afro-brasileiros, vol. 7, Trabalhos apresentados ao 10 Congresso Afro-Brasileiro realizado no Recife, em 1934. Preface by Arthur Ramos. Recife: Massangana, 1988.

Fernandes, Aníbal. *Um senhor de engenho pernambucano*. Rio de Janeiro: Cruzeiro, 1959.

Fernandes, Bernardo Mançano. *Questão agrária, pesquisa e MST*. São Paulo: Cortez, 2001.

Fernandes, Eliane Moury. *O movimento político-militar de 1964 no Nordeste: Catálogo da história oral*. Recife: Massangana, 2004.

Ferreira, Darlene Aparecida de Oliveira. *Mundo rural e geografia; Geografia agrária no Brasil: 1930–1990*. São Paulo: Editora UNESP, 2002.

Figueiredo, Vilma de Mendonça. "O campo histórico político da tecnologia e os trabalhadores rurais sindicalizados." *Temas rurais* 2, no. 3 (1989): 27–42.

Fletcher, James C., and Daniel P. Kidder. *Brazil and the Brazilians Portrayed in Historical and Descriptive Sketches*. Boston: Little, Brown, 1868.

Fontes, Paulo. *Trabalhadores e cidadãos—Nitroquímica: A fábrica e as lutas operárias nos anos 50*. São Paulo: Annablume, 1997.

———. *Um Nordeste em São Paulo: Trabalhadores migrantes em São Miguel Paulista (1945–1966)*. Rio de Janeiro: Fundação Getúlio Vargas, 2008.

Fortes, Alexandre. *Nós do Quarto Distrito: A classe trabalhadora porto-alegrense e a era Vargas*. Caxias do Sul: Educs; Rio de Janeiro: Garamond, 2004.

Freire, Paulo. *Pedagogy of Hope: Reliving Pedagogy of the Oppressed*. Translated by Robert R. Barr. New York: Continuum, 1994.

French, John D. *Drowning in Laws: Labor Law and Brazilian Political Culture*. Chapel Hill: University of North Carolina Press, 2004.

———. "A história latino-americana do trabalho hoje: Uma reflexão auto-crítica." *Revista de historia* (UNISINOS, Rio Grande do Sul), no. 6 (2002): 11–28.

———. "The Latin American Labor Studies Boom." *International Review of Social History* 45 (2000): 279–308.

French, John D., and Thomas D. Rogers. "Slavery as a 'Sinister Principle' of Authority: Continuities between Slavery and Freedom in the Making of Modern Brazil." Unpub. ms., 2007.

Freyre, Fernando de Mello, ed. "Simpósio sobre as enchentes de Pernambuco (informe final)." Recife: IJNPS, 1975.

Freyre, Gilberto. *Casa-grande e senzala: Formação da família brasileira sob o regime de economia patriarcal*. Recife: Imprensa Oficial, 1966 [1933].

———. "Homens, terras e águas na formação agrária do Brasil: Sugestões para um estudo de interrelações." *Boletim do Instituto Joaquim Nabuco de Pesquisas Sociais* 3 (1954): 3–12.

———. *Interpretação do Brasil*. Translated by Olívio Montenegro. São Paulo: Companhia das Letras, 2001.

———. "Introdução à 9a edição." In *Minha formação*, by Joaquim Nabuco, 4–13. Brasília: Editora Universidade de Brasília, 1981 [1900].

———. "Introdução à 10a edição." In *Minha formação*, by Joaquim Nabuco, 14–24. Brasília: Editora Universidade de Brasília, 1981 [1900].

———. "Joaquim Nabuco e as reformas sociais." Introduction to *O abolicionismo*, by Joaquim Nabuco, 15–27. 4th ed. Petrópolis: Vozes, 1977.

———. *The Mansions and the Shanties: The Making of Modern Brazil*. Translated by Harriet de Onís. New York: Alfred A. Knopf, 1963.

———. *The Masters and the Slaves: A Study in the Development of Brazilian Civilization*. Translated by Samuel Putnam. 2d ed. New York: Alfred A. Knopf, 1963.

———. *Nordeste: Aspectos da influência da cana sobre a vida e a paisagem do Nordeste do Brasil*. 6th ed. Rio de Janeiro: Record, 1989 [1937].

———. *A presença do açúcar na formação brasileira*. Rio de Janeiro: IAA, 1975.

———. *Região e tradição*. Rio de Janeiro: José Olympio, 1941.

———. *Tempo de aprendiz: Artigos publicados em jornais na adolescência e na primeira mocidade do autor, 1918–1926*. São Paulo: Instituição Brasileira de Difusão Cultural, 1979.

———. *Tempo morto e outros tempos: Trechos de um diário de adolescência e primeira mocidade, 1915–1930*. Rio de Janeiro: José Olympio, 1975.

———, ed. *Diário íntimo do engenheiro Vauthier, 1840–1846*. Rio de Janeiro: Ministério da Educação e Saude, 1940.

Freyre, Gilberto, et al. *Cana e reforma agrária*. Recife: IJNPS, 1970.

———. *Novos estudos afro-brasileiros: Trabalhos apresentados ao 1o Congresso Afro-brasileiro do Recife*. Rio de Janeiro: Civilização Brasileira, 1937.

Funes Monzote, Reinaldo. *From Rainforest to Cane Field in Cuba: An Environmental History since 1492*. Translated by Alex Martin. Chapel Hill: University of North Carolina Press, 2008.

Furriela, Rachel Biderman. *Democracia, cidadania e proteção do meio ambiente*. São Paulo: Anablume, FAPESP, 2002.

Galloway, Joey H. "Agricultural Reform and the Enlightenment in Late Colonial Brazil." *Agricultural History* 53, no. 4 (October 1979): 763–79.

———. "The Last Years of Slavery on the Sugar Plantations of Northeastern Brazil." *Hispanic American Historical Review* 51, no. 4 (1971): 586–605.

———. "Northeast Brazil, 1700–1750: The Agricultural Crisis Reexamined." *Journal of Historical Geography* 1, no. 1 (1975): 21–38.

———. *The Sugar Cane Industry: An Historical Geography from Its Origins to 1914*. Cambridge: Cambridge University Press, 1989.

———. "The Sugar Industry of Pernambuco during the Nineteenth Century." *Annals of the Association of American Geographers* 58, no. 2 (2005): 285–303.

Garcia, Afrânio. *Terra de trabalho: Trabalho familiar de pequenas produtores*. Rio de Janeiro: Paz e Terra, 1983.

Gardner, George. *Travels in the Interior of Brazil, Principally through the Northern Provinces, and the Gold and Diamond Districts, during the Years 1836–1841*. 2d ed. London: Reeve, Benham, and Reeve, 1849.

Gaspari, Elio. *A ditadura derrotada*. São Paulo: Companhia das Letras, 2003.

GERAN. *Programa de melhoria das condições de vida dos trabalhadores com base em projeto de educação vocacional agrícola*. Publication no. 24/70. Recife: Grupo Especial para Racionalização da Agroindústria Canavieira do Nordeste, 1970.

Giblin, James. "Land Tenure, Traditions of Thought about Land in Tanzania." In *Land,*

Property and the Environment, edited by John Richards, 138–78. Oakland, Calif.: ICS Press, 2002.

Gidwani, Vinay. "Labored Landscapes: Agro-ecological Change in Central Gujarat, India." In *Agrarian Environments: Resources, Representations, and Rule in India*, edited by Arun Agrawal and K. Sivaramakrishnan, 216–47. Durham, N.C.: Duke University Press, 2000.

Giusti-Cordero, Juan A. "Labor, Ecology and History in a Caribbean Sugar Plantation Region: Piñones (Loíza), Puerto Rico, 1770–1950." Ph.D. diss., SUNY-Binghamton, 1994.

Gonçalves, Fernando Antônio. "Condições de vida do trabalhador rural da zona da mata do estado de Pernambuco—1964." *Boletim do Instituto Joaquim Nabuco de Pesquisas Sociais* 15 (1966): 117–74.

Gonçalves Neto, Wenceslau. *Estado e agricultura no Brasil: Política agrícola e modernização econômica brasileira, 1960–1980*. São Paulo: Hucitec, 1995.

Gonsalves de Mello, J. A. Introduction to *Diálogos das grandezas do Brasil*, by Ambrósio Fernandes, xiii–liii. Recife: Massangana, 1997 [1618].

———. "Notas acêrca da introdução de vegetais exóticos em Pernambuco." *Boletim do Instituto Joaquim Nabuco de Pesquisas Sociais* 3 (1954): 33–64.

González, Luis Arnaldo. "Law, Hegemony, and the Politics of Sugarcane Growers under Getúlio Vargas: Campos, Rio de Janeiro, Brazil, 1930–1950." Ph.D. diss., University of Minnesota, 1998.

Gorenstein, Ossir. "O trabalhador rural e o acesso à propriedade da terra." *Cadernos do CEAS* 84 (March/April 1983): 48–53.

Grant, Andrew. *History of Brazil, Comprising a Geographical Account of That Country, Together with a Narrative of the Most Remarkable Events Which Have Occurred There since Its Discovery; a Description of the Manners, Customs, Religion, &C. of the Natives and Colonists; Interspersed with Remarks on the Nature of Its Soil, Climate, Productions, and Foreign and Internal Commerce. To Which Are Subjoined Cautions to New Settlers for the Preservation of Health*. London: Henry Colburn, 1809.

Greenfield, Gerald. *The Realities of Images: The Great Drought*. Philadelphia: American Philosophical Society, 2001.

Greider, Thomas, and Lorraine Garkovich. "Landscapes: The Social Construction of Nature and the Environment." *Rural Sociology* 59, no. 1 (1994): 1–24.

Grove, Richard H. *Green Imperialism: Colonial Expansion, Tropical Island Edens and the Origins of Environmentalism, 1600–1860*. Cambridge: Cambridge University Press, 1995.

Grzybowski, Cândido. *Caminhos e descaminhos dos movimentos sociais no campo*. Petrópolis: Vozes, 1987.

Hall, Anthony. "Innovation and Social Structure: The Sugar Industry of Northeast Brazil." In *The Logic of Poverty: The Case of the Brazilian Northeast*, edited by Simon Mitchell, 157–69. London: Routledge and Kegan Paul, 1981.

Hamasaki, Cláudia Satie. "O setor sucroalcooleiro e seus trabalhadores: Emprego e pobreza na zona da mata de Pernambuco." M.A. thesis, Universidade Federal de Pernambuco, 1997.

Hartshorne, Richard. *The Nature of Geography*. Lancaster, Penn.: Association of American Geographers, 1939.

Harvey, David. *The Condition of Postmodernity: An Enquiry into the Origins of Cultural Change*. Oxford: Blackwell, 1989.

Hayden, Dolores. "Urban Landscape History: The Sense of Place and the Politics of Space."

In *Understanding Ordinary Landscapes,* edited by Paul Groth and Todd Bressi, 111–33. New Haven, Conn.: Yale University Press, 1997.

Heidegger, Martin. "Building Dwelling Thinking." In *Martin Heidegger: Basic Writings,* edited by David Farrell Krell, 319–39. San Francisco: Harper San Francisco, 1993.

Henderson, James. *A History of Brazil; Comprising Its Geography, Commerce, Colonization, Aboriginal Inhabitants, &C. &C. &C.* London: Longman, Hurst, Rees, Orme, and Brown, 1821.

Heredia, Beatriz Maria Alásia de. *A morada da vida: Trabalho familiar de pequenos produtores do Nordeste do Brasil.* Rio de Janeiro: Paz e Terra, 1979.

Hewitt, Cynthia. "Brazil: The Peasant Movement of Pernambuco, 1961–1964." In *Latin American Peasant Movements,* edited by Henry A. Landsberger, 374–98. Ithaca, N.Y.: Cornell University Press, 1969.

Holloway, Thomas H. *Immigrants on the Land: Coffee and Society in São Paulo, 1886–1934.* Chapel Hill: University of North Carolina Press, 1980.

Holt, Thomas C. *The Problem of Freedom: Race, Labor, and Politics in Jamaica and Britain, 1832–1938.* Baltimore: Johns Hopkins University Press, 1992.

Honório Rodrigues, José. *História da história do Brasil,* part 1, *Historiografia colonial.* São Paulo: Companhia Editora Nacional, 1979.

Horowitz, Irving Louis. *Revolution in Brazil: Politics and Society in a Developing Nation.* New York: E. P. Dutton, 1964.

Houaiss, Antônio. *Dicionário Houaiss da língua portuguesa.* Rio de Janeiro: Objetiva, 2001.

Houtzager, Peter. "State and Unions in the Transformation of the Brazilian Countryside, 1964–1979." *Latin American Research Review* 33, no. 2 (1998): 103–42.

Howe, Gary, and David Goodman, eds. *Smallholders and Structural Change in the Brazilian Economy: Opportunities in Rural Poverty Alleviation.* San José, Costa Rica: IFAD and IICA, 1992.

Huntington, Ellsworth. *Civilization and Climate.* New Haven, Conn.: Yale University Press, 1915.

Igler, David. *Industrial Cowboys: Miller and Lux and the Transformation of the Far West, 1850–1920.* Berkeley: University of California Press, 2001.

Ingold, Tim. "The Temporality of the Landscape." *World Archaeology* 25, no. 2 (1993): 152–74.

Instituto do Açúcar e do Alcool (IAA). *10 congresso açucareiro nacional: Anais,* vol. 1. Rio de Janeiro: IAA, 1950.

Instituto Joaquim Nabuco de Pesquisas Sociais (IJNPS). "O trabalhador rural volante na zona da mata de Pernambuco (Relatório preliminar)." Recife: IJNPS, 1978.

Jaccoud, Luciana de Barros. *Movimentos sociais e crise política em Pernambuco, 1955–1968.* Recife: Massangana, 1990.

Julião, Francisco. *O Cambão—The Yoke: The Hidden Face of Brazil.* Translated by John Butt. Middlesex, U.K.: Penguin, 1972.

Junqueira, Antônio Augusto B., and Bento Dantas. "A cana-de-açúcar no Brasil." In *Cultura e adubação de cana-de-açúcar,* by E. Malavolta et al., 27–60. São Paulo: Instituto Brasileiro de Potassa, 1964.

Kalleberg, Arne L. "Nonstandard Employment Relations: Part-Time, Temporary and Contract Work." *Annual Review of Sociology* 26 (2000): 341–65.

Kenny, Judith T. "Climate, Race, and Imperial Authority: The Symbolic Landscape of the British Hill Station in India." *Annals of the Association of American Geographers* 85, no. 4 (1995): 694–714.

Kidder, Daniel P. *Sketches of Residence and Travels in Brazil, Embracing Historical and Geographical Notices of the Empire and Its Several Provinces*, vol. 2. Philadelphia: Sorin and Ball, 1845.

Kirkendall, Andrew. "Entering History: Paulo Freire and the Politics of the Brazilian Northeast, 1958–1964." *Luso-Brazilian Review* 41, no. 1 (2004): 168–89.

Kittrell, Edward. "Wakefield's Scheme of Systematic Colonization and Classical Economics." *American Journal of Economics and Sociology* 32, no. 1 (1973): 87–111.

Koster, Henry. *Travels in Brazil*, 2 vols. 2d ed. London: Longman, Hurst, Rees, Orme, and Brown, 1817.

Koury, Mauro Guilherme Pinheiro. "Movimentos sociais no campo (Estudos)." *Texto de debate* 9 (1986): 1–55.

———. "O trabalhador da cana: Luta comum e concorrência." *Cadernos de CEAS* 92 (1984): 23–34.

Lara, Silvia Hunold. *Campos da violência: Escravos e senhores na capitania do Rio de Janeiro, 1750–1808*. Rio de Janeiro: Paz e Terra, 1988.

Leeds, Anthony. "Brazil and the Myth of Francisco Julião." In *Politics of Change in Latin America*, edited by Joseph Maier and Richard Weatherhead, 190–204. New York: Praeger, 1974.

Lefebvre, Henri. *The Production of Space*. Translated by Donald Nicholson-Smith. London: Blackwell, 1991.

Leff, Nathaniel H. "Economic Development in Brazil, 1822–1913." In *How Latin America Fell Behind: Essays on the Economic Histories of Brazil and Mexico, 1800–1914*, edited by Stephen Haber, 34–64. Stanford, Calif.: Stanford University Press, 1997.

Leite Lopes, José Sérgio. *A tecelagem dos conflitos de classe na cidade das chaminés*. São Paulo: Marco Zero, 1988.

———. *O vapor do diabo: O trabalho dos operários do açúcar*. 2d ed. Rio de Janeiro: Paz e Terra, 1978.

Leite Lopes, José Sérgio, Celso Brandão, and Rosilene Alvim. *Tecido memória*. DVD. Rio de Janeiro: Museu Nacional de Anthopologia, 2009.

Levantamento das condições ambientais referentes à poluição atmosférica (Diagnóstico das condições ambientais do estado de Pernambuco). Recife: CPRH, 1977.

Levantamento sócio-econômico dos trabalhadores rurais. Recife: CONTAG/FETAPE/SORPE, 1968.

Levinson, Jerome, and Juan de Onís. *The Alliance That Lost Its Way: A Critical Report on the Alliance for Progress*. Chicago: Quadrangle, 1970.

Lévi-Strauss, Claude. *Tristes Tropiques: An Anthropological Study of Primitive Societies in Brazil*. Translated by John Russell. New York: Atheneum, 1968.

Levy, Henrique. "Maintaining Class Domination in the Sugar Area of Northeast Brazil through State Intervention, U.S. Congress, and Ethanol Program." Ph.D. diss., University of Maryland, 1985.

Lima, Dárdano de A. *Estudos fitogeográficos de Pernambuco*. Recife: IPA, 1957.

Linhart, Robert. *O açúcar e a fome: Pesquisa nas regiões açucareiras do Nordeste brasileiro*. Translated by J. Silveira. Rio de Janeiro: Paz e Terra, 1981.

Lins, Rachel Caldas. "Efeitos sociais da degradação dos rios do açúcar do Nordeste do Brasil." *Ciência e trópico* 8, no. 2 (1980): 193–202.

———. "Natureza e limites do conhecimento geográfico." *Ciência e trópico* 6, no. 2 (1978): 271–786.

Lins do Rego, José. *Banguê*. Rio de Janeiro: José Olympio, 1935.

———. *Fogo morto*. 6th ed. Rio de Janeiro: José Olympio, 1965 [1943].

———. *Menino de engenho*. 36th ed. Rio de Janeiro: Nova Fronteira, 1986.

———. *Meus verdes anos (Memórias)*. 2d ed. Rio de Janeiro: José Olympio, 1957 [1956].

———. *O moleque Ricardo: Romance*. Rio de Janeiro: José Olympio, 1935.

———. "Notas sobre Gilberto Freyre." In *Região e tradição*, by Gilberto Freyre, 9–21. Rio de Janeiro: José Olympio, 1941.

———. *Plantation Boy*. Translated by Emmi Baum. New York: Alfred A. Knopf, 1966.

———. Preface to *Memórias de um senhor de engenho*, by Júlio Bello, 3d ed, xix–xx. Recife: FUNDARPE, 1985 [1935].

———. *Pureza*. Translated by Lucie Marion. New York: Hutchinson International, 1948.

———. *Usina*. 2d ed. Rio de Janeiro: José Olympio, 1940.

Lockhart, James, and Stuart Schwartz. *Early Latin America: A History of Colonial Spanish America and Brazil*. Cambridge: Cambridge University Press, 1983.

Lopes, Luiz Simões. "Entrevista." *A lavoura* 79 (1976): 10–11.

Lowenthal, David. "Environmental History: From the Conquest to the Rescue of Nature." In *Cultural Encounters with the Environment: Enduring and Evolving Geographic Themes*, edited by Alexander Murphy and Douglas Johnson, 177–200. New York: Rowman and Littlefield, 2000.

Maciel, Telmo Frederico do Rêgo. *Nível de vida do trabalhador rural da zona da mata—1961*. Recife: Instituto Joaquim Nabuco de Pesquisas Sociais, 1964.

Magalhães, Juraci Perez. *A evolução do direito ambiental no Brasil*. 2d ed. São Paulo: Juarez de Oliveira, 2002.

Malavolta, E. "Tendências no uso de fertilizantes na América Latina—princípios e perspectivas." *Anais da Escola Superior de Agricultura "Luiz de Queiroz"* 32 (1975): 657–70.

Malavolta, E., and M. Rocha. "Recent Brazilian Experience on Farmer Reaction and Crop Response to Fertilizer Use." *Anais da Escola Superior de Agricultura 'Luiz de Queiroz'* 35 (1978): 297–313.

Mallon, Florencia. "Peasants and Rural Laborers in Pernambuco, 1955–1964." *Latin American Perspectives* 5, no. 4 (1978): 49–70.

Manzo, Lynne C. "Beyond House and Haven: Toward a Revisioning of Emotional Relationships with Places." *Journal of Environmental Psychology* 23 (2003): 47–61.

Mão de obra volante na agricultura, IV reunião nacional: Anais. Universidade Estadual Paulista, December 7–8, 1978.

Marcondes, Maria José de Azevedo. *Cidade e natureza: Proteção dos mananciais e exclusão social*. São Paulo: FAPESP, 1999.

Marquardt, Steve. "'Green Havoc': Panama Disease, Environmental Change, and Labor Process in the Central American Banana Industry." *American Historical Review* 106, no. 1 (2001): 49–80.

———. "Pesticides, Parakeets, and Unions in the Costa Rican Banana Industry, 1938–1962." *Latin American Research Review* 37, no. 2 (2001): 3–36.

Martins, José de Souza. *Os camponeses e a política o Brasil: As lutas sociais no campo e seu lugar no processo político*. 5th ed. Petrópolis: Vozes, 1995.

Massey, Doreen. "Double Articulation: A Place in the World." In *Displacements: Cultural Identities in Question*, edited by Angelika Bammer, 110–21. Bloomington: Indiana University Press, 1994.

Massey, Doreen, John Allen, and Philip Sarre, eds. *Human Geography Today*. Cambridge: Polity, 1999.

Mattos de Castro, Hebe Maria. *Das cores do silêncio: Os significados da liberdade no sudeste escravista, Brasil século XIX*. Rio de Janeiro: Arquivo Nacional, 1995.

May, Peter Herman. "A Modern Tragedy of the Non-commons: Agro-Industrial Change and Equity in Brazil's Babassu Palm Zone." Ph.D. diss., Cornell University, 1986.

Maybury-Lewis, Biorn. *The Politics of the Possible: The Brazilian Rural Workers' Trade Union Movement, 1964–1985*. Philadelphia: Temple University Press, 1994.

McCook, Stuart. "Promoting the 'Practical': Science and Agricultural Modernization in Puerto Rico and Colombia, 1920–1940." *Agricultural History* 75, no. 1 (Winter 2001): 52–82.

McGillivray, Gillian. *Blazing Cane: Sugar Communities, Class, and State Formation in Cuba, 1868–1959*. Durham, N.C.: Duke University Press, 2009.

McWilliams, Carey. *Factories in the Fields: The Story of Migratory Farm Labor in California*. Boston: Little, Brown, 1939.

Medeiros, Leonilde Sérvolo de. *História dos movimentos sociais no campo*. Rio de Janeiro: FASE, 1989.

Mello, Evaldo Cabral de. "Mata norte e mata sul: O surto algodoeiro rompeu trezentos anos de hegemonia açucareira e aprofundou diferenças seculares entre as duas zonas da mata de Pernambuco." *Continente multicultural* (July 2002): 84–91.

Mello Filho, Luiz Emydio de. "Conservação e agricultura a visão histórica e a visão prospectiva." *A lavoura* 87 (1985): 31–32.

Melo, Mário. "Ensaio sobre alguns topônimos pernambucanos." *Revista do Instituto Arqueológico, Histórico e Geográfico de Pernambuco* 39, nos. 135–42 (1928–29): 140.

Melo, Mário Lacerda de. *O açúcar e o homem: Problemas sociais e econômicos do Nordeste canavieiro*. Recife: IJNPS, 1975.

———. *Paisagens do Nordeste em Pernambuco e Paraíba*. Rio de Janeiro: Conselho Nacional de Geografia, 1958.

Melo Neto, José Francisco de. *Usina Catende: Entre a doçura e a harmonia*. Catende: mimeo, 2002.

Meneses, Joedna Reis de. "A indústria do atraso ou o atraso da indústria? O discurso da industrialização no Nordeste, 1950–1970." M.A. thesis, Universidade Federal de Pernambuco, 1999.

Menezes, José Luiz Mota. *Nova Iorque nasceu em Pernambuco: Duas estrelas—o mesmo sonho*. Recife: Instituto Cultural Bandepe, 2002.

Mesorregião da mata pernambucana: Microrregião da mata setentrional da mata meridional e de Vitória de Santo Antão. Recife: CONDEPE, 2001.

Milet, Henrique Augusto. *A lavoura da cana de açúcar*. Edited by Manuel Correia de Andrade. Recife: Massangana, 1989 [1881].

Miller, Joseph. "The Numbers, Origins, and Destinations of Slaves in the Eighteenth-Century Angolan Slave Trade." *Social Science History* 13, no. 4 (1989): 421–37.

Miller, Shawn W. *Fruitless Trees: Portuguese Conservation and Brazil's Colonial Timber.* Stanford, Calif.: Stanford University Press, 2000.

———. "Fuelwood in Colonial Brazil: The Economic and Social Consequences of Fuel Depletion for the Bahian Recôncavo, 1549–1820." *Forest and Conservation History* 38, no. 4 (Oct. 1994): 181–92.

———. "Stilt-Root Subsistence: Colonial Mangroves and Brazil's Landless Poor." *Hispanic American Historical Review* 83, no. 2 (2003): 223–53.

Milliet, Sérgio. "A obra de José Lins do Rego." In *Fogo morto*, by José Lins do Rêgo, 6th ed, xviii–xxv. Rio de Janeiro: José Olympio, 1965 [1943].

Mintz, Sidney. "The Rural Proletariat and the Problem of Rural Proletarian Consciousness." In *The Emergence of Worker Consciousness*, vol. 1 of *Ideology and Social Change in Latin America*, edited by June Nash and Juan Corradi, 34–51. New York: mimeo, 1975.

———. *Sweetness and Power: The Place of Sugar in Modern History.* London: Penguin, 1985.

———. *Worker in the Cane: A Puerto Rican Life History.* New Haven, Conn.: Yale University Press, 1960.

Miranda, Edson. *Chapéu de palha: O segundo governo Arraes.* São Paulo: Alfa-Omega, 1991.

Mitchell, Don. *The Lie of the Land: Migrant Workers and the California Landscape.* Minneapolis: University of Minnesota Press, 1996.

Montenegro, Antônio Torres. *História oral e memória: A cultura popular revisitada.* São Paulo: Contexto, 1994.

———. "Labirintos do medo: O comunismo (1950–1964)." *Clio: Revista de pesquisa histórica* 22 (2004): 215–35.

Montgomery, David. *Workers' Control in America: Studies in the History of Work, Technology, and Labor Struggles.* Cambridge: Cambridge University Press, 1983.

Montrie, Chad. *Making a Living: Work and Environment in the United States.* Chapel Hill: University of North Carolina Press, 2008.

Moraes, José Geraldo Vinci de, and José Marcio Rego. *Conversas com historiadores brasileiros.* São Paulo: Editora 34, 2002.

Moreira, Roberto José. "Tecnologia, relações sociais e condições de vida dos trabalhadores rurais." *Temas rurais* 2, no. 3 (1989): 17–21.

Mosher, Jeffrey. "Political Mobilization, Party Ideology, and Lusophobia in Nineteenth-Century Brazil: Pernambuco, 1822–1850." *Hispanic American Historical Review* 80, no. 4 (2000): 881, 901.

Mota, Carlos Guilherme. *Ideologia da cultura brasileira, 1933–1974.* São Paulo: Atica, 1994.

Mota, Mauro. *Os bichos na fala da gente.* Rio de Janeiro: Tempo Brasileiro, 1978.

Moura, Denise A. Soares de. *Saindo das sombras: Homens livres no declínio do escravismo.* Campinas: FAPESP, 1998.

Moura, Margarida Maria. *Camponeses.* São Paulo: Atica, 1986.

Moura, Severino Rodrigues de. *Memórias de um camponês.* Recife: Editora Universitária, 1978.

Movimento dos Trabalhadores Rurais Sem Terra. *Assassinatos no campo: Crime e impunidade, 1964–1986.* São Paulo: Global, 1987.

Nabuco, Carolina. *The Life of Joaquim Nabuco.* Translated by Ronald Hilton. Stanford, Calif.: Stanford University Press, 1950.

Nabuco, Joaquim. *O abolicionismo.* 4th ed. Petrópolis: Vozes, 1977.

———. *Abolitionism: The Brazilian Antislavery Struggle.* Edited and translated by Robert Conrad. Urbana: University of Illinois Press, 1977.

———. *Discursos parlamentares (1879–1889)*, vol. 11 of *Obras completas*. São Paulo: Instituto Progresso Editorial, 1949.

———. *A escravidão*. Edited by Leonardo Dantas Silva. Rio de Janeiro: Nova Fronteira, 1999.

———. *Minha formação*. Brasília: Editora Universidade de Brasília, 1981 [1900].

Nascimento, Cláudio. "Do 'beco dos sapos' aos canaviais de Catende." Ministério do Trabalho e Emprego. <http://www.mte.gov.br/ecosolidaria/prog_becosapos.pdf>. Accessed October 30, 2009.

Negro, Antonio Luigi, and Paulo Fontes. "Using Police Records in Labor History: A Case Study of the Brazilian DOPS." *LABOR: Studies in Working-Class History of the Americas* 5, no. 1 (Spring 2008): 15–22.

Neiman, Guillermo, and Germán Quaranta. "Restructuring and Functional Flexibilization of Agricultural Labor in Argentina." Translated by Carlos Pérez. *Latin American Perspectives* 31, no. 4 (2004): 45–61.

Nieuhof, Joan. *Memorável viagem marítima e terrestre ao Brasil*. Translated by Moacir N. Vasconcelos. São Paulo: Martins, 1942 [1682].

Nunberg, Barbara. "Structural Change and State Policy: The Politics of Sugar in Brazil since 1964." *Latin American Research Review* 21, no. 2 (1986): 53–92.

Ó, Manoel do. *100 anos de suor e sangue: Homens e jornadas da luta operária do Nordeste*. Petrópolis: Vozes, 1984.

O homem e a terra na Usina Catende. n.p.: 1941.

Oliveira, Ariovaldo U. *A geografia das lutas no campo*. São Paulo: Contexto, 2001.

Oliveira, Aurenéa Maria de. "A reconstrução do personagem mítico-político de Miguel Arraes na região da Zona da Mata pernambucana (1986–1990)." M.A. thesis, Universidade Federal de Pernambuco, 2001.

Oliveira, Francisco de. *Elegia para uma re(li)gião: SUDENE, Nordeste, planejamento e conflito de classes*. 6th ed. Rio de Janeiro: Paz e Terra, 1977.

Olwig, Kenneth R. "Recovering the Substantive Nature of Landscape." *Annals of the Association of American Geographers* 86, no. 4 (1996): 630–53.

Oramas, Luis Pérez. "Frans Post, invenção e 'aura' da paisagem." In *O Brasil e os holandeses, 1630–1654*, edited by Paulo Kerkenhoff, 218–37. Rio de Janeiro: GMT, 1999.

Outtes, Joel. *O Recife: Gênese do urbanismo, 1927–1943*. Recife: Massangana, 1997.

Pádua, José Augusto. "'Cultura esgotadora': Agricultura e destruição ambiental nas ultimas décadas do Brasil Império." *Estudos de sociedade e agricultura* 11 (1998): 134.

———. "Natureza e projeto nacional: As origens da ecologia política no Brasil." Estudo no. 54, Instituto Universitário de Pesquisas do Rio de Janeiro, December 1986.

———. *Um sopro de destruição: Pensamento político e crítica ambiental no Brasil escravista (1786–1888)*. Rio de Janeiro: Jorge Zahar, 2002.

Page, Joseph A. *The Revolution That Never Was: Northeast Brazil, 1955–1964*. New York: Grossman, 1972.

Paixão, Marcelo Jorge de Paula. "No coração do canavial: Estudo crítico da evolução do complexo agroindustrial sucro-alcooleiro e das relações de trabalho na lavoura canavieira (estudo comparativo em 12 estados do Brasil)." M.A. thesis, Universidade Federal do Rio de Janeiro, 1994.

Palmeira, Moacir. "The Aftermath of Peasant Mobilization: Rural Conflicts in the Brazilian Northeast since 1964." In *The Structure of Brazilian Development*, edited by Neuma Aguiar, 71–98. New Brunswick, N.J.: Transaction, 1979.

———. "Casa e trabalho: Nota sobre as relações sociais na '*plantation*' traditional." In *Actes du XLIIe Congrès International des Américanistes*, vol. 1, 305–15. Paris: Société des Américanistes, 1976.
Pandeiro, Jackson do. "Boi tungão." Audio recording. *Raízes nordestinos*. 1999 [1956].
Paraíso, Rostand. *A Esquina do Lafayette e outros tempos do Recife*. Recife: R. Paraíso, 2001.
Passmore, John. *Man's Responsibility for Nature*. New York: Charles Scribner and Sons, 1974.
Passos, J. Brito P. *Cronistórias de minhas atividades profissionais*. Recife: Editora Universitária UFPE, 1985.
Pearson, Neale J. "Small Farmer and Rural Worker Pressure Groups in Brazil." Ph.D. diss., University of Florida, 1967.
Peck, Gunther. "The Nature of Labor: Fault Lines and Common Ground in Environmental and Labor History." *Environmental History* 11, no. 2 (2006): 212–38.
Pereira, Anthony. *The End of the Peasantry: The Rural Labor Movement in Northeast Brazil, 1961–1988*. Pittsburgh: University of Pittsburgh Press, 1997.
Pereira, Luiz. *Trabalho e desenvolvimento no Brasil*. São Paulo: Difel, 1965.
Pernambuco, problemática de suas atividades econômicas: Agro indústria açucareira. Recife: GERAN, 1970.
Perruci, Gadiel. "O canto do cisne dos barões do açúcar (Um congresso de classe)." In *Trabalhos do Congresso Agrícola do Recife, outubro de 1878*, by Sociedade Auxiliadora da Agricultura de Pernambuco, i–xlii. Recife: CEPA-PE, 1979.
———. *A República das usinas: Um estudo de história social e econômica do Nordeste, 1889–1930*. Rio de Janeiro: Paz e Terra, 1978.
Pessoa, Dirceu. *Espaço rural e pobreza no Nordeste do Brasil*. Recife: Massangana, 1990.
———, ed. *Política fundiária no Nordeste: Caminhos e descaminhos*. Recife: Massangana, 1990.
Peters, Edson Luiz, and Paulo de Tarso de Lara Pires, eds. *Legislação ambiental federal*. 2d ed. Curitiba: Juruá, 2002.
Pickersgill, Barbara, Spencer C. H. Barrett, and Dardano de Andrade-Lima. "Wild Cotton in Northeast Brazil." *Biotropica* 7, no. 1 (1975): 42–54.
Pinto, Estevão, ed. *O problema agrário na zona canavieira de Pernambuco*. Recife: Imprensa Universitária, 1965.
Pinto, Eudes de Souza Leão. "Considerações agronônicas objetivando o aproveitamento do potencial econômico de Pernambuco." *Estudos brasileiros* 20. Rio de Janeiro: Ministério de Agricultura, 1963.
Por, Francis Dov. *Sooretama: The Atlantic Rain Forest of Brazil*. The Hague: SPB Academic, 1992.
Portelli, Alessandro. *The Death of Luigi Trastulli and Other Stories: Form and Meaning in Oral History*. Albany: SUNY Press, 1991.
Prado, Caio, Jr. *The Colonial Background of Modern Brazil*. Berkeley: University of California Press, 1967.
Pred, Allan. *Making Histories and Constructing Human Geographies: The Local Transformation of Practice, Power Relations, and Consciousness*. Boulder: Westview, 1990.
Price, Robert. "The Brazilian Land Reform Statute." Land Tenure Center of University of Wisconsin Working Paper 15 (April 1965).
Programa de ação para o desenvolvimento da zona da mata do Nordeste. Recife: SUDENE, 1997.

Programa de apoio ao desenvolvimento sustentável da zona da mata, hierarquização dos municípios da zona da mata pernambucana. Recife: CONDEPE, 1997.

Programa governo nos municípios: Região de desenvolvimento mata norte. Plano de Ação Regional, 2000–2003. Recife: FIDEM, 1999.

Programa governo nos municípios: Região de desenvolvimento mata sul. Plano de Ação Regional, 2000–2003. Recife: FIDEM, 1999.

Rabello, Sylvio. "O pobre Rio Siriji." *Boletim do Instituto Joaquim Nabuco de Pesquisas Sociais* 13/14 (1964–65): 243–46.

Radding, Cynthia. *Landscapes of Power and Identity: Comparative Histories in the Sonoran Desert and the Forests of Amazonia from Colony to Republic.* Durham, N.C.: Duke University Press, 2005.

———. *Wandering Peoples: Colonialism, Ethnic Spaces, and Ecological Fronteirs in Northwestern Mexico, 1700–1850.* Durham, N.C.: Duke University Press, 1997.

Raffles, Hugh. *In Amazonia: A Natural History.* Princeton, N.J.: Princeton University Press, 2002.

Raiz, Jovino da. "O trabalhador negro no tempo do banguê comparado com o trabalhador negro no tempo das uzinas de assucar." In *Estudos afro-brasileiros*, 191–94. Recife: Massangana, 1988.

Ramos, Hélio. *Nordeste: Nação espoliada.* Rio de Janeiro: Civilização Brasileira, 1982.

Rask, Kevin. "The Social Costs of Ethanol Production in Brazil: 1978–1987." *Economic Development and Cultural Change* 43, no. 3 (April 1995): 628–31.

Real, Katarina. *O folclore no carnaval do Recife.* Recife: Massangana, 1990.

Rebouças, André. *Agricultura nacional, estudos econômicos; propaganda abolicionista e democrática.* Recife: Massangana, 1988 [1883].

Reis, Jaime. "From Banguê to Usina." In *Land and Labour in Latin America: Essays on the Development of Agrarian Capitalism in the Nineteenth and Twentieth Centuries*, edited by Kenneth Duncan and Ian Rutledge, 369–96. Cambridge: Cambridge University Press, 1977.

Reis, João José. "'The Revolution of the *Ganhadores*': Urban Labour, Ethnicity and the African Strike of 1857 in Bahia, Brazil." *Journal of Latin American Studies* 29 (1997): 455–93.

Repsold, Carlos Arthur. "Uma resposta ao Brasil." *A lavoura* 79 (1976): 3–9.

Resende, Alexander S. de, et al. "Long-Term Effects of Pre-harvest Burning and Nitrogen and Vinasse Applications on Yield of Sugar Cane and Soil Carbon and Nitrogen Stocks on a Plantation in Pernambuco, N.E. Brazil." *Plant and Soil* 281, nos. 1–2 (March 2006): 339–51.

Reynolds, Maura. "Brazil, U.S. to Promote Biofuels; Bush and Lula Tout Ethanol as the Right Path Economically." *Los Angeles Times*, March 10, 2007.

Ribeiro, Joaquim. *Folclore do açúcar.* Rio de Janeiro: Campanha de Defesa do Folclore Brasileiro, 1977.

Richards, John. *Unending Frontier: An Environmental History of the Early Modern World.* Berkeley: University of California Press, 2003.

Richardson, Bonham C. *Igniting the Caribbean's Past: Fire in British West Indian History.* Chapel Hill: University of North Carolina Press, 2004.

Riechmann, Deb. "Bush, Brazil Leader Talk Trade, Ethanol." *Washington Post*, April 2, 2007.

Robock, Stefan H. *Brazil's Developing Northeast: A Study of Regional Planning and Foreign Aid*. Washington, D.C.: Brookings Institution, 1963.

Rodney, Walter. *A History of the Guyanese Working People, 1881–1905*. Baltimore: Johns Hopkins University Press, 1981.

Rodrigue, John C. *Reconstruction in the Cane Fields: From Slavery to Free Labor in Louisiana's Sugar Parishes, 1862–1880*. Baton Rouge: Louisiana State University Press, 2001.

Rodrígues, Mônica dos Santos, and Paula de Andrade Rollo. *Estudo de caso: O mercado de terras rurais na região da zona da mata de Pernambuco, Brasil*, vols. 1 and 2. Santiago de Chile: CEPAL/Sociedad Alemana de Cooperación Técnica, 2000.

Rogers, Thomas D. "Laboring Landscapes: The Environmental, Racial, and Class Worldview of the Brazilian Northeast's Sugar Elite, 1880s-1930s." *Luso-Brazilian Review* 46, no. 2 (2009): 22–53.

———. "Geneticistas da gramínea doce em campos decadentes: Variedades de cana-de-açúcar, agrônomos e plantadores na abordagem da modernização agrícola (1930–1964)." *Clio* 26, no. 2 (2009): 161–88.

———. "José Lins do Rego, *Moleque Ricardo* (1935)." *Labor: Studies in Working-Class History of the Americas* 4, no. 4 (2007): 26–27.

Rosa, Marcelo. "As novas faces do sindicalismo rural brasileiro: A reforma agrária e as tradições sindicais na zona da mata de Pernambuco." *Dados* 47, no. 3 (2004): 473–503.

Rosa e Silva Neto, J. M. da. *Contribuição ao estudo da zona da mata em Pernambuco: Aspectos estruturais e econômicos da área de influência das usinas de açúcar*. Recife: IJNPS, 1966.

——— "Sugestões para uma nova orientação a ser tomada pelos proprietários da zona sul do estado de Pernambuco." *Separata do Boletím da Secretaria da Agricultura, Indústria e Comércio* 3–4 (1948): 4–7.

Roseberry, William. *Coffee and Capitalism in the Venezuelan Andes*. Austin: University of Texas Press, 1983.

———. "Potatoes, Sacks, and Enclosures in Early Modern England." In *Golden Ages, Dark Ages: Imagining the Past in Anthropology and History*, edited by William Roseberry and Jay O'Brien, 19–47. Berkeley: University of California Press, 1991.

Rottenberg, Simon. "Negotiated Wage Payments in British West Indian Agriculture." *Journal of Farm Economics* 33, no. 3 (1951): 402–5.

Rufino de Araújo, Espedito. *O trator e o 'burro sem rabo': Consequências da modernização agrícola sobre a mão-de-obra na região canavieira de Pernambuco, Brasil*. Geneva: IUED, mimeo, 1990.

Rufino de Araújo, Espedito, and Christine Rufino Dabat. "'Se não fosse o sindicato . . .': Papel do sindicato de trabalhadores rurais na vida dos canavieiros de Pernambuco." In *O novo mapa do mundo: Problemas geográficos de um mundo novo*, edited by Milton Santos et al., 180–92. São Paulo: Hucitec, 1995.

Rutenberg, Jim. "Bush, Following Up on Trip, Meets with Brazilian Leader." *New York Times*, April 1, 2007.

Sackman, Douglas. "'Nature's Workshop': The Work Environment and Workers' Bodies in California's Citrus Industry, 1900–1940." *Environmental History* 5, no. 1 (2000): 27–53.

Sales, Apolônio. "Em torno da Lei 178." *Boletim da Secretaria de Agricultura, Indústria e Comércio* 7, no. 2 (1941): 107–17.

———. *Hawaii açucareiro*. Recife: IPA, SAIC, 1937.

———. *O Ministério da Agricultura no govêrno Getúlio Vargas (1930–1944)*. Rio de Janeiro: Serviço de Documentação Ministério da Agricultura, 1945.

Sampaio, Cid Feijó. *Mensagem apresentada pelo Exmo. Sr. Governador Cid Feijó Sampaio à Assembléia Legislativo do Estado de Pernambuco, em 15 de março de 1959*. Recife: Imprensa Oficial, 1959.

———. *O problema açucareiro em Pernambuco (Carta ao Presidente da República)*. Recife: Grupo de Estudos do Açúcar, 1965.

Sartre, Jean-Paul. *The Philosophy of Jean-Paul Sartre*. Edited by R. Cumming. New York: Vintage, 1965.

Sauer, Carl Ortwin. "The Morphology of Landscape." In *Land and Life: A Selection from the Writings of Carl Ortwin Sauer*, edited by John Leighly, 315–50. Berkeley: University of California Press, 1967.

Schama, Simon. *Landscape and Memory*. New York: Vintage, 1995.

Scheper-Hughes, Nancy. *Death without Weeping: The Violence of Everyday Life in Brazil*. Berkeley: University of California Press, 1992.

Schwarcz, Lilia Moritz. *The Spectacle of the Races: Scientists, Institutions, and the Race Question in Brazil, 1870–1930*. New York: Hill and Wang, 1999.

Schwartz, Stuart B. "Indian Labor and New World Plantations: European Demands and Indian Responses in Northeastern Brazil." *American Historical Review* 83, no. 1 (Feb. 1978): 43–79.

———. *Slaves, Peasants, and Rebels: Reconsidering Brazilian Slavery*. Urbana: University of Illinois Press, 1995.

———. *Sugar Plantations in the Formation of Brazilian Society: Bahia, 1550–1835*. Cambridge: Cambridge University Press, 1985.

Scott, James C. *Seeing Like a State: How Certain Schemes to Improve the Human Condition Have Failed*. New Haven, Conn.: Yale University Press, 1998.

Scott, Rebecca J. *Degrees of Freedom: Louisiana and Cuba after Slavery*. Cambridge: Harvard University Press, 2005.

———. *Slave Emancipation in Cuba: The Transition to Free Labor, 1860–1899*. Princeton, N.J.: Princeton University Press, 1985.

Scott, Rebecca J., and Michael Zeuske. "Property in Writing, Property on the Ground: Pigs, Horses, Land, and Citizenship in the Aftermath of Slavery, Cuba, 1880–1909." *Comparative Studies of Society and History* 44, no. 4 (2002): 669–99.

Semple, Ellen Churchill. *Influences of Geographic Environment on the Basis of Ratzel's System of Anthropo-geography*. New York: Russell and Russell, 1911.

Sgrecia, Alexandre. "O assalariado da cana: Resistência do trabalhador 'fichado' à dominação do capital." Paper presented at conference "Movimentos sociais no campo," November 24–26, 1979, Recife.

Short, John Rennie. *Imagined Country: Society, Culture and the Environment*. New York: Routledge, 1991.

Sicsú, Abraham Benzaquen, and Lúcia Carvalho Pinto de Melo. "Desenvolvimento sustentável para a zona da mata de Pernambuco: A dimensão tecnologica." Recife: Instituto Inter-americano de Cooperativa para a Agricultura, 1993.

Sigaud, Lygia. "Armadilhas da honra e do perdão: Usos sociais do direito na mata pernambucana." *Mana* 10, no. 1 (2004): 131–63.

———. *Os clandestinos e os direitos: Estudo sobre trabalhadores da cana-de-açucar de Pernambuco*. São Paulo: Duas Cidades, 1979.

———. "A Collective Ethnographer: Fieldwork Experience in the Brazilian Northeast." *Social Science Information* 47, no. 1 (2008): 71–97.

———. *Greve nos engenhos*. Rio de Janeiro: Paz e Terra, 1980.

———. "The Idealization of the Past in a Plantation Area: The Northeast of Brazil." In *Emergence of Worker Consciousness*, vol. 1 of *Ideology and Social Change in Latin America*, edited by June Nash and Juan Corradi, 167–82. New York: mimeo, 1975.

———. "A luta de classes em dois atos: Notas sobre um ciclo de greves camponesas." *Revista de ciências sociais* 29, no. 3 (1986): 319–43.

———. "Luta política e luta pela terra no Nordeste." *Revista de ciências sociais* 26, no. 1 (1983): 77–95.

———. "A morte do caboclo: Um exercício sobre sistemas classificatórios." *Boletim do Museu Nacional* 30 (1978): 1–29.

———. "A percepção do salário entre trabalhadores rurais no Nordeste do Brasil." In *Actes du XLIIe Congrès International des Américanistes*, vol. 1, 316–30. Paris: Société des Américanistes, 1976.

Silva, Gerson Victor, and João Policarpo R. Lima. "Desenvolvimento sustentável na zona da mata: Econômica." Recife: Instituto Inter-americano de Cooperativa para a Agricultura, 1993.

Silva, José Graziano da. *O novo rural brasileiro*. Campinas: UNICAMP, 1999.

———. *De bóias-frias a empregados rurais: As greves dos canavieiros paulistas de Guariba e de Leme*. Maceió: Edufal, 1997.

Simas, Paula. *Açúcar bruto*. Brasília: Editora Universidade de Brasília, 1997.

Singer, Paulo. "Economic Evolution and the International Connection." In *Brazil: A Century of Change*, edited by Ignacy Sachs, Jorge Wilheim, and Paulo Sérgio Pinheiro, 55–100. Translated by Robert N. Anderson. Chapel Hill: University of North Carolina Press, 2009.

Sistemas de produção florestal para a zona da mata pernambucana. Recife: IBAMA, 1996.

Skidmore, Thomas E. *Black into White: Race and Nationality in Brazilian Thought*. Durham, N.C.: Duke University Press, 1993.

———. *Brazil: Five Centuries of Change*. 2d ed. New York: Oxford University Press, 2010.

———. *Politics in Brazil, 1930–1964: An Experiment in Democracy*. London: Oxford University Press, 1967.

———. *The Politics of Military Rule in Brazil, 1964–1985*. New York: Oxford University Press, 1990.

———. "Raízes de Gilberto Freyre." *Journal of Latin American Studies* 34 (2002): 1–20.

Soares, José Arlindo. *A frente do Recife e o governo do Arraes: Nacionalismo em crise—1955/1964*. Rio de Janeiro: Paz e Terra, 1982.

Sociedade Auxiliadora da Agricultura de Pernambuco. *Trabalhos do Congresso Agrícola do Recife, outubro de 1878*. Recife: CEPA-PE, 1979.

Soluri, John. *Banana Cultures: Agriculture, Consumption, and Environmental Change in Honduras and the United States*. Austin: University of Texas Press, 2004.

———. "People, Plants, and Pathogens: The Eco-social Dynamics of Export Banana Production in Honduras, 1875–1950." *Hispanic American Historical Review* 80, no. 3 (2000): 463–501.

Souto Maior, Mario, and Leonardo Dantas Silva, eds. *A paisagem pernambucana*. Recife: Massangana, 1993.

Souza, Francisco de Assis Lemos de. *Nordeste: O Vietnã que não houve; Ligas camponesas e o golpe de 64*. Londrina: UEL/Ed. da Universidade Federal da Paraíba, 1996.

Souza, Gabriel Soares de. *Tratado descritivo do Brasil em 1587*. Belo Horizonte: Itatiaia, 2000.

Staden, Hans. *Primeiros registros escritos e ilustrados sobre o Brasil e seus habitantes*. Edited by Mary Lou Paris and Ricardo Ohtake. São Paulo: Terceiro Nome, 1999.

Steinberg, Ted. "Down to Earth: Nature, Agency, and Power in History." *American Historical Review* 107, no. 3 (2002): 798–820.

Stewart, Mart A. *"What Nature Suffers to Groe": Life, Labor, and Landscape on the Georgia Coast, 1680–1920*. Athens: University of Georgia Press, 1996.

Strauss, E. "Cana de açúcar e fertilidade do solo." Special issue of *Boletim da Secretaria de Agricultura, Indústria e Comércio* 7, no. 4 (1946): 323.

Striffler, Steve. *In the Shadows of State and Capital: The United Fruit Company, Popular Struggle, and Agrarian Restructuring in Ecuador, 1900–1995*. Durham, N.C.: Duke University Press, 2002.

Strohmeier, Gerhard. "Wild West Imagery: Landscape Perception in Nineteenth-Century America." In *Nature and Society in Historical Context*, edited by Mikuláš Teich, Roy Porter, and Bo Gustafsson, 257–73. Cambridge: Cambridge University Press, 1997.

Sugarcane Breeding Institute. "Breeding." <http://sugarcane-breeding.tn.nic.in/breeding.htm>. Accessed April 23, 2005.

Szmrecsányi, Tamás, and Oriowaldo Queda, eds. *Vida rural e mudança social: Leituras básicas de sociologia rural*. São Paulo: Companhia Editora Nacional, 1976.

Taunay, Affonso de E. *André João Antonil (João Antonio Andreoni, S.J.) e sua obra: Estudo bio-bibliographico*. São Paulo: Companhia Melhoramentos de S. Paulo, 1923.

———. "Antonil e sua obra." In *Cultura e opulência do Brasil por suas drogas e minas*, by André João Antonil, 7–59. São Paulo: Companhia Melhoramentos de São Paulo, 1923 [1711].

Taunay, Carlos Augusto. *Manual do agricultor brasileiro*. Edited by Rafael de Bivar Marquese. São Paulo: Companhia das Letras, 2001 [1839].

Taylor, Thomas Griffith. *Environment and Race: A Study of the Evolution, Migration, Settlement and Status of the Races of Man*. London: Oxford University Press, 1927.

Teixeira, João Gabriel L. C. "Estado, sindicatos e as transformações tecnológicas na agricultura brasileira." *Cadernos de estudos sociais* 5, no. 1 (1989): 113–28.

Teixeira, Luís. "Roteiro de todos os sinais, conhecimentos, fundos, baixos, alturas e derrotas que há na costa do Brasil desde o cabo de Santo Agostinho até ao Estreito de Fernão de Magalhães." In *A paisagem pernambucana*, edited by Mário Souto Maior and Leonardo Dantas Silva, 9–12. Recife: Massangana, 1993.

Thomas, Julian. "Comments on Part I: Intersecting Landscapes." In *Contested Landscapes: Movement, Exile, and Place*, edited by Barbara Bender and Margot Winder, 181–88. Oxford: Berg, 2001.

Thrift, Nigel. "Steps to an Ecology of Place." In *Human Geography Today*, edited by Doreen Massey, John Allen, and Philip Sarre, 295–323. Cambridge: Polity, 1999.

Tollenare, L. F. *Notas dominicais, tomadas durante uma viagem em Portugal e no Brasil em 1816, 1817 e 1818*. Salvador: Progresso, 1956.

Topik, Steven. *The Political Economy of the Brazilian State, 1889–1930*. Austin: University of Texas Press, 1987.

Torres, Vasconcelos. *Condições de vida do trabalhador na agroindústria do açúcar*. Rio de Janeiro: Instituto do Açúcar e do Alcool, 1945.

———. *Oliveira Vianna, sua vida e sua posição nos estudos brasileiros de sociologia*. Rio de Janeiro: Freitas Bastos, 1956.

Trigo, Luciano. *Engenho e memória: O Nordeste do açúcar na ficção de José Lins do Rêgo*. Rio de Janeiro: Topbooks, 2002.

Trouillot, Michel-Rolph. *Silencing the Past: Power and the Production of History*. Boston: Beacon, 1995.

Tuan, Yi-Fu. "Place: An Experiential Perspective." *Geographical Review* 65, no. 2 (1975): 151–65.

Varnhagen, Francisco Adolfo de. *História geral do Brasil antes da sua separação e independência de Portugal, vol. 1*. São Paulo: Companhia Melhoramentos, 1956.

Várzea, Affonso. *Geografia do açúcar no leste do Brasil*. Rio de Janeiro: IAA, 1944.

Vasconcelos Sobrinho, João de. *As regiões naturais do Nordeste: O meio e a civilização*. Recife: Conselho do Desenvolvimento de Pernambuco, 1970.

Vianna, Hermano. *The Mystery of Samba: Popular Music and National Identity in Brazil*. Translated by John C. Chasteen. Chapel Hill: University of North Carolina Press, 1999.

Vicente, Ana Valéria. *Maracatu rural: O espetáculo como espaço social*. Recife: Editora Associação Reviva, 2005.

Vidal e Souza, Candice. *A pátria geográfica: Sertão e litoral no pensamento social brasileiro*. Goiânia: Editora UFG, 1997.

Viola, Eduardo J. "O movimento ecológico no Brasil (1974–1986): Do ambientalismo à ecopolítica." Working Paper no. 93, Kellogg Institute for International Studies, University of Notre Dame, April 1987.

Viotti da Costa, Emilia. *The Brazilian Empire: Myths and Histories*. Chapel Hill: University of North Carolina Press, 2000.

Waldman, Maurício. *Ecologia e lutas sociais no Brasil*. São Paulo: Contexto, 2001.

Wanderley, Maria de Nazareth Baudel. "Algumas reflexões sobre o campesinato do Nordeste: Conceito e realidade." *Ciência e cultura* 29, no. 5 (1977): 537–44.

———. *Capital e propriedade fundiária: Suas articulações na economia açucareira de Pernambuco*. Rio de Janeiro: Paz e Terra, 1979.

———. "O 'lugar' dos rurais: O meio rural no Brasil moderno." Paper presented at the XXI Encontro Annual da ANPOCS, Caxambu, October 1997.

Watts, David. *The West Indies: Patterns of Development, Culture and Environmental Change since 1492*. Cambridge: Cambridge University Press, 1987.

Welch, Cliff. *The Seed Was Planted: The São Paulo Roots of Brazil's Rural Labor Movement, 1924–1964*. University Park: Pennsylvania State University Press, 1974.

Wells, Miriam. *Strawberry Fields: Politics, Class, and Work in California Agriculture*. Ithaca, N.Y.: Cornell University Press, 1996.

White, Richard. "Are You an Environmentalist or Do You Work for a Living? Work and Nature." In *Uncommon Ground: Rethinking the Human Place in Nature*, edited by William Cronon, 171–85. New York: W. W. Norton, 1995.

———. *The Organic Machine*. New York: Hill and Wang, 1996.

Wilkie, Mary E. "A Report on Rural Syndicates in Pernambuco." Rio de Janeiro: Centro Latino-americano de Pesquisas em Ciências Sociais, 1964.

Williams, Raymond. *The Country and the City*. New York: Oxford University Press, 1973.

Wolford, Wendy. "Of Land and Labor: Agrarian Reform on the Sugarcane Plantations of Northeast Brazil." *Latin American Perspectives* 31, no. 2 (2004): 147–70.

———. *This Land Is Ours Now: Social Mobilization and the Meanings of Land in Brazil*. Durham, N.C.: Duke University Press, 2010.

Woortmann, Klaas. "'Com parente não se neguceia': O campesinato como ordem moral." *Anuário antropológico* (1987): 11–65.

Worster, Donald. *Dust Bowl: The Southern Plains in the 1930s*. New York: Oxford University Press, 1979.

———. *Nature's Economy: A History of Ecological Ideas*. 2d ed. Cambridge: Cambridge University Press, 1994.

———. "Transformations of the Earth: Toward an Agroecological Perspective in History." *Journal of American History* 76, no. 4 (1990): 1087–106.

Zaidan Filho, Michel. *O fim do Nordeste e outros mitos*. São Paulo: Cortez, 2001.

Zoneamento agroecológico do Nordeste: Diagnóstico do quadro natural e agrosocioeconômico. 2 vols. Petrolina: EMBRAPA, 1993.

Zoneamento pedoclimático do estado de Pernambuco. Recife: CONDEPE, 1987.

INDEX

Abolitionism, 43, 47–49, 51–53, 65
Accord of the Fields, 160
Agrarian reform, 3, 134, 163–64, 174–76, 247 (n. 138), 260 (n. 85)
Agriculture: plantation-based, 2, 8; and social change, 9, 95, 153, 160; modernization of, 10, 17, 51–52, 83, 100, 103, 108, 113, 153–54, 180–84, 208; non-Brazilian sugarcane, 12, 142; indigenous, 24, 30, 33; in colonial Brazil, 27, 33, 36, 37; in imperial Brazil, 40, 42, 69; critiques of sugarcane, 45–46, 60, 187, 204; state and, 158, 179, 188, 200. *See also* Agronomists; Agronomy; *Engenhos*; Estatuto da Lavoura Canavieira; Sugarcane; Tasks; *Usinas*
Agronomists, 3, 14, 15, 81, 83, 126, 128, 244 (n. 77); and agricultural modernization, 103, 105, 109, 133, 150, 154, 177, 205–6, 261 (n. 18); and rural workers, 106, 118–19, 121, 177; and CO 331 (3X), 111, 113; and 1970s sugarcane boom, 180–83, 190, 193, 201, 205, 206
Agronomy, 100, 103, 154
Alagôas, 41, 52, 216
Alcohol. *See* Ethanol
Aliança da Renovação Nacional (ARENA), 180
Aliança Nacional Libertadora (ANL), 130
Alliance for Progress, 3, 136
Andrade, Manuel Correia de Oliveira, 135
Anthropology, 6, 4, 11, 220 (n. 11), 230 (n. 38), 233 (n. 88)
Antonil, João, 33–34, 83, 142
Arraes, Miguel: electoral victories, 134, 139; as cane technician, 134, 141; importance to cane workers, 137, 140–41, 159–60, 194, 197, 206, 212, 251 (n. 56); as polarizing figure, 141–42, 146, 157, 251 (n. 67)
Associação Pernambucana de Defesa de Natureza (ASPAN), 186, 262 (n. 33)
Atlantic Forest, 12, 22–23, 29, 47, 222 (n. 4); destruction of, 33, 36, 39–40, 44, 203, 226 (n. 70), 262 (n. 33)
Azevedo, Antônio da Costa "Tenente," 106–8, 114, 128, 180, 213

Bagasse, 34, 152, 185–87, 262 (n. 34)
Bananas, 10–11, 85, 90, 91–92, 207, 216
Barracão (*engenho* store), 90, 238 (n. 85)
Bello, Júlio, 46, 100, 114; commentary on modernization and change, 52–64 passim, 76, 83, 101, 213; commentary on slavery, 57, 68, 74, 234 (n. 121)
Bezerra, Gregório: childhood and rural labor, 66–67, 78, 81–83, 89, 188, 209, 233 (n. 110); and union organizing, 130, 138, 148, 153, 157, 160, 209, 248 (n. 22), 252 (n. 75), 255 (nn. 1, 5)
Boletim canavieiro, 144, 145
Braça (unit of measure), 118–19, 121–22, 161, 246 (nn. 112, 126), 247 (n. 131), 251 (n. 64), 256 (n. 20)
Brasil açucareiro, 89–90, 108, 110; social research and commentary in, 90, 105, 115, 245–46 (n. 110); agricultural research in, 103–4, 107, 150, 152, 180

Cabo de Santo Agostinho, 22–24, 27, 30, 159, 255 (n. 7)
Callado, Antônio, 133–36
Caminha Filho, Adrião, 104, 107–9, 241 (n. 25), 242 (n. 47)
Canal Pointe: CP sugarcane varieties, 110, 112, 243 (n. 63)
Capibaribe River, 27, 88, 203
Captivity: concept among workers, 6, 8, 72, 75, 85, 92, 94, 162, 171–74, 177–78, 189, 206, 216; and freedom, 6, 78, 92, 94–95, 124, 172, 178, 205, 216; landscape of, 15, 71, 89–96 passim, 141, 180, 205. *See also* Freedom
Cardim, Padre Fernão, 24, 26, 31, 225 (nn. 60, 63), 226 (n. 75), 227 (n. 93)
Caribbean islands, 10, 32, 34–36, 38, 64, 73–74, 77, 101, 103, 109, 252 (n. 79). *See also* Cuba; Puerto Rico
Castelo Branco, General Humberto, 123, 174
Castro, Josué de, 86–88, 125, 129–30, 134–35, 174, 237 (n. 71), 238 (nn. 73, 74, 76)
Catende, 128
Catende, Usina, 80, 242 (nn. 38, 39), 245 (n. 100); and agricultural innovation,

106–8, 115, 141, 180, 182, 262 (n. 39); and labor disputes, 166, 171, 173, 190–92, 194, 263 (nn. 55, 57); as workers' cooperative, 213–16, 266 (n. 37)
Catholic Church, 4, 16, 135, 138–40, 196, 226 (n. 75), 250 (n. 47)
Cavalcanti, Carlos de Lima, 86, 91, 231 (n. 59)
Clandestinos, 165, 168, 171–72, 194, 257–58 (n. 44)
Class, 10, 13, 67–68, 142; planter, 3, 8, 10, 14, 15, 31, 45–67 passim, 175, 177, 207, 230 (n. 43); class relations, 5, 59, 69, 123, 141, 159; working, 12, 85–88, 130, 132, 207–8; middle, 164
Coffee: in southeast Brazil, 2, 11, 14, 30, 39, 51; in *zona da mata*, 5, 216; regulation of, 11, 103; economic importance of, 30, 37, 40–42, 53, 73, 75; and labor, 63, 74, 76, 116
Coimbatore: CO sugarcane varieties, 110–12, 241 (n. 17), 242 (n. 47), 243 (n. 63). *See also* CO 331 (3X)
Coimbra, Estácio, 58, 86, 231 (n. 65)
Commodities, 10–12, 39, 40, 48, 103, 116, 198
Communist Party, 4, 16, 66, 130–32, 135, 138–40, 144, 146–48, 157, 160, 196
Companhia Pernambucana de Controle da Poluição Ambiental e de Administração de Recursos Hídricos (CPRH, now Agência Estadual de Meio Ambiente in Pernambuco), 152, 186, 265 (n. 1)
Conferência Nacional dos Trabalhadores na Agricultura (CONTAG), 195, 219–20 (n. 9)
Consolidação das Leis do Trabalho (CLT), 130–31, 138, 263 (n. 61)
Contractors, 92, 166–67, 171–72, 175, 207, 258 (n. 58), 266 (n. 13)
CO 331 (3X), 15; rapid adoption of in Pernambuco, 109–14, 123–24, 129, 140; consequences of use, 158, 163, 185, 205–6; science of, 180, 182, 243 (nn. 64, 65), 244 (n. 83)
Cotton, 21–22, 37, 39–42, 51, 72–73, 79, 116, 127, 204
Coup of 1964. *See* Dictatorship
Crespo, Padre Paulo, 138–39, 146, 149, 174–75, 211, 259 (n. 76)
Cuba, 10, 12, 36, 65, 215–16; sugar production in, 42, 75–76, 102, 109, 181, 215–16, 228 (n. 125); and politics, 130, 135, 249 (n. 32), 250 (n. 49)
Curado. *See* Instituto Agrícola do Nordeste

Dantas, Bento, 109, 112, 121, 151–52, 177, 184
Deforestation, 14, 34, 36, 44, 105, 185, 187, 189
Departamento de Ordem Político e Social (DOPS), 12, 125, 143, 146–48, 195–96, 247–48 (n. 1), 262 (n. 38)
Dictatorship: coup of 1964, 3, 140, 157, 159, 196; influence in *zona da mata*, 4, 16, 158, 179; landscape discourse of, 6, 187, 193, 200, 204–6, 217; labor policies of, 11, 158, 160, 177, 198, 207–8, 212, 256 (n. 16); and transition to democracy, 12, 195, 212; management of sugarcane industry, 17, 158, 161, 164, 177, 179–80, 192, 206; and strikes, 17, 193–94, 204, 207; and agrarian reform, 175–77, 209, 210. *See also* Estado Novo
Donkeys, 37, 53, 57, 79–80, 90, 190, 192
Dutch, the, 10, 27, 32–33, 225 (n. 60)

Education, 7, 82, 84, 103, 129, 135, 141, 182, 250 (n. 46), 260 (n. 85)
Emancipation. *See* Slavery
Empreiteiros. *See* Contractors
Engenhos (sugar plantations), 15, 16, 40, 143, 176, 187, 214, 216; description and analysis of, 29–30, 64, 66, 68–69, 73, 90–95, 159, 172–73, 178; historic impact and distribution of, 31–36, 42, 51, 73, 75–76, 79, 100–107, 128, 184; and transition to *usinas*, 42, 43, 182; planters' perceptions of, 45, 50, 52 55, 57, 59, 66, 67, 68, 200; workers and, 72, 74, 77, 78, 87, 90–95, 114–22 passim, 129, 148, 162, 167, 171–75, 207; conflict on, 125–39 passim, 165, 166, 168, 208–9; labor cases against, 191; organizing on, 194, 196–97; and subsistence crops, 210, 213, 215. *See also* Expulsion; *Senhores de engenho*
Erosion, 25, 30, 34, 44, 187, 228 (n. 134)
Estado Novo, 86, 104, 131, 158
Estatuto da Lavoura Canavieira (ELC), 104, 114–15, 124, 158, 175
Estatuto da Terra (ET). *See* Agrarian reform
Estatuto do Trabalhador Rural (ETR): effects of on workers, 138, 158–59, 162–63, 172–73, 177, 193, 197; passage and content of, 138, 165, 167, 176; effects of on planters, 153, 158, 173, 206; and compliance, 162–63, 167–68, 172–73, 176, 206
Ethanol, 3, 17, 210–12, 215–16
Expulsion, 132, 138, 168, 172–74, 207, 209, 212, 258 (nn. 54, 62)

Federação dos Trabalhadores na Agricultura do Estado de Pernambuco (FETAPE), 162, 168, 174–76, 193–95, 208–9, 212, 257 (n. 43)
Fertilizer, 38, 42, 106–7, 111, 114, 152, 181–84, 215, 261 (n. 21)

Fichas (work cards), 165–71, 173, 175, 191, 206–7
Fires, 9, 16, 30; and sugarcane harvest, 121, 149–53, 206, 254 (n. 132); politicization and investigation of, 125–26, 141–49, 206, 210, 252 (nn. 79, 86), 253 (nn. 95, 99)
Floods, 17, 186–87, 196, 200–201, 208
Folklore, 7, 15, 72, 93–94, 205, 236 (n. 47), 239 (n. 104)
Forests, 1, 10, 12, 18, 90, 189; and *zona da mata*, 2, 22, 24, 28; in Cuba, 10, 12, 36; historical accounts of in Pernambuco, 21, 23, 26, 28, 29, 38, 60; physical description of in Pernambuco, 23, 25, 34; and fires, 24, 30, 35, 149; impact of sugar industry on, 31–35, 43, 49, 80, 105–6, 184, 201; regeneration of, 39–40, 87, 185; preservation of, 55, 185. *See also* Atlantic Forest; Deforestation
Freedom: and slavery, 2, 74–75, 247 (n. 134); and sense of captivity, 6, 44, 71, 78, 92, 94–95, 124, 172, 178, 205, 216; and use of land, 76, 92, 178, 211–12, 216; and mobility, 172, 174, 178, 210
Freyre, Gilberto: and analysis of *zona da mata*, 1–2, 17–18, 46, 59–68, 85–87, 189, 203–4, 213; and Joaquim Nabuco, 14, 46, 50, 59–60, 65, 68, 89; and generation of 1930s, 15, 45, 52, 54, 58–59, 76, 88, 204, 230 (n. 40); and politics, 58–59, 88, 135, 231 (n. 65); and engagement with academia, 60–63, 109, 185, 232 (nn. 77, 80, 87, 88). *See also* Landscape: laboring
Fundação Joaquim Nabuco. *See* Joaquim Nabuco Foundation
Fundo de Assistência ao Trabalhador Rural (FUNRURAL), 163, 256 (n. 23), 263 (n. 61)

Galiléia, Engenho, 132, 134, 136, 174, 240 (n. 108)
Gardens. *See* Subsistence crops
Geography, 6, 61, 88, 103, 105, 146, 220 (n. 19), 230 (n. 38), 232 (n. 77), 232–33 (n. 88)
Goitá Grande, Engenho, 208–11, 213
Goulart, João, 3, 133–34, 136–37, 139, 157, 177, 206
Green Revolution, 11, 181
Grupo de Estudo do Açúcar (GEA), 163–64, 183, 185
Grupo de Trabalho para a Indústria Açucareira (GTIA), 164, 259–60 (n. 85)
Grupo Especial para a Racionalização Açucareira do Nordeste (GERAN), 164, 177, 257 (n. 30), 259 (n. 79), 259–60 (n. 85)

Habitus, 14, 46, 64, 67–68, 205
Hawaii, 103, 107, 111, 119, 256 (n. 26)

Instituto Agrícola do Nordeste (IANE), 103, 105, 110–12, 241 (n. 23), 243 (n. 71)
Instituto de Pesquisas Agronômicos (IPA), 105–6, 108, 242 (n. 36)
Instituto do Açúcar e do Álcool (IAA): creation and structure of, 76, 103–4, 199, 205, 214; and rural workers, 90, 115, 179–80, 238 (n. 80), 262 (n. 38); and support for planters, 106, 108–9, 112–13, 151, 158, 213, 242 (n. 39), 244 (n. 83), 251 (n. 71); and connection to political figures, 132, 134, 141, 205, 213–14
Instituto Nacional de Colonização e Reforma Agrária (INCRA), 208, 265 (n. 9)
Instituto Nacional de Previdência Social, 192
Irrigation, 9, 38, 106–8, 111, 183, 186, 200, 262 (n. 39)

Joaquim Nabuco Foundation, 89, 134, 146, 174, 185
Julião, Francisco, 17, 58, 132, 135–37, 153, 174, 249 (n. 32), 256 (n. 15)
Juntas de Conciliação e Julgamento (JCJ), 131, 138, 165–73 passim, 190–92, 246–47 (n. 128), 263 (n. 55)

Kennedy, John F., 3, 136, 250 (n. 52), 259–60 (n. 85)
Kidder, Daniel, 37, 39, 227 (n. 93)
Koster, Henry, 37, 39
Kubitschek, Juscelino, 133

Labor: and exploitation, 1–2, 8, 75, 94, 96, 200; and environment, 1–5, 9, 17, 48, 54, 57, 60, 71, 89, 152–53, 204; labor movements, 4, 16, 126, 130–40, 193–98, 204, 211; and landscape, 7–8, 14, 45, 50, 54, 64–65, 72, 205–6; and law, 12, 47, 129–31, 137–38, 159, 163, 165–71, 190–92, 207, 212; and state, 11, 15–16, 137, 158, 177, 188, 206–7; and tenantry, 17, 73–77, 91–92, 116; and slavery, 29, 41, 46–48; management and labor relations, 50, 91, 95, 114, 160–62, 164, 189–90, 214; and race, 57, 63, 67; foremen, 81, 84, 99, 139, 168, 176, 194, 213, 216; in sugarcane, 83, 105, 116–23; workers' views of, 84, 89; urban, 86–87, 130, 138; and workers without contracts, 172, 175; industrial, 192. *See also Clandestinos*; Consolidação das Leis do Trabalho; Contractors; Estatuto do Trabalhador Rural; *Fichas*; Juntas de Conciliação e Julgamento; Labor

history; Labor mobility; Landscape: laboring; Slavery; Strikes; Tasks; Tenantry; Tribunal Regional de Trabalho; Tribunal Superior de Trabalho
Labor Court. *See* Juntas de Conciliação e Julgamento; Tribunal Regional de Trabalho; Tribunal Superior de Trabalho
Labor history, 2, 8, 12–13, 15, 265 (n. 5)
Labor mobility, 105, 115, 129, 168, 174, 178, 245 (n. 95)
Land reform. *See* Agrarian reform
Landscape, 2, 4, 13, 17, 18, 46, 93, 95; of *zona da mata*, 1, 18, 21–42 passim, 50, 59, 60, 88, 126, 136; definitions of, 5–10, 61, 178, 216–17, 220 (n. 19), 232 (n. 77), 234 (n. 124), 240 (n. 109); laboring, 14, 45–46, 50–55, 57–58, 64–69 passim, 114, 124, 141, 162, 180, 193, 204–6; of captivity, 15, 71–72, 89, 91, 93–96, 141, 180, 204–6; of state, 162, 180, 187, 193, 200, 204–6
Liberty. *See* Freedom
Ligas Camponesas. *See* Peasant Leagues
Lima Sobrinho, Alexandre Barbosa, 104, 106, 114–15, 124, 132, 143, 175, 245 (n. 96)
Lins do Rego, José: contributions to elite discourse about land and labor, 2, 14, 45, 46, 52, 54–59, 66–68, 217; relationship to Freyre, 54, 230 (n. 40); thoughts on social change, 64, 87–88; thoughts on slavery, 74; and environmental change, 200
Literacy. *See* Education

Magalhães, Agamenon, 86–87
Mandar (to order, to rule), 57, 64, 66, 137
Manioc, 91, 128, 210, 216, 225 (n. 60)
Massapê. *See* Soil
Mata Atlântica. *See* Atlantic Forest
Maurice of Nassau, 32, 225 (n. 60)
Migration. *See* Labor mobility; Recife; Rio de Janeiro; São Paulo
Military. *See* Dictatorship
Monoculture, 1–2, 10, 17, 31–32, 60, 61, 63, 126, 189, 213. *See also* Bananas; Coffee; Sugarcane
Morada. *See* Tenantry
Movimento Sem-Terra, 212

Nabuco, Joaquim, 2, 14, 46–68 passim, 89, 229 (n. 3), 230 (n. 39), 232 (n. 87)
Nascimento, Euclides Almeida do, 162, 257 (n. 43)
Nazaré da Mata, 72, 119, 150, 168, 169, 171, 258 (n. 50), 263 (n. 55)
New York Times, 3, 126, 135, 138, 140
Nieuhof, Johan, 26, 27, 32, 33

Nutrition, 61, 62, 76, 88, 158–59, 238 (n. 76), 239 (n. 101)

Ó, Manoel do, 44, 71–75, 78, 93, 94
Oil, 3, 11, 181, 216
Oral history, 7, 12, 15, 44, 94
Oxen, 79, 90; and ox carts, 33, 37, 45, 81, 95, 236 (n. 47); relationship to workers, 53, 57, 72, 79, 81–82, 95

Palmares, 43, 128, 138, 140, 166–67, 168, 196
Paraíba, 41, 54, 59, 216
Partido do Movimento Democrático Brasileiro (PMDB), 198
Peasant Leagues, 5, 13, 16, 17, 131–32, 135–36, 144, 147, 211, 249 (n. 25)
Peasants, 3, 11, 79, 92, 115, 131, 146, 176, 178, 207, 212
Pedrosa, Usina, 78, 86, 91, 187, 231 (n. 59), 239 (n. 88)
Pernambuco: and sugar agriculture, 1–2, 8, 13, 15, 28, 32, 36, 46, 65, 100, 104, 145; and historiography, 2, 4–5, 9, 10, 77, 125; and labor movements, 4, 117, 119, 120, 131, 137, 162, 175, 193–99, 207; and police, 5, 14, 143, 147; and sugar industry, 10, 34–35, 40–43, 52, 54, 66, 75–76, 106, 110–14, 123, 180–84, 198–99, 216; and politics, 16, 40, 53, 58, 83, 86, 104, 130, 133–37, 157, 174, 206, 212; environment of, 21–27 passim, 31, 37, 126; and slavery, 30, 41, 46–47, 74–75; and cotton, 39–40; and *usinas*, 51, 73, 102, 106, 108; and modernization, 101, 105, 163, 182; and state producer organizations, 103, 144, 215; and secretary of agriculture, 113, 115, 147–48, 164; and secretary of state, 141; and fires, 142–51; Pernambuco Development Council (CONDEPE), 185, 188. *See also* Police
Piecework. *See* Tasks
Pindobal, Engenho, 5, 8, 13
Pinto, Eudes de Souza Leão, 147–48, 164, 245 (n. 97)
Piracicaba Research Station, 103, 150
POJ sugarcane varieties, 102, 104, 107, 110, 112, 242 (n. 47)
Police: and investigations on *engenhos*, 5, 13, 121, 143, 146–47, 149, 168; documents of, 12, 148, 195, 204, 249 (n. 25); relations with planters, 121, 131, 144–46, 158, 179; Arraes's interactions with, 137, 142; relations with unions, 176, 179, 195–96, 204, 208–9, 256 (n. 17). *See also* Departamento de Ordem Político e Social
Pollution. *See* Rivers; Water

Prestes, Luiz Carlos, 130, 131
Programa de Apoio à Pequena Produção Familiar Rural Organizada (PRORURAL), 192
Programa de Redistribuição de Terras e Estimulo à Agroindústria do Norte e do *Nordeste* (PROTERRA), 177, 260 (n. 87)
Programa Especial de Apoio às Populações Pobres das Zonas Canavieiras do Nordeste (PROCANOR), 208–9
Programa Nacional de Álcool (Proálcool), 181, 185, 186, 203, 204, 210
Programa Nacional de Melhoramento da Cana-de-Açucár (PLANALSUCAR), 180–81, 182, 199, 260 (n. 6)
Puerto Rico, 75, 102, 109, 120, 150

Quadros, Jânio, 133

Race, 1, 49–68 passim, 88, 231 (n. 59)
Railroads, 14, 22, 37, 42–44, 49, 71, 74, 90, 102, 142, 203, 205
Rainfall, 25, 39, 105, 126, 128, 196
Rationalization: of sugarcane industry, 99, 100, 104, 114, 141, 157, 164, 180, 204, 205, 207; of labor, 117, 121, 137; definition of, 119, 260 (n. 1)
Recife, 27, 32, 58, 59, 75; as reference point for discussing cane region, 5, 22–44 passim, 107, 125, 130, 131, 139, 173; regional importance of, 42, 108, 133; migration to, 83, 85–88, 169; as site of institutional or government activity, 103, 133, 143, 147, 157, 174, 192, 196–97, 209; and politics, 132–34, 139; flooding of, 187
Recife Front, 132
Rio de Janeiro, 88, 105, 133; and coffee, 30, 51; and sugar industry, 34, 36, 102, 104, 112, 113, 216; as political power, 40, 46, 58; as migration destination, 86, 169
Rivers: and pollution, 1, 14, 17, 32, 44, 60, 105, 184–86, 188, 203, 208; and sedimentation, 18, 187; river systems of Pernambuco, 24, 25, 27, 30, 37, 38, 50, 52, 88, 187, 196; historical accounts of, 26–27, 31, 60, 63, 200; use by workers, 188–89. *See also* Floods; Water
Roça. See Subsistence crops
Rural workers' unions. *See* Unions

Sales, Apolonio, 106–7
Sampaio, Cid F., 99–100, 122–23, 132–34, 136, 139, 145, 174, 210, 253 (n. 98)
Santos, José Lopes de Siqueira, 138, 142, 157
São Paulo, 4, 39, 58, 131, 198; coffee industry of, 11; economic and political power of, 30, 40, 51, 53, 65, 163, 216; as destination for migration, 86, 169; agricultural research in, 103, 150–51; sugarcane industry of, 104, 108, 111, 119, 137, 150–51, 163, 198–99, 216
Senhores de engenho, 31, 35, 40, 100, 125; and management of *engenhos*, 42, 76, 77, 78, 81, 83, 101, 137, 148, 160, 209; social role of, 52, 55, 57, 64–66, 69, 73, 76, 90, 92–93, 104, 172, 174, 177, 209, 213, 230 (n. 43); in labor disputes, 166, 169
Serviço de Orientação Rural de Pernambuco (SORPE), 139
Sigaud, Lygia, 78, 83–84, 195, 197–98, 220 (n. 11)
Silveira, Pelopidas, 132
Sítio. See Subsistence crops
Slavery, 1–2, 29, 31; attitudes of slaveowners, 8, 46, 48, 50, 52, 57, 59, 65–68, 73, 77; abolition of, 14, 15, 41–43, 48, 51, 50, 74, 76, 105, 205; and connection to landscape, 48–49, 53, 54, 60, 63, 71; and workers' perspectives, 72–73, 75, 82, 85
Soares de Souza, Gabriel, 29, 37
Social science, 85–89 passim, 104–5, 134
Sociedade Agro-Pecuária de Pernambuco (SAPP), 132, 136. *See also* Galiléia, Engenho
Soil, 9, 17, 24, 34, 48, 49, 50, 69; *massapê*, 24, 31, 127, 223 (n. 17); and natural fertility, 26, 34, 43, 63, 100, 225 (n. 57); treatment of, 42, 80–81, 107; degradation of, 49, 63, 105, 152, 153, 184; and sugarcane varieties, 102, 110. *See also* Erosion
Strikes, 10, 87, 159, 198; November 1963 strike, 3, 16–17, 140, 148, 160, 207; September 1979 strike, 3, 17, 193–97, 206, 207, 208, 210–11, 251 (n. 71); "for lower wages," 99, 122–24, 137, 206; 1963 strikes, 125, 139, 140, 147, 159; Strike Law, 159, 193–94; 1980s strikes, 195–98, 204, 208, 211
Subsistence crops: and worker gardening, 10, 77, 80, 87, 90, 91, 123–24, 176, 190, 210; slave plots, 32, 34, 235 (n. 25); workers' inability to plant, 76, 90, 92, 114, 173–74, 196, 216; policies to promote, 115, 174–75, 213; and peasants, 132
Sugarcane: impact on *zona da mata*, 1, 9, 21, 22, 28, 29–31, 36, 38, 40, 45–46, 60, 188, 203; planters of, 5; landscape of, 8, 17, 18, 58, 96; in other countries, 10, 12, 116, 120, 215; varieties of, 15, 28, 101, 112–13; expansion of, 90, 179, 181, 192, 193; research on, 100, 103, 105, 109–10; fertilization of, 107,

182; regional distribution of, 126, 128; and fires, 142–54 passim; and future, 215–16

Sugar Cycle, 54–55, 230 (n. 44)

Sugar market, 40, 75, 163, 181, 256 (n. 25)

Sugar mills. See *Usinas*

Superintendência de Desenvolvimento do Nordeste (SUDENE), 133

Tasks, 85, 176, 191; transformation of, 99, 116–23, 139, 140, 152, 206, 214, 245 (n. 103); Task Table (*Tabela de Tarefas*), 159–64, 168–71, 177, 190, 195, 204, 211–12. *See also* Agriculture; *Braça*; Labor

Tenantry: transformation of, 3, 17, 114–16, 123, 129, 138, 140, 168, 172, 209; norms of, 5, 76–78, 84, 90–93, 235 (n. 27); establishment of, 15, 72, 75, 77; and legal system, 169–70, 209

Tenentes (lieutenants), 86, 158

Textiles, 86–88

Tollenare, L. F., 21, 23, 36–37, 39, 44, 73, 227 (n. 93)

Tribunal Regional de Trabalho (TRT), 131, 138, 263 (n. 60)

Tribunal Superior de Trabalho (TST), 131, 138, 190–92

Two-hectare law, 175–76, 211, 259 (nn. 76, 80)

União de Lavradores e Trabalhadores Agrícolas do Brasil (ULTAB), 131

Unions, 2, 120; and strikes, 3, 16, 17, 140, 159, 193–98, 204; unionization, 3, 16, 123, 130–31, 136–41, 144, 148, 153, 158, 193, 249 (n. 27); union officials, 14, 17, 73, 146, 157, 159, 162, 165, 168, 170, 173, 209, 212, 214–15, 256 (n. 17); and conflict with planters, 84, 160; and politics, 104, 130–31, 136, 147, 208; and agrarian reform, 163, 172, 174–77, 211; and labor courts, 165–66, 192, 207

United States: Freyre travels to, 58; during slavery and emancipation, 62, 116; labor history of, 121, 245–46 (n. 110); perspective on Pernambuco, 130, 135; Recife consulate in, 147, 162–63, 250 (n. 52), 259–60 (n. 85); agriculture in, 164; as sugar market, 181

Usinas (sugar mills): transition to, 43, 51, 52, 54–55, 65, 73, 75–76, 79, 83, 87, 104, 199, 203–5, 248 (n. 7); impacts of, 43, 60, 67, 94, 184–85, 187–88; distribution of in *zona da mata*, 101–2, 128, 140, 163, 184; modernization of, 106–9, 113, 114–15, 183, 200; and politics, 132, 149, 158, 212; and burning sugarcane, 150–52; and labor, 166–68, 188, 189, 191, 213–14, 216, 238 (n. 80)

Vargas, Getúlio: and modernization, 53, 104; and populist politics, 58, 133; and economic planning, 76, 103–4, 158; and Pernambuco, 86, 104, 106, 130; and labor, 115, 118, 131, 137, 165

Vicência, 79, 128, 145, 211, 263 (n. 55)

Vinhoto. See Water: wastewater

Water: and early sugarcane cultivation, 23, 29, 33, 34, 37, 44; and geography of *zona da mata*, 24, 26–28; and human consumption, 81, 129, 188; and protection, 105; and soils, 110, 127; and modern sugarcane industry, 152, 185–86, 188, 199–200; wastewater, 185–88, 203. *See also* Floods; Irrigation; Rivers

Zona da mata (forest zone): definition and characteristics of, 2, 24, 25–27, 126, 130, 169; agro-environmental change in, 3, 9, 13, 22, 36, 38, 39, 43–44, 106, 182; research on, 4–5, 15, 85; landscape of, 6, 17, 18, 42, 50, 89, 136, 137; sugarcane and, 10, 17, 27, 29, 35, 40, 129, 183, 203, 210, 213, 216–17; politics in, 15–16, 126, 134, 141, 145; forests and, 24, 31, 33, 38; state and, 158, 160, 161, 174, 195; workers and, 168, 175–76, 179, 188, 196–97, 204–5